DESIGN ENGINEERING: Inventiveness, Analysis, and Decision Making

DESIGN ENGINEERING: Inventiveness, Analysis, and Decision Making

JOHN R. DIXON

ASSOCIATE PROFESSOR OF ENGINEERING
SWARTHMORE COLLEGE

McGraw-Hill Book Company
NEW YORK ST. LOUIS SAN FRANCISCO TORONTO
LONDON SYDNEY

DESIGN ENGINEERING:
Inventiveness, Analysis, and Decision Making

Library of Congress Catalog Card Number 66–20003

ISBN 07-017071-1 4567890 KPKP 79876543

PREFACE

This book is dedicated to the idea that engineering design is problem solving—science-based problem solving with social-human awareness. It is also written in accordance with my conviction that engineering design is *not* essentially an "art," but rather an activity that can be researched, analyzed, and taught. I believe that problem solving (engineering design is a particular kind of problem solving) is a high-level intellectual activity. The application of knowledge is every bit as respectable an intellectual challenge as is the acquisition of knowledge. The application of knowledge requires understanding and purpose.

My book has fallen considerably short of my aims and hopes. If you use it as a text, there is much left for you to do. As it stands, it makes a crowded one-semester course or a sparse two-semester course, depending upon how much you supplement the book with projects of your own. It is intended, in fact, that you will use problems and projects of your own invention and from your own experience to supplement the text. How else can a design problem-solving course be kept stimulating for both student and teacher?

I have abstracted three general skills from an analysis of the design problem-solving process: inventiveness, analysis, and decision making. Each of these has been made the subject of special attention in its own part of the book.

The book is the result of teaching design to seniors in mechanical engineering at both Purdue University and Swarthmore College.

Experience with these courses led me more and more to the conclusion that my students knew more than they understood or could use. I took it as my task to help them learn how to use purposefully what they had been taught in their science and engineering science courses. There is little doubt in my mind that the *way* they had been taught was interfering with their ability to use what they had "learned." Hence I set out to discover the general skills used in design problem solving so that each could be explicitly taught.

The concept of design as problem solving is somewhat broader than the traditional association of the word "design" with machine design. Consistent with this broader concept, I have included example problems from the subject matter areas of heat transfer, fluid mechanics, and thermodynamics, as well as from the more traditional design areas.

During the decade or so following World War II, engineering education began to stress the teaching of science and engineering science. The idea, and it was a good one, was to teach, not specifics of practice, but the principles and fundamentals of the science upon which the practice, rapidly changing was based. The change is virtually complete. Now it is time that we begin to discover and teach, not the practice of design, but the principles and fundamentals of design problem solving. It is to this end that this book is contributed as a beginning.

JOHN R. DIXON

CONTENTS

1
INTRODUCTION TO DESIGN ENGINEERING

1.1 ENGINEERING AND SOCIETY

Consider for a moment the role that engineering activity is playing in the various societies of the world. In America, Western Europe, England, Scandinavia, Japan, and parts of the Soviet Union, there exist advanced technical societies. Poverty, where it is present in these places, exists for political or social reasons and is not due to lack of scientific, engineering, or technical knowledge. A few nations, notably Sweden and Denmark, have effectively eliminated serious poverty altogether. The success of engineering activity in all these advanced societies is obvious. Great bridges and highways carry reliable and rapid trucks, trains, and automobiles. Ships and planes are abundant. Factories turn out more and more goods with fewer and fewer men. Tractors plow. Mechanical harvesters reap. Rapid communications systems and computers serve business, industry, science, and government.

Now consider the nontechnical societies—Africa; most of India, China, and the Middle East; Central America; and many parts of South America. Farming is done primarily by manual and animal labor. Manufacturing is primitive and inefficient. Transportation and communication are slow and unreliable. Poverty is commonplace.

What is the difference? The gap between the advanced technical societies and the poorer societies is the process called the industrial revolution. This process, though painful in many ways, has lifted a great deal of brute labor from man's shoulders. It has enabled societies to reach a point where one man can do the farming for many while another produces the goods for many. As a result, working hours for all have been reduced and more and more luxury items are produced.

It is not necessary to indulge in a great debate on which of man's many successful activities—engineering, science, medicine, law, etc.—contributed most to the success of the industrial revolution. It is self-evident that engineering was one integral and indispensable part of the process. Furthermore, it is equally evident that the transformed societies are now utterly dependent upon their engineers and technicians. Through the years, engineers have solved many problems associated with industrialization, and these solutions have become commonplace. More important, the advanced and expanding societies have made an economic habit of getting more and more new solutions from engineers. Not only is society dependent on the old solutions continuing to function, it is perpetually in need of new working ideas from its engineers. The economic fat would really be in the fire if engineers quit inventing new and better ways to do things.

At present, especially in the United States, a new kind of revolution called automation is in progress. Whereas the industrial revolution largely eliminated the burden of heavy labor, the new revolution is reducing the burden of routine work. Automation is also reducing mental labor since large and rapid computers have relieved men of endless calculations. As the new revolution continues, more delicate mental tasks are being automated. Even a field so personalized as teaching has begun to feel the influence of automation. As with the industrial revolution, it is again quite obvious that engineering is a prime contributor to the revolutionary forces. Engineering activity has played a great and important role in the history of modern man, is now playing a significant role, and has an important obligation to, and stake in, the future.

1.2 ENGINEERS AND SOCIETY

Engineering activity is clearly vital in an advanced modern society. But what about the engineers themselves? The people who carry on this important activity have not generally become significant in public affairs, despite the dependence of society on engineering. This is often recognized and deplored in editorials and articles found in the publications of the various professional engineering societies. Unfortunately, the seriousness of the lack of engineering representation in public affairs is seldom stressed in such articles. Social, economic, and political decisions in an industrial society involve scientific and engineering considerations which are, for the most part, beyond the knowledge and understanding of the general public and their elected representatives. In order to have these technical factors properly considered, it follows that the views of knowledgeable and competent engineers should be included in the public decision-making processes. It is not fame that is sought, but without proper technical and engineering counsel modern industrial societies are quite likely to make serious errors in their social, political, and economic planning. Hence, the relatively small number of publicly prominent engineers is unfortunate, even dangerous.

The argument that public decisions must take technology and science into account was made popular by the British novelist–scientist–public servant C. P. Snow, in his book "Science and Government." Scientists, in contrast to engineers, have become quite interested and influential in public affairs in recent years. The exciting debate stirred by another C. P. Snow book, "The Two Cultures and the Scientific Revolution," is evidence of the concern of scientists for their role in public affairs. Snow argues that there exist two intellectual "cultures," science and literature, that do not understand or communicate with each other. He also argues that public affairs are influenced largely by the literary culture, with the voice of science insufficiently heard. This, he says, is intolerable for a modern society in a world so greatly influenced by scientific and technical developments. He closes his book by raising the question of how a democratic society can incorporate the viewpoint of science, a body of knowledge and an activity it little understands, into its political, economic, and social decision making, and especially into its educational planning.

Though a few engineers have gained public importance in recent years, it is scientists generally and not engineers who have shown a great deal of concern for, knowledge of, and interest in, public affairs as they are affected by scientific and technical matters. C. P. Snow also speaks of this subject in his novel, "The New Men," where his central character says:

> It struck me that all the top scientists . . . were present, but none of the engineers. As an outsider it had taken me years to understand this rift in technical

> society. . . . I had expected scientists and engineers to share the same response to life. . . . The engineers, . . . the people who made the hardware, who used existing knowledge to make something go, were in nine cases out of ten, conservatives in politics, acceptant of any regime in which they found themselves, interested in making their machine work, indifferent to long term social guesses.
>
> Whereas the physicists, whose whole intellectual life was spent in seeking new truths, found it uncongenial to stop seeking when they had a look at society. They were rebellious, questioning, protestant, curious for the future and unable to resist shaping it. The engineers buckled to their jobs and gave no trouble, in America, in Russia, in Germany. It was not from them, but from the scientists, that came the heretics, forerunners, martyrs, traitors.

The thing that is so strange about the "rift in technical society" is that it is the scientists who have emerged as the branch most involved in bringing technical information and viewpoints to the social, economic, and political spheres. Engineering activity, after all, has always been much closer to, and more involved with, the real and practical economic, social, and political problems facing people. Engineering is science-based problem solving with social-human awareness, and one would therefore expect that engineers would be closer to, more informed about, and more interested in, technical effects on society. However, it has not been so. Why?

1.3 THE NEW CHALLENGE OF ENGINEERING

It is not difficult to trace in general terms the various roles of engineering in the industrial and automation revolutions. Civil engineers were needed first to pioneer the highways, bridges, and railways and to build the ports and cities. In our engineering schools, civil-engineering departments were the first and for many years were the largest. Next the development of a complex industrial society from a simple, rural one required increasing numbers of mechanical engineers to build machines, develop and operate factories, supply power, and design new products. As expected, university departments of mechanical engineering grew while civil-engineering departments declined numerically. Now the need for better controls, automation, the great increase in communication problems, and many new materials and scientific discoveries have increased the need for electrical engineers. This is again reflected by the rising enrollments in electrical-engineering departments. Projecting a little, one can predict the relative leveling off of all engineering, a rise in the physical sciences and mathematics, followed by a rise of the behavioral and biological sciences combined. In the light of the *needs* of a developing society, all these changes seem quite natural.

Actual activity tends to lag need, however. Consider the needs of the advanced nations now. An essential problem is not how to span great rivers, but how to deal with river pollution. An essential problem is not how to build great cities, but how to clean up slums. An essential problem is not how to find new weapons, but how to keep from using the ones we have. When there was a physical frontier to overcome, a nation to build, and an industrial revolution in progress, then engineering had a clear and evident mission. But there is little satisfaction in providing more and more luxuries to people already dying of overweight.

The mission, however, continues to exist if we engineers will only look to see. The important problems are still important, but they now require a degree of social concern and knowledge not previously required of engineers. *Food*, for example, is still in critical shortage in two-thirds of the world. With the ever-increasing population, this shortage is bound to become more critical as time goes on. High-productivity low-cost compact food manufacturing systems are needed. *Water* is a problem in both advanced and developing societies. More fresh water is needed in more places. *Air pollution* is a vitally significant problem in and near many huge population centers throughout the world. *Communication* and *transportation* problems are tremendously important to the operation of the complex systems being developed. *Recreation* and *old age* also promise to become huge social problems with engineering implications.

The mission of engineering continues, but it presents a new challenge to the people who make engineering their work. The mission now calls for a great deal more social, political, and economic knowledge and consideration than in the past. Solving pollution problems involves dealing with the industries doing the polluting, the public doing the complaining, and the government doing the legislating. Solutions to modern transportation problems may dislocate or otherwise affect the lives of thousands of people. Similar statements can be made about most aspects of the modern mission of engineering.

There is another relatively new development that presents a new challenge to engineers. The rise of science and the great increase in scientific discovery require a knowledge of basic science not formerly essential for engineers. It is to be remembered that the close association of engineering with science is new. Engineering developed historically out of the crafts. Early engineers were primarily craftsmen and artisans who had learned a little science. Now, however, it is necessary for the engineer to know and understand a great deal of science. In fact, basic science has become the very foundation of an engineering education. The engineer, therefore, is now challenged with the need to know and to be able to apply modern science to practical matters.

Though the new challenging problems facing engineering involve social and political factors to a larger degree than ever before, and though solutions require scientific understanding to a larger degree than every before, it is a mistake to assume that all the remaining problems are social, political, or scientific. Many essential problems are still *design engineering problems.* Solutions will use science, and solutions will consider social and political factors, but solutions will *be* new designs for things and processes. Writing in the *Saturday Review*, August 29, 1964, R. Buckminster Fuller, inventor of the geodesic dome and the Dymaxion three-wheeled automobile, says:

> Parading in multitudes, students demand that their political leaders take steps to bring about peace and plenty. The fallacy of this lies in their mistaken, age-old assumption that the problem is one of political reform. The fact is that the politicians are faced with a vacuum and you can't reform a vacuum. The vacuum is the apparent world condition of not enough to go around—not enough for even a majority of mankind to survive more than half of its potential life span. It is a "you or me to the death" situation that leads from impasse to ultimate showdown by arms. Thus more and more students around the world are learning of the new and surprising alternative to politics—the design science revolution, which alone can solve the problem.
>
> The students are thrilled to realize that it is themselves they must turn to in order to make the world work, through practical use of their university science and technology resources and their laboratory-supported design science capabilities. The students know that they need no more license to invent the tools that will make the world work than the Wright brothers needed a license to invent one of the most needed more-with-less tools—the airplane. The student's task is clear-cut. It is to increase the over-all efficiency of the world's mechanical devices from their present 4 per cent to an over-all efficiency of 12 per cent. This is easy, since over-all efficiencies up to 80 per cent are now feasible. The students know that if they invent the right tools, the tools will be used, given the right emergency. And they know that their design science revolution is bound to work because the emergencies to foster its realization are already here. Their revolution is a bloodless revolution that brings peace in the only way it may ever become effective—by elimination of the physical wants that always underlie war.

The new challenge of engineering thus involves attaining the ability to integrate three factors, the theoretical, the practical, and the social, for the purpose of finding new solutions to old problems. Theoretical, experimental, and mathematical science must be applied to real, useful, and practical needs, taking fully into account the social, political, and economic aspects involved. It is a unique challenge, and those who choose to accept it on these terms can expect to be rewarded with an exciting and significant career.

1.4 SOCIAL RESPONSIBILITY AND ENGINEERING

Before going on to a more detailed discussion of the role of engineering, especially engineering design, it is appropriate that consideration be given to the ethical questions which confront an engineer. Though much attention is given by the professional societies to those knotty economic-type ethical problems which arise in client relations—when changing employers, for instance, or with respect to patents and inventions—the fundamental ethical question is of a more personal nature. It is this: Should you contribute yourself—your time, energy, talent—to the project in question? Closely related to this prime question are those follow-up questions: Do you have a responsibility for the use others make of the results of your engineering efforts? How can you discharge that responsibility?

The scientist has a similar problem, of course, but he is theoretically in a somewhat less critical position because, in the purest sense, he does not know ahead of time what it is he is going to discover. The engineer, on the other hand, knows full well what he is attempting to build or design before he begins. Notice, for example, that a moral standard among engineers condemning, say, artificial food storage would have prevented development of the refrigerator and food freezer. And a moral standard among engineers condemning weapons of war would certainly have reduced the scope of warfare. Each of us, of course, has personal limitations and economic constraints which prevent us from being completely the persons we would like to be, but there is no one who is not free to begin by admitting the truth to himself.

One can ignore this basically ethical aspect of engineering activity, as many unfortunately do, but to do so is to make an ethical-moral decision by default. For those who wish to pursue this matter, there are some questions for discussion purposes at the close of the chapter.

1.5 ENGINEERING IN CONTEXT

There are several ways of organizing the constructive activities of man in a scientific-technical society. One dimension shows the relationship of engineering to science on the one hand and to production on the other, as shown in Fig. 1.1. As noted earlier, engineers cannot now escape the need to know, understand, and utilize basic physical and mathematical science. On the other hand, engineers must relate their work to the conditions of production. That which cannot be built is not useful.

In addition to the science-production dimension, the engineer is linked with the arts and crafts in what might be called the functional-aesthetic dimension. The practical thing is seldom ugly. This is true not only because we tend to see beauty in the practical but also because we seldom tolerate for long that which is not reasonably attractive. The first working model of a new useful item may be aesthetically horrible, but you can be sure that the first revision will include a great deal of attention to beauty.

In still another dimension, the engineer's activity is linked to economics, sociology, and politics. The connection with economics should be obvious. No government or company can afford to do everything it may wish at any given time. Economic resources must be available and the effective design engineer, if he is to have any significant decision responsibility at all, must be able to comprehend the economics of his department, company, community, society, and world. Comprehension here means more than the ability to coin phrases about the virtues of profits and free enterprise and includes an understanding of such things as the public costs of free enterprise (e.g., river pollution) and of the human and other resources which made a society wealthy and an economy vital.

The engineer has a personal life, too—a family, home, friends, recre-

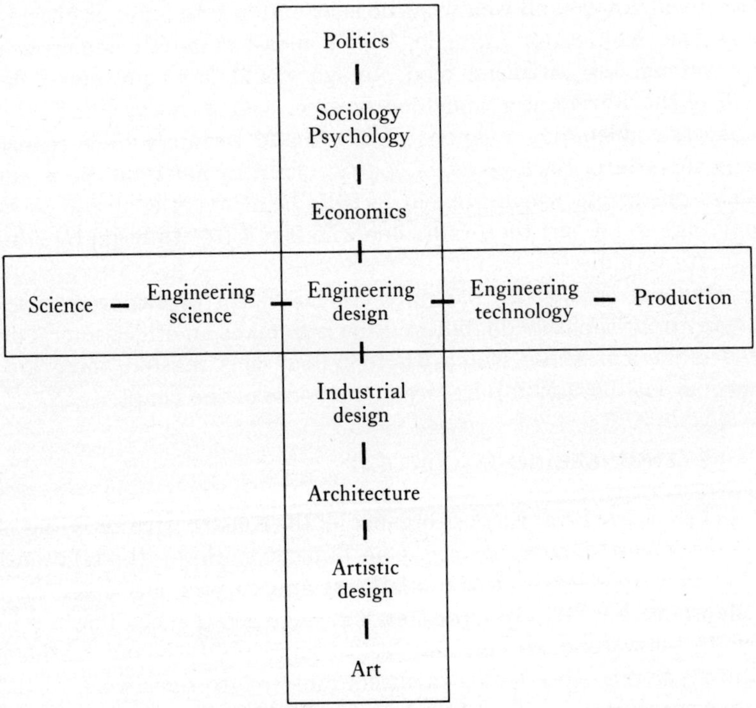

Fig. 1.1 Engineering design in several dimensions.

ation, a philosophy, and perhaps a religion. Because he is a human being he cannot help being interested in the human predicament of life and death, the history of mankind and his world, and the various expressions of man through literature, art, and music.

Because of the broad range of activities which are related to engineering, the person who starts out to train to be an engineer is confronted with the broadest spectrum of subject matter imaginable. He must be well-grounded in the basic physical sciences, engineering sciences, and mathematics. He must become familiar with the various means of production.

He must be knowledgeable in economics and other social sciences as they relate to his engineering work. He is human and a social being and hence his education will necessarily need to include the liberal arts and humanities. Finally his education must include some training in the *techniques or methods of making purposeful use of his knowledge*. This latter training is the purpose of courses in engineering design.

Engineering curricula reflect the breadth of engineering. Courses in the humanities, economics, and social sciences are generally required. Courses in basic sciences, modern sciences, engineering sciences, and mathematics are required. Courses in manufacturing processes or industrial engineering are usually included. And there are, of course, courses in engineering design.

1.6 OBJECTIVES OF COURSES IN ENGINEERING DESIGN

The objective of courses in engineering design is to help the student develop skill in applying what he has learned in his science, engineering science, social studies, and humanities courses to the solution of practical engineering problems. The object is not to teach new engineering science or specific technical subject matter, but rather to help the student learn how to use purposefully and effectively what he already knows. The scientific or engineering subject matter of design problems comes from many different areas of civil, chemical, electrical, and mechanical engineering. Thus design is not limited to its traditional area of "machine" design but includes heat transfer, fluid mechanics, electronics, and other areas as well.

An engineering design problem is almost always stated as follows: *Devise, subject to certain problem-solving constraints, a component, system, or process to accomplish a specified task optimally, subject to certain solution constraints.* The phrase "subject to certain constraints" has been included in this statement twice to emphasize an important point of view taken in this book. It is that there are *two* kinds of constraints on the engineering problem solver. One set of constraints applies to his *problem-solving procedure* and consists of such items as his own knowledge and the time, laboratory facilities, or computational facilities available. The other set of constraints applies to his *problem solution* and consists of such items as the limitations of nature, cost, availability of materials, equipment, or manufacturing skills. Our object in this book is to help the student attain certain skills and viewpoints which will improve his chances of successfully solving engineering problems, considering the implications of both kinds of constraints.

This brings up an important premise of this book: It is here *taken on faith that conscious awareness and study of the design and analysis problem-solving process (its successes and failures) will improve the probability of future success.* Students who wish to say dogmatically that engineering design is completely an "art," that there is nothing about it for a given individual that can be learned, analyzed, or improved, should stop reading this book now.

1.7 THE ENGINEERING DESIGN PROCESS

Since you are reading on, you (or your slave-driving instructor!) must at least *tentatively* agree that awareness or study of the design and analysis process *might* produce better results. Let us then begin to become aware of the designing process. Remember, the problem is this: Devise, subject to certain problem-solving constraints, a component, system, or process to accomplish a specified task optimally, subject to certain solution constraints. The first step is the recognition and understanding of the goal or objective. It may be given as an assignment or it may be a self-imposed goal (e.g., a patentable item, a workbench for the basement, etc.). Note how the fundamental ethical question for engineers comes into play right here. Should a moral and ethical person contribute himself—his time, energy, talent—to this objective? In any case, the first step in the design process is to articulate the goal, need, or objective to be fulfilled. This is shown in Fig. 1.2 as the step labeled "goal recognition."

The second step in the design process is the description of a more specific task to be done which will accomplish the general goal. For example, if fresh water from the sea is the general goal, a nuclear-powered-boiling seawater-conversion plant might be the task specification. Alternatively, the task specification might be a chemical conversion plant, or a solar plant. Notice that the choice of a task involves a *decision.*

The next stage of design usually requires that the engineering designer get an idea—a new idea, or an old idea applied in a new way to his problem. He needs to formulate a way, a method, or a concept of how to get the task done. Sometimes this requires a great deal of imagination, ingenuity, and inventiveness. Sometimes it is quite a routine application or revision of an existing idea. But the quality of the solution is determined to a large extent by the *quality* of the idea or concept used in this step. As shown in Fig. 1.2, it is in a sense the heart of the design process and is called concept formation.

Once an idea for a way of solving the problem has been found and selected (another decision), then the engineer must analyze his idea. Such engineering analysis calls for a clear definition of the problem or question to be solved. It requires the formation of a model (either on paper or in the laboratory or both) which is simultaneously simple enough to solve in reasonable time and yet complex enough to give meaningful results. Analysis requires the application of physical principles (science or engineering science) to this model and the computation of numerical results. It also involves checking, evaluating, generalizing, and optimizing the results. A great deal of space in this book will be devoted to the process of engineering analysis because it is so very important in the overall design process.

With the analysis completed, if the results are favorable, the engineer

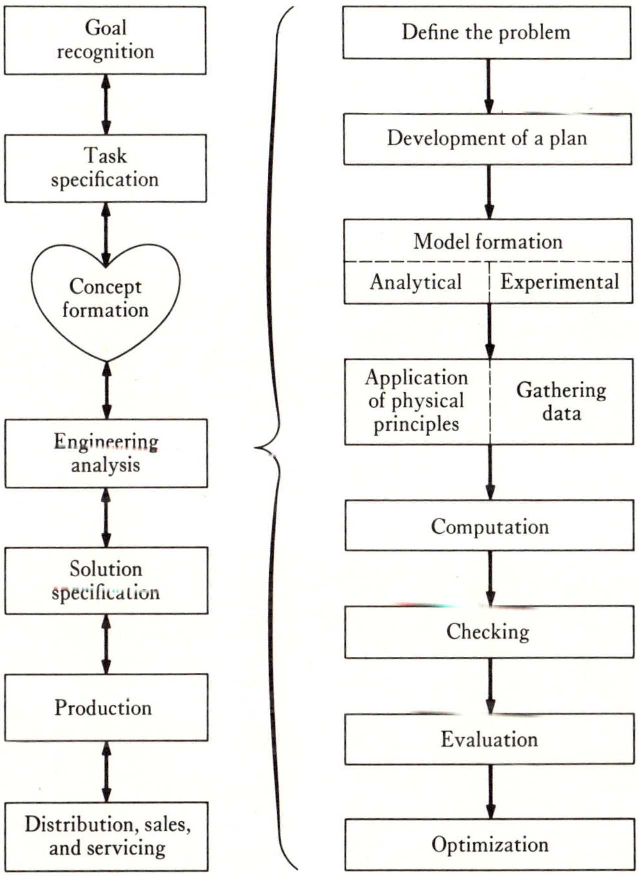

Fig. 1.2 Block diagram of the design process.

next must translate his solution into production terms. We call this *solution specification*, and questions here have more to do with "How shall it be built?" than with "How will it work?" Finally, then, it must actually be built, distributed, sold, and serviced.

Notice that in Fig. 1.2, where this complete design process is shown in block-diagram form, we have used two-way arrows to indicate that in actual practice the design process is not a straight step-by-step path. Any one solution may require a number of iterations through various parts of the process and much going back and forth. Seldom is a problem so simple or a design engineer so lucky that the first idea he has (1) works, (2) works well, (3) works optimally, (4) can be produced at low cost, (5) can be sold, (6) will require little service, etc., etc., etc.

Another way to look at the design process is shown in Fig. 1.3. In the terms of the figure, the engineering design task is to find the system or component which will produce a specified output from a given input. Other activities, science and analysis, can be described in similar terms,

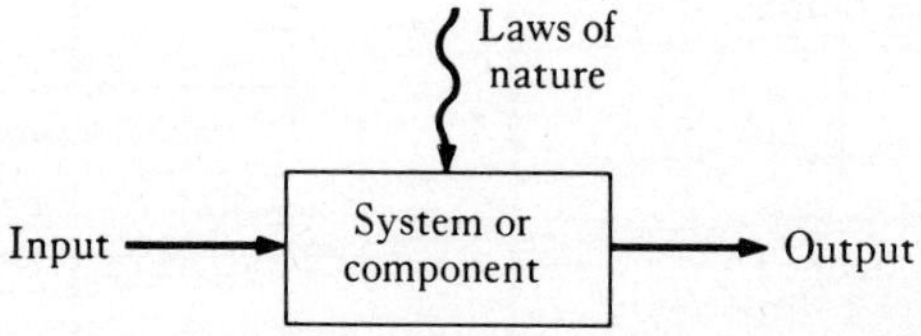

Given	Find	Process
Input, laws of nature, system or component	Output	Analysis (deduction)
Output, laws of nature, system or component	Input	Inverse analysis
Input, output, system or component	Laws of nature	Science (induction)
Input, output, laws of nature	System or component	Engineering design

Fig. 1.3 Schematic of the design process.

as the diagram shows. The task of science is to discover the laws, whereas the task of analysis is to determine either the input or output as the case may be. Because most analysis is concerned with outputs, the task of solving for the inputs given the desired output is called *inverse analysis*. This little schematic representation is a neat way of illustrating the fundamental difference between design and other scientific-technical processes.

1.8 THE SKILLS OF ENGINEERING DESIGN

What are the skills and knowledge which enable one to be a successful engineer? There are, of course, the attributes that all successful people have in common, such as the personal traits of energy, perseverance, the pursuit of excellence, a pleasant and interesting personality, imagination, etc. There are also luck, and a sense of history, and a sense of the future, none of which can be taught overtly. There are skills and knowledge, however, which are more closely identified with one's engineering education and which can be and are taught. These are the factors which help a person do a successful job of solving an engineering problem:

1. *Inventiveness*—the ability to think of or discover valuable, useful ideas or concepts for things or processes to accomplish given objectives
2. *Engineering analysis*—the ability to analyze a given component, system, or process using engineering or scientific *principles* in order to arrive quickly at meaningful answers
3. *Engineering science*—thorough knowledge and training in depth in an engineering science specialty
4. *Interdisciplinary ability*—the ability to deal competently and confidently with basic problems or ideas from disciplines outside one's specialty
5. *Mathematical skill*—the ability to bring powerful mathematical and computational skills to a problem when appropriate
6. *Decision making*—the ability to make decisions in the face of uncertainty but with a full and balanced grasp of all the factors involved
7. *Manufacturing processes*—a knowledge of and an appreciation for the potential and limitations of both old and new manufacturing processes
8. *Communication skill*—the ability to express oneself clearly and persuasively orally, graphically, and in writing

In this book, we shall be concerned primarily with items 1, 2, and 6: inventiveness, analysis, and decision making. Your science, engineering science, social science, and other specialized courses deal with the other items.

1.9 INVENTIVENESS VS. ANALYSIS

Both inventiveness and analysis are integral parts of engineering design. There is reason to believe, however, that the skills, talents, and temperaments which tend to make a top inventive engineer are opposite in a sense to those skills, talents, and temperaments which tend to make a top analytical engineer. This is a problem which has been discussed and studied by engineering educators, industrialists, and psychologists. Though a great deal is still to be learned about the question, present indications strongly suggest a real conflict between the inventive and analytic roles expected of modern design engineers. It is important, therefore, that engineers be aware of this conflict and take steps to resolve it.

Consider again the design problem and the steps in its solution. The problem is to devise, subject to certain problem-solving constraints, a component, system, or process to accomplish a specified task optimally, subject to certain solution constraints. Though the solution involves the several more detailed steps shown in Fig. 1.2, in general there are two primary things to be done: (1) the engineer must first conceive of a method, scheme, or idea that he thinks will work; (2) then he must analyze his method, scheme, or idea quantitatively to ensure that it can be made to work satisfactorily, subject to the given constraints. The first of these general steps is a creative, searching, innovative step. It requires breadth of knowledge and experience, the relating of diverse elements, and an open-ended type of thinking which diverges to and ranges among many possible solutions. The analytic step, on the other hand, requires depth of specialized knowledge, the recognizing and remembering of specific facts, mathematical skill, and thinking which converges to the single correct answer. The two steps are quite different and it is not difficult to imagine how people skilled at one may not be generally skilled at the other.

Resolution of the conflict is not to be found by positioning oneself permanently at some halfway point between the poles of inventor and analyst. That would produce very little. Rather, resolution is to be found by learning to do both well *and* by being able to change rather completely from one to the other as is appropriate for the situation.

The specialized knowledge of engineering science and the associated analytical skills must be learned. There is discipline involved in the learning which would certainly interfere with and inhibit creative faculties. The solution is to learn also what is known about inventiveness and creativity. Having an awareness of the differing nature of analysis and inventiveness will enable the careful engineer to move from analysis to inventiveness and back without a great deal of interference. What is called for is almost a Jekyll-and-Hyde type of switching from one to the other.

Almost all your college work is of the analytical nature. Inventiveness, however, is an important skill in engineering design, and therefore two full chapters in this book will be devoted exclusively to it.

1.10 SUMMARY AND PREVIEW

This chapter has contained two major parts. First, there was presented an overview of engineering and engineers in the total social context. The relationship of engineering to other activities—science, production, art, economics, etc.—was briefly discussed. The basic ethical question for engineers—"Should I contribute myself to this assignment?"—was mentioned. This is not the kind of subject matter which should be dogmatically presented by either an author or teacher. For this reason, many of these topics are continued as questions at the end of the chapter. The questions are intended to stimulate discussion and study, but they also contain some new material which is of interest per se. You are strongly urged to at least read over these questions, for they are intended to be an integral part of the chapter.

The second major part of this chapter was a presentation and introductory discussion of the engineering design process. The skills involved in engineering design were listed and the possible conflict between two of them—inventiveness and analysis—was discussed. It was pointed out that three of the design skills—inventiveness, analysis, and decision making—were the special subject matter of courses in engineering design and would be the subject matter of this book.

Now that the design process as a whole has been discussed, it is appropriate that attention be given to these three topics in more detail. Inventiveness, analysis, and decision making are three quite different mental processes. Inventiveness is generating or creating alternatives; analysis is detailed study of one alternative; decision making is choosing among the alternatives. Because they are different mental processes, they are considered separately in this book. Inventiveness is taken first and is the topic of Chaps. 2 and 3. Next analysis is considered (Chaps. 4 to 10), and finally decision making is considered (Chaps. 11 to 17).

SUGGESTED READING

Huxley, Aldous L.: "Literature and Science," Harper & Row, Publishers, Incorporated, New York, 1963. Huxley's last book before his death takes up the two-cultures debate. This is an excellent analysis of the reasons for the split Snow describes.

Industrial Research, Beverly Shores, Ind., 46301. Another of the newer periodicals that publish good articles of interest to engineers who are designers in the broadest sense.

International Science and Technology, Conover-Mast Publications, 205 East 42d St., New York, N.Y., 10017. This periodical contains general technical articles on a variety of modern technical topics.

Johnson, Lee H.: "Engineering Principles and Problems," McGraw-Hill Book Company, New York, 1960. An excellent introduction to engineering for freshmen.

Krick, Edward V.: "An Introduction to Engineering and Engineering Design," John Wiley & Sons, Inc., New York, 1965. This is a good book for you to recommend to your high school or freshman friends who are thinking about engineering as a career.

Machine Design, Penton Publishing Co., Penton Building, Cleveland, Ohio, 44113. This is the traditional magazine of the mechanical engineering field, and it carries many articles of current interest.

Science, and *Scientific American*, American Association for the Advancement of Science, 1515 Massachusetts Ave., NW, Washington, D.C., 20005. These publications carry excellent news coverage of events in Washington that are important to science, engineering, and education. There are also technical papers of general interest to engineers.

Smith, Ralph J.: "Engineering as a Career," 2d ed., McGraw-Hill Book Company, New York, 1962. A good description of engineering for high school students and freshmen.

Snow, Charles P.: "The New Men," Macmillan & Co., Ltd., London, 1954. A novel about the British scientists who developed Britain's atomic energy project.

———: "The Two Cultures and the Scientific Revolution," Cambridge University Press, New York, 1959. Sir Charles Snow, British novelist–scientist–public servant, describes the communication gap between science and literature in the modern world. Also see "The Two Cultures—A Second Look" by Snow, Mentor Books, New American Library of World Literature, Inc., New York, 1963.

Society for Social Responsibility in Science, 719 Stradone Road, Bala Cynwyd, Pa. This organization welcomes engineers and publishes a monthly newsletter on subjects related to the social concerns of technical people.

Whinnery, J. R. (ed.): "The World of Engineering," McGraw-Hill Book Company, New York, 1965. A good book for you to recommend to your nontechnical friends who wonder what engineers do. Also recommended for high school students exploring the possibilities of an engineering career.

Wiesner, Jerome B.: "Where Science and Politics Meet," McGraw-Hill Book Company, New York, 1965. The author of this collection of speeches is an engineering scientist who was President Kennedy's Special Assistant for Science and Technology.

Wilson, Warren E.: "Concepts of Engineering System Design," McGraw-Hill Book Company, New York, 1965. This is a very good introductory book for systems design in the general sense.

PROBLEMS

1.1 What is your answer to the basic ethical question for engineers? Consider, for example, the responsibility of engineers in various nations working on such products as nuclear missiles, television, antipersonnel land mines, a 310-hp automobile engine, switch blades, etc.

1.2 As an inhabitant of Mars you note that engineers on Earth in nation A and in nation B are both designing nuclear missiles for use against each other. Comment on this fact from the viewpoint of the Martian. Also try commenting from the viewpoint of the engineers in each nation.

1.3 A definition of engineering recommended recently by the Engineer's Council for Professional Development (ECPD) is: "Engineering is the profession in which a knowledge of mathematics and natural science gained by study, experience, and practice is applied, with judgment, to develop ways to utilize, economically, the materials and forces of nature." The committee considered adding the phrase "for the progressive well-being of mankind" but did not do so. Do you think this phrase should have been added? Discuss.

1.4 Discuss the role of the engineer as it relates to science, production, art, economics, sociology, psychology, and politics, in the case of (*a*) the automobile, (*b*) the nuclear missiles, (*c*) the computer, and (*d*) you name it.

1.5 Study your curriculum and see how it meets the various needs for training engineers.

1.6 Do you think being aware of the engineering design process will make you a better designer? Why? (Note: Can you find any analogies in other fields? Do painters just paint, musicians just play, authors just write, researchers just research, etc.?)

1.7 Discuss the argument "If I don't do it, somebody else will" from an ethical viewpoint.

1.8 Review your knowledge of world history and see if you can find a correlation between poverty and political unrest.

1.9 In a recent speech a prominent engineering educator predicted the major tasks facing engineers in the near future as "the conquest of space and the advancement of underdeveloped nations." Do you see any conflict between the educational needs of engineers conquering space and those advancing underdeveloped nations? How can we manage and balance our resources of people, money, and educational facilities to meet both these objectives?

1.10 Discuss the following editorial which appeared in *Mechanical Engineering:*

ENGINEERS IN GOVERNMENT

President Kennedy has continued to stress the importance of science and scientists in the Executive Branch with the establishment of the Office of Science and Technology and the increased emphasis on national scientific programs. But, asks Glenn T. Seaborg, Chairman, U.S. Atomic Energy Commission, is this enough? Have we done everything necessary to give science and technology their proper places in Government? These complex questions were posed by Dr. Seaborg at a recent American Chemical Society meeting in New York. Statistics, which can be deceiving, indicate that only about 3 per cent of our Congressmen have a scientific or technical background, compared to about 30 per cent of the Soviet leaders.

"I hesitate to make any value judgment based on this comparison," he said, "because of the great difference between our national goals and our governmental organization and those of the Soviet Union. I do believe," he added, "that the participation of scientists and engineers in our Government is bound to increase. The scientist and engineer now are as necessary in active governmental service as the businessman or lawyer. The knowledge of the scientist or the engineer and his reasoning judgment are indispensable to the formulation of national programs.

"Scientists and engineers should increasingly assume positions of responsibility comparable to the importance of scientific and technological undertakings in the Government, and consistent with the influence of science on Government and society. The important national decisions to be made, the programs to be administered, the funds to be allotted, the proposals to be judged, all depend increasingly upon scientific and technological factors."

1.11 Criticize the description and outline of the engineering design and analysis process given in this chapter. Is anything omitted? Could it be done better?

1.12 Criticize the description and outline of the skills of engineering design as given in this chapter. Is anything omitted? Could it be done better?

1.13 What is the nature of the conflict between inventiveness and analysis? What do you think of the suggestion made for resolving it in this chapter? Can you make alternative or additional suggestions?

1.14 To what extent are problems of poverty amenable to political as distinguished from engineering design solutions in (*a*) America, (*b*) India, (*c*) Africa, and (*d*) you name it?

1.15 Repeat Prob. 1.14 but apply it to problems resulting from affluence.

1.16 Do you see any possible natural conflict between the skills of decision making and those of either inventiveness or analysis? Discuss.

I
INVENTIVENESS

2 FACTORS AND METHODS OF ENGINEERING INVENTIVENESS

2.1 OBJECTIVES AND LIMITATIONS

A great deal of effort has been expended by psychologists in an attempt to isolate and measure a unique quality of people called *creativity*. The word "creativity" is often used to connote some inherent, abstract property or characteristic which people have. In order to avoid the assumption (and the arguments which follow the assumption) of the existence of such a property, the word "inventiveness" will be used in this book. By inventiveness is meant simply the demonstrable ability to get new and useful ideas for accomplishing engineering goals. The objective of this chapter is to describe explicit ways that an engineer can improve his success in getting such ideas. This is a pragmatic, no-nonsense objective. Concrete results are wanted.

Some limitations must be admitted, however. First, it must be recognized that heredity, environment, and past training impose some very real limitations on the inventive ability of every person. On the other hand, it is also quite likely that few people are working up to their full inborn imaginative potential. Our objective is to describe specific ways that students can bring their inventive performance closer to their inherent potential.

Second, it must be recognized that for modern engineers, inventiveness depends to some extent on scientific and technical knowledge and understanding. Except for a very few intuitive geniuses, thorough knowledge and understanding of a scientific or technical field are now a prerequisite to inventiveness in that field.

There is an implied premise underlying this chapter and the next of which the student should be aware. It is that inventive ability—that is, the ability to get many new, useful ideas for accomplishing engineering goals—can be improved by a study of the creative process, by practice, and by use of certain methodologies presented later in this chapter.

As a rule, most engineers and engineering students do not like the subject of inventiveness. It is too vague. Engineers generally like things that are specific and real and problems that are well defined. Hence they feel uncomfortable with the vagueness surrounding inventiveness. This is an unfortunate characteristic and it has hurt the engineering profession. It has resulted in an unwillingness to become involved in poorly defined problems, hence in *new* problems, which are always poorly defined. Inventiveness is an important skill in engineering, and though there is a lot of foolishness being written and spoken about gimmicks for creating creativity, there are also some valuable and useful things known about ways to get new ideas. The techniques presented later in the chapter are proven, but their successful use requires effort, practice, and an appreciation of their limitations. In order better to understand the techniques, therefore, it is useful first to summarize what responsible psychologists know about inventiveness and inventive people, to discuss the inventive process itself, and to consider some of the important mental blocks to inventiveness. The next three sections will deal with these topics.

2.2 DEFINITION OF INVENTIVE

For the purposes of introduction, inventiveness was defined above as the ability to get new and useful ideas for accomplishing engineering goals.

It is desirable now, however, to have a somewhat more precise and complete definition. What are the distinguishing qualities of those things, processes, problem solutions, ideas, or works of art that people generally agree to call inventive?

First of all, of course, there is the quality of *newness* or *uniqueness*. This is an obvious requirement and needs no elaboration.

Second, those things are called inventive that are either *useful*, as with a product, or *appreciated*, as with art work. Even though an idea, thing, process, or picture may be new and unique, if it is never—past, present, or future—useful or appreciated by someone, it does not deserve or get the label inventive. Those things that are called inventive are more than just different; they are also at some time either purposeful or beautiful or both.

A third quality of those things called inventive is that they bring *simplicity* where formerly there had been complexity. This is a quality often referred to as *elegance*. A new and simple solution to a complex problem is elegant, and would be called inventive. A new but complex solution to a simple problem would not qualify to be called inventive.

It is also characteristic of inventive solutions to create *new relationships*. Formerly unrelated elements are often put together to get a new, unique effect or solution. This is not always a recognizable feature of an inventive solution, but it is quite common.

Thus inventive things—whether they are physical things, ideas, theories, paintings, etc.—are called inventive when they have three essential features: (1) they must be new and unique; (2) they must be useful or appreciated; and (3) they must be elegant. The reader will recognize that this definition has a qualitative nature. It is sometimes a matter of opinion whether a particular thing is inventive. Nevertheless, the definition is useful because most people will agree most of the time about the particular use of the word. It is well to remember, however, that this is an area where subjective judgments are present.

2.3 CHARACTERISTICS OF INVENTIVE PEOPLE

What kind of people are especially inventive? Do they differ from the general population in significant ways? These are important questions because the answers may aid in the search for specific ways to improve inventiveness. Also, industry is interested in the answers, so that the more potentially inventive people can be given appropriate assignments.

Much of the work done to determine the characteristics of inventive people has been an attempt to find correlations between actual inventive performance and various standard psychological measurements. For example, a group of engineers might be divided into two groups, the more inventive and the less inventive, based on their actual job performance.

Then a battery of tests on such things as IQ, values, likes and dislikes, etc., could be given to both groups. Finally, statistical analysis of the results could be made to determine whether inventiveness correlates significantly with any of various personality factors. Such experiments have been performed and some of the results are as follows. See item 6 in the Suggested Reading for this chapter.

The tests indicate that there is no significant correlation between IQ and inventiveness. There are two qualifications to this statement which must be made. First, there is no correlation between creativity and IQ *above a certain minimum IQ* required to do the job in question. That is, there are no creative engineers with IQ = 90. But, above the necessary minimum, say about 120, there is no noticeable correlation of intelligence scales with inventive performance.

The other qualification to the statement applies at the very high end of the IQ scale. At *very* high IQ ratings ("gifted" people), there appears to be a negative correlation between IQ and inventiveness. These results have been obtained by several investigators (see, for example, "Intelligence and Creativity," by Getzels and Jackson), but you should bear in mind that "correlation" and "no correlation" are statistical generalizations; they do not mean "everybody" or "nobody"! For most persons, these results simply mean that our intelligence limitations need not also limit our inventiveness, or vice versa.

Tests have also been made to attempt to relate value systems to inventive performance. A person's values are those things he thinks important, that he is interested in, and on which he tends naturally to spend his time and energy. It appears that the more inventive individuals (the tests being reported were done with scientists, engineers, and architects) valued highly things *theoretical* and things *aesthetic*, placing less value on things economic and social.

There are also psychological scales that categorize people into certain so-called psychological types suggested by Dr. Carl Jung (see "Psychological Types," by Jung). One of the scales is a *judging-perceptive scale*. On one extreme of this scale, the *judging person* is one who is full of "ought to be's." Confronted with a new situation, such a person quickly judges (prejudges?) it as good, bad, or how it ought to be. He is less interested in or curious about what the new thing or situation *is* or how it works than he is with his own judgment of the thing. On the other end of this scale, the *perceptive person* is one who is fundamentally curious about things, how they work, etc. He is more concerned with simply finding out how things *are* than he is with making judgments of how they *ought to be.*

On this judging-perceptive scale, those who are the more perceptive tend to be the more inventive. The more perceptive people are more open to experience. They *see* more, *hear* more, and find out about and remember more things. Hence they have more to relate when new relationships are needed. The more judging person allows his judgment to get in the way of his perception and he has less experience because of it. There is an old saying that "Seeing is believing." *It is false!* The true version is "Believing is seeing," and many good experiments have been performed that prove it. We often see things or hear things or interpret things very much as we believe them to be. Judgment gets in the way of perception, limits our experience, and hence inhibits our ability to be inventive.

Another aspect of personality has to do with the *way* people judge. Living requires many decisions and judgments, of course. Even the most perceptive of people must judge when necessary. Two kinds of judgment can be discussed: *feeling judgment* and *thinking judgment.* Engineers, as you might expect, tend to be thinking types as compared with the general population. However, among engineers, those who are the more feeling tend to be the more inventive.

There are also personality types based on the kind of perception which predominates: *sense perception* or *intuitive perception.* Sense perceivers are more oriented to the physical world and their five physical senses. Intuitive perceivers are less oriented to the physical world and tend to rely more on their intuition in "seeing" things. Engineers, with their natural orientation to the physical world, tend generally to be sense perceivers. Among engineers, however, those who are the more intuitive tend to be the more inventive. Note how this ties in with the data on theoretical values. Intuitive means looking behind the obvious or the physical for meaning or implications. Theoretical means what "might be" or what "could be." The similarity is clear. Also, of course, the feeling (vs. thinking) judgment ties in here in much the same way. In sum, inventiveness in people seems to run stronger in those who are perceptive, who are open to experience, who are curious, who value the theoretical and the beautiful, and who tend to be more feeling or intuitive than their colleagues.

Though people cannot alter their personalities very much, there is nevertheless at least one practical course of action which the experiments suggest. Something can be done with the data on perceptiveness. It is possible to practice being more perceptive and less judging. A course in art, especially painting or drawing, is very helpful in training visual perception. Faced with the necessity for making a picture of an object, one must first see it as it really is. Many nonartists comment on how much a painting course helped them to become more aware of their environment.

Improving perceptiveness generally requires continual reminding at first, especially to prevent the habit of judgment from interfering with accurate and complete observation. But it can be done. Look carefully at things. See how they work. Look for hidden elements or functions. In short, be more *perceptive.*

2.4 THE CREATIVE PROCESS

How do inventive people behave when they are being inventive? (One physicist says he gets his best ideas while shaving!) Several eminently inventive people were asked to review their own behavior and to describe the steps in the inventive process. One of the best descriptions is the following:

1. Preparation: gathering skills, formulating the problem
2. Concentrated effort: working hard for a solution
3. Withdrawal: a period of mental rest away from the problem
4. Insight: the idea or reorganization which is the solution
5. Follow-through: generalizing, evaluating

The steps are not necessarily gone through linearly. There may be alternating periods of work and withdrawal or of work and preparation (gathering more skills). Usually, though, insight follows closely on withdrawal. Also, inventive people tend to follow through on their ideas to some fruition. Just as a useless or unappreciated idea is not inventive, neither is an unfulfilled one. The image of the inventive person as one who starts much but finishes nothing is a myth.

There is a lesson in this list of steps in the creative process. It is that if inventive solutions are expected, the problem-solving circumstances had better be arranged so that the inventive process is at least possible. If one comes unprepared to a problem, or does not allow time for concentrated effort *and withdrawal*, one can hardly expect a great deal of inventiveness to result. In planning work, it is helpful to keep the inventive process in view so that it may be allowed to work.

2.5 PSYCHOLOGICAL "SET"

Figure 2.1 shows three pitchers, each a different size, marked A, B, and C. The pitchers, which have no partial volumes marked on them, are to be used to place certain exact quantities of water in an accompanying large bucket of unknown size. The three columns labeled A, B, and C give the capacities of three pitchers respectively for a series of nine problems. As an example, the solution to the first problem can be obtained by filling pitcher A and then removing one pitcherful of B and adding one of C. The solution is therefore $A - B + C$. The object of the problem is to find ways to obtain the desired final total quantity with as little pouring of water back and forth as possible. It is also desired that the nine problems be solved as quickly as possible. To see how good you are, time yourself while you do the remaining eight problems. The correct answers and some

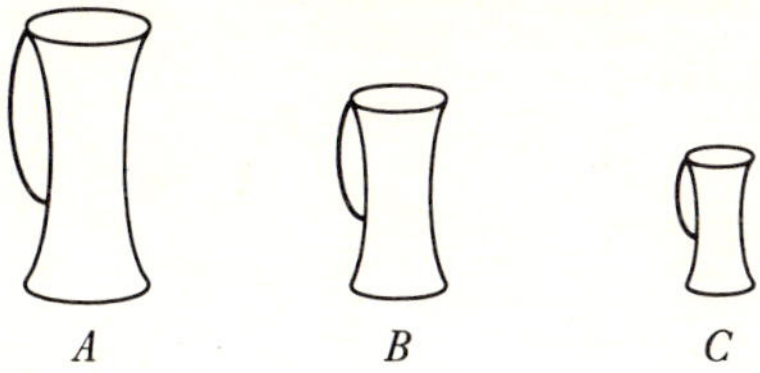

Problem	A	B	C	Desired	Answer
2.1	10	7	5	8	$A - B + C$
2.2	20	25	11	6	$A - B + C$
2.3	14	3	2	13	
2.4	18	10	7	15	
2.5	11	8	6	9	
2.6	12	9	7	10	
2.7	18	23	9	4	
2.8	13	10	8	11	
2.9	23	28	14	9	

Fig. 2.1 For each problem, what is the simplest way to obtain the desired quantity, using only full pitchers of the capacity shown?

Problem	A	B	C	Desired	Answer
2.1	10	7	5	8	$A - B + C$
2.2	20	25	11	6	$A - B + C$
2.3	14	3	2	13	$A - B + C$
2.4	18	10	7	15	$A - B + C$
2.5	11	8	6	9	$A - B + C$
2.6	12	9	7	10	$A - B + C$
2.7	18	23	9	4	$A - B + C$
2.8	13	10	8	11	$A - B + C$
2.9	23	28	14	9	$A - C$

Fig. 2.2 Solution to the pitcher problem.

grades based on how quickly you worked are given in Figs. 2.2 and 2.3. Write your answers in the space below.

Set is a word used by psychologists to mean a predisposition to a particular method or way of thought in solving a problem. "Being in a rut" is a common expression which connotes set. It is probably the single most important obstacle to inventiveness. If, in the pitcher problem, you got the correct answers to problem 9, you are to be congratulated. You were not *set* by the first eight solutions.

This pitcher problem is one often used to illustrate set. The answer $A - B + C$ to the last example is not wrong. It works, but is not the best or simplest answer. Incidentally, for some reason, a few students will start

*Time for nine correct answers**	*Grade*
Less than 2 min	*A*
2–5 min	*B*
Over 5 min	*C*

* It is the author's experience that approximately 25 percent of engineering students get the nine answers correct (see Fig. 2.2).

Fig. 2.3 Results of the pitcher problem.

at the bottom when given this problem. They invariably get the desired answer to problem 9 because they are not set by the string of $A - B + C$ answers above.

Set is not to be considered always "bad" just because it is often a potential obstacle to inventive solutions. Set is simply learning. Having learned a particular method, one quite naturally wants to use it again. In breaking set, do not say, "Forget what is known," or "Never use an old method." Rather say, "Remember, there are a lot of methods—not just one." Set is ignoring all the possibilities but the one set upon.

Set on a particular method or solution is developed by habit, or it can be part of one's personality. Everyone has experienced reluctance to drop a pet idea. "Stick-to-itiveness" is considered a universal virtue. Perseverance, however, can become stubbornness if carried too far. When a person's object has become to find a way to make a particular method work in spite of everything, rather than simply to find the best method, then he has become set.

Certain circumstances increase or promote set—for example, anxiety or tension. Fewer people will get the "right" answers to the pitcher problem if they are made to hurry or are otherwise placed under tension. Other experiments have shown the same effect. In critical or tense situations, people cling harder and harder to the old solutions, old methods, and old habits. People seldom venture out for something new in a crisis.

Familiarity also increases set. Familiar objects, processes, or ideas are seldom used in novel ways. There is also an effect called *functional set.* It has been shown that objects performing a conventional function in the given situation are seldom used in novel ways. A piece of string lying on the table, for example, will be used by many to solve a problem. A piece of string that is used functionally to hold a picture on the wall will seldom be used in the same problem. A hammer just used to drive a nail can be expected to be less frequently used in some novel way (e.g., as a doorstop) than one which is just lying there. Set due to familiarity and function is quite common.

Set also affects learning. Believing is seeing! Experimenters have shown that people learn more facts which support their own prejudices than they do facts which are contrary to their beliefs. Thus set is a great obstacle to inventiveness in human, social, and political affairs, as well as in engineering situations.

Fortunately, there is something that can be done about set. It is relatively easily corrected simply *by being aware of it!* In the pitcher problem, if people are warned not to get fixed on one solution, many more will get the answers desired. In other problem situations, calling attention to possible set has been shown to reduce its effect markedly. One of the biggest obstacles to inventive solutions is thus one of the most easily corrected. Watch out for set!

2.6 SET-BREAKING EXPERIENCES

The set-breaking experience is one which is designed to shake a person's view of things out of the conventional pattern. The idea is to provide an experience which, at least for a time, will break the sets of the individuals participating. Most set-breaking experiences begin with the definition of a totally strange physical environment. One plan, for example, is to describe the environment of a new and imaginary planet. (This procedure was initiated by Prof. John Arnold of Stanford University. He called his planet Arcturus IV.) Engineering projects are then assigned in this environment, usually in groups so that a variety of ideas and views will be presented. Notice how set *must* be discarded in this situation because conventional solutions just do not apply. Gravity, if there is any, may be up, or circular, or go on and off. The inhabitants may be one-legged, or blind, or have some superhuman abilities. Problems of design in buildings, communications, transportations, etc., are boundless on the new planet. After working on them, one does not return to earth problems with the belief that there is only one way to do things.

A variation of the above project is to have the group itself invent a new planet, its science, environment, inhabitants, etc. Then problems in politics, sociology, and literature, as well as in engineering and architectural design, can be considered by groups. One's conventional set is shaken, if not broken, by the need to deal with a weird world and its people, customs, and science.

These set-breaking experiences are rather long, complex projects, and are best done under experienced leadership. There is no harm, of course, in a leaderless, inexperienced group trying them out. Plan carefully, however, allowing enough time for idea gathering and decision making. Also, for a first project, use only willing and enthusiastic people. A well-planned project of this nature can be both exciting and valuable.

2.7 MIDWAY SUMMARY AND PREVIEW

Up to this point in this chapter the suggestions for improving inventiveness have been of a long-range type. Specifically, the following suggestions have been made: (1) practice perceptiveness in observing the world; (2) plan time and circumstances so that the inventive process has a chance; (3) be aware of set as the major factor inhibiting inventiveness; and (4) take advantage of opportunities to take part in specially designed set-breaking experiences. Now it is time to turn attention to shorter-range techniques which can be applied to specific and current problems. Practice at being perceptive will gradually improve a person's idea-getting capacity but is of little help if a new idea is needed right *now*. The following techniques are intended to provide direct help when a new idea is needed.

2.8 BRAINSTORMING

Criticism and fear of criticism are known to inhibit inventive thinking. Any new idea may, of course, turn out bad. If the originator fears the criticism which may result from his poor idea, he will not wish to try out or express unproven thoughts. In this way, many potentially good ideas—or thoughts which may suggest good ideas to others—are lost. The technique known as *brainstorming* was developed to remove the obstacle caused by fear of criticism from idea sessions.

Brainstorming has had a great deal of attention in the public press in the past, so that most people have heard about it. It is not a technique which should be thought of as a panacea. It has had some successes and some failures. The effective use of brainstorming requires a knowledge of its limitations, how it works, and when it might be appropriately applied.

Used properly in the right situations by people who know how, it can be a powerful idea generator.

Contrary to popular belief, brainstorming is quite difficult to do and takes considerable practice. It can often be a frustrating experience. There are rules to the game of brainstorming. The first rule is that criticism and judgment, whether favorable or unfavorable, are not allowed. Because people are habitual critics, this postponement of judgment is the hardest and most important rule of brainstorming.

The second rule of brainstorming is that quantity is desired. The object is to generate a *lot* of ideas. Also, one idea can lead to another. Again, it isn't easy just to blurt out ideas that come to us without first turning them over (i.e., judging them!) in our own minds first. However, with no criticism allowed, one idea is as good as any other. In brainstorming, the more ideas the better.

The third rule of brainstorming is to *think wild.* Far-out ideas are wanted. In the final judgment which comes later, many of the ideas will not be useful, but the process is one of letting the ideas flow as freely and as wildly as possible. A sort of group chain of consciousness (or unconsciousness) is to be encouraged in brainstorming.

The use of brainstorming on engineering design problems also suggests that for best results the members of a brainstorming group should not have close personal relationships and should not be too intimately involved with the project under consideration. Obviously, they should have a general appreciation for, and knowledge and understanding of, the problem, but for brainstorming purposes they need not be experts. The experts can judge and develop the ideas in detail later.

Brainstorming is best used on problems which are not extremely precise or specialized. Highly sophisticated and technically detailed problems are less likely to meet with brainstorming success than are more generalized situations. For example, "What new products can be made in this factory?" is a question more amenable to brainstorming than "How can we reduce the noise interference on the 0/2 oscilloscope used to test the Type ABC tubes?"

In the right circumstances, as noted above, and with people who know how to follow the rules, brainstorming can be a valuable method for getting new ideas. In the next chapter, two actual brainstorming sessions are recorded so that you can see how the process works in action. There are also some suggested problems for you to try at the end of that chapter.

2.9 INVERSION

It is very difficult to think up a new problem. New solutions are much more common. Many new solutions come from a new viewpoint. One way of getting a new viewpoint is by what might be called the inversion method.

The method simply calls for conscious breaking of the set, the old way of looking at the problem, for the purpose of looking at it in some new or inverted fashion. If the usual way of viewing a problem is from the outside, then inverting it means to look at it from the inside. If a part in the problem situation is always standing up, then inverting it means to turn it upside down, or perhaps horizontal, or perhaps at some angle. If one part of a system moves and another is stationary, inversion means switching them around. Upside down, inside out, reversed, etc.: these are the words which describe the inversion method of getting new ideas.

An example will help illustrate this method. The problem is to get the meat out of walnuts on an assembly line. All efforts at crushing, sawing, cutting, etc., have failed for one reason or another. It is customary to think of this problem from *outside* the walnut. Suppose, however, that we invert the problem and consider it from the *inside* instead. Couldn't the shell be pushed out and away from the meat? By drilling a small hole in the shell and applying air pressure, one can push the shell away and solve the problem. Inversion is a very simple and very powerful technique for getting a new viewpoint.

Another example of inversion used to solve an actual problem is included in the next chapter.

2.10 ANALOGY

A great deal of original thinking is done by analogy, and this process can be consciously used to stimulate new ideas. Solutions to problems are often suggested by analogous situations in other problems, in nature, or even in literature. The use of analogous engineering solutions, especially from other areas, to get new ideas is fairly straightforward, but it is appropriate here to point out how being perceptive about how things work will provide one with a much broader experience which can be used in this way.

Getting ideas for engineering solutions by use of analogy with nature is also straightforward but is hard for most engineers to practice because they lack even basic training in biology, especially physiology. Nature has found so many different ways of doing things, however, that it provides a wealth of possibly useful new ideas for solutions to engineering problems. There are, for example, literally hundreds of different kinds of "pumps" represented by the hearts of various animals.

Using analogy to literature requires a good knowledge of literature, folklore, myth, poetry, and science fiction. It means taking a real or fancied idea from the literature and applying it, or some variation of it, or something suggested by it, to the engineering problem at hand. Alice's looking glass, for example, might provide an idea, and the Lilliputians of "Gulliver's Travels" were quite clever engineers.

Using other engineering solutions to get new ideas by analogy should come easily to the perceptive engineer. Ideas by analogy with nature will be more difficult, but a medical doctor or biologist might be called in on a special problem for this purpose, or a course in physiology and anatomy would help. Ideas suggested by the literature will be hardest of all, partly because many engineers have not had the time or inclination to read widely, but also partly because it is not common to look for new *engineering* ideas in literature. Analogy provides a tremendous source of new ideas, however, and should be consciously used. There are several examples in the next chapter to show how this can be done, and item 5 in the suggested reading discusses the use of analogy in more detail.

2.11 EMPATHY

Empathy means personal identification and feeling for another. It is usually used in the context of human relations and means, roughly, putting oneself "in the other man's shoes." The term can also be applied to the act of identifying personally with a thing, part, or process being devised. The object is to "become" the part, to see from *its* position and viewpoint what it might do.

As an example, reconsider the problem of getting the meat out of walnuts. In Sec. 2.9, it was shown how inversion led to a satisfactory solution. Empathy could also have been used to obtain that solution. Identify personally with meat inside the shell. (Dark in there, isn't it?) To get out, you need help in *pushing* the shell away. It's only a short step from this realization to the air-pressure scheme. The essential step in solving this problem is thinking from inside the shell, and both inversion and empathy suggest this step immediately.

Empathy requires a willingness to do some playacting. An uninhibited and extroverted nature helps with this, but most people can do it with practice. It's a very useful technique for getting new ideas.

Both empathy and the use of analogy are useful to people working alone or in groups. Brainstorming, of course, is generally a group activity, though some claim it can be used alone. In the discussion of brainstorming, it was pointed out that group interaction on a personal level is not desirable. That is, comment on the last idea presented in the group is not wanted; all that is wanted is the next idea. When using analogy and empathy, however, a more conscious problem-solving attitude is probably best. Group interaction is wanted. Highly sensitive, insecure, or greatly inhibited persons are not best suited to this kind of activity. Tact is called for at all times, of course, and it is usually best to be positive rather than negative, but all idea sessions do not have to be brainstorming sessions. As was pointed out, brainstorming is a specialized technique with limitations. The use of analogy and empathy (and fantasy, which is described next) are appropriate for conscious use any time an idea is needed. Also see item 5 in the Suggested Reading.

2.12 FANTASY

Fantasy is make-believe—wishing it were so. The use of fantasy to stimulate new ideas involves dreaming of some fanciful solutions which use unreal things or supernatural processes if necessary. Consideration of the ideal solution is often useful, even if it involves some fantasy. The hope is, of course, that the wishful thinking may bring up a new idea or viewpoint which ultimately leads to a new, practical solution. Examples of the use of fantasy are also included in the next chapter and are discussed in item 5 of the Suggested Reading.

2.13 SYSTEMATIC SEARCH FOR NEW COMBINATIONS

It was noted earlier in the chapter that sometimes it is characteristic of those things called creative that they are formed of or by new relationships among diverse or previously unrelated parameters. That is, creative solutions often are found by forming new combinations of things, processes, or ideas. It follows that a *systematic* search for new combinations may be a fruitful way of aiding the inventive process. This process is discussed at some length in item 1 in the Suggested Reading for this chapter, where it is called "morphological analysis."

Basically, the method is one of breaking the problem into two or more important dimensions based on the required functions of the system or component to be designed. Then each of the functional dimensions is

	Distribution					
Fuel	*Air*	*Water*	*Oil*	*Stones*	*Electric wires*	*Animals*
Gas						
Oil						
Coal		x				
Nuclear						
Earth						
Sun						
Biological						
Wood						

Fig. 2.4 Matrix for a systematic search for new alternatives in a home heating system.

essentially brainstormed to compile as long a list as possible of ways of meeting each one. Finally, the lists are placed into a kind of matrix form so that each and every combination can be seen.

As a very simple example, a home heating system can be used. Let the "functions" or dimensions be *source of heat* and *distribution of heat*. Possible sources of heat are gas, electricity, fuel oil, wood, coal, nuclear fuel, the sun, the earth, a biological colony, etc. Possible methods of distribution are air, water, steam, electric wires, stones or other solids, oil, other liquids, etc. A two-dimensional matrix of these can now be made, as shown in Fig. 2.4. Each block in the matrix suggests a possible solution combination, such as a coal source with water distribution (the one marked x). Note how this procedure calls attention to combinations that may not have been thought of otherwise. When three or more functions are involved, a matrix with three or more dimensions results, but the procedure is the same.

2.14 SUMMARY AND PREVIEW

In the first part of this chapter, the subject of inventiveness was discussed in a general way. The meaning of inventiveness, the personal qualities of inventive people, and the process of inventiveness were all discussed. In the second half of the chapter, some more specific idea-stimulating methods were introduced. In the beginning, the objective was described as a no-nonsense one. Concrete results were wanted. Here they are. To improve your success at getting new, useful ideas:

1. Practice being perceptive. Learn more from experiences. Be curious about how things work. Don't let judgment spoil the view. Remember that believing is seeing.
2. Plan problem-solving activity with the inventive process in mind. Especially important is providing for both hard work and withdrawal, in alternating fashion if necessary.
3. Be aware of set. Take advantage of opportunities to participate in set-breaking experiences.
4. Learn *how* to brainstorm and *when* to brainstorm. Remember the rules: no criticism, quantity is desired, think wild. Brainstorming is best used on larger-range, general problems.
5. Use inversion. Look for a new viewpoint, inside out, upside down, stop the moving and move the stopped, etc.
6. Use analogy to other engineering solutions, to nature (especially biology), and to things found in literature.
7. Use empathy. Identifying with the thing being designed often leads to a new viewpoint.
8. Use fantasy. Considering the ideal solution, even if it involves wishing, can stimulate new ideas.
9. Try a systematic search for a new combination.

All these techniques take practice. Most of us have had a good share of our inventiveness disciplined out of us by parents, teachers, and other well-meaning souls. Now that we are civilized, however, we need some of it back. It won't come easy. Don't expect anything like "instant creativity." Getting new ideas is often hard, frustrating work. Practice with the above suggestions will help, however.

Before we go on to Chap. 3, there is one additional point made early in the chapter that should be reiterated. Inventiveness in the advanced technical world of today and tomorrow requires a solid scientific background and depth of knowledge in an engineering science specialty. Everything said in this chapter is based on the assumption that the reader has or will have such a solid background and depth of knowledge and understanding. With those things as prerequisites, the ideas presented here, if practiced, can be expected to lead to improved inventiveness.

In the next chapter, a number of examples of creativity are presented, mostly in dialogue form. Where possible, the use of specific techniques, such as empathy, are pointed out.

SUGGESTED READING

Alger, John R. M., and Carl V. Hays: "Creative Synthesis in Design," Prentice-Hall, Inc., Englewood Cliffs, N.J., 1964. This paperbook on creativity in engineering has an especially useful explanation of how to conduct a systematic search for new combinations. The process is called "morphological analysis" in the book.

Buhl, Harold R.: "Creative Engineering Design," The Iowa State University Press, Ames, Iowa, 1960. A very good book on creative engineering which stresses habit (set) as an inhibiting factor. Also discusses some interesting examples.

Design in America, *Saturday Review*, May 23, 1964, pp. 12–29. This issue of *Saturday Review* magazine contains articles on creativity in design. See especially The Twenty Best Industrial Designs since World War II.

Gardner, John W.: "Self-renewal: The Individual and the Innovative Society," Harper & Row, Publishers, Incorporated, New York, 1964. This is a most useful and interesting discussion of the problem of maintaining a creative personality or a creative organization.

Gordon, W. J. J.: "Synectics," Harper and Brothers, New York, 1961. An excellent book on the use of analogy in stimulating new ideas. Good examples are included. Highly recommended.

MacKinnon, D. W.: Fostering Creativity in Engineering, *J. Eng. Educ.*, vol. 52, no. 3, pp. 129–142, December, 1961. A good presentation of experimental results concerning the creativity of scientists, engineers, and architects.

Marks, Robert W.: "The Dymaxion World of Buckminster Fuller," Reinhold Publishing Corporation, New York, 1960. A well-done, illustrated description of the work and philosophy of one of the most inventive designers. Very stimulating.

Osborn, Alexander: "Applied Imagination," Charles Scribner's Sons, New York, 1957. The originator of brainstorming discusses many ways to develop creative ability.

PROBLEMS

2.1 What are the factors you consider in determining whether one thing is more, or less, inventive than some other thing?

2.2 What are the characteristics of the more inventive people? Do the characteristics listed in the chapter check with your own observation?

2.3 What are the steps of inventive problem solving? Think back over instances when you got good, new ideas. Do you agree with the steps given in the chapter?

2.4 What is meant by *set?* What can be done about it? Think back over instances when set interfered with your getting good, new ideas. How could you have broken your set?

2.5 Get a good, diverse group of students, preferably from several departments in your university (not just engineering and sciences, if possible), and invent a world. Start by inventing the physical world. Brainstorm it for a specified period. Then decide on one idea and stick with it. Next invent inhabitants for this world in the same way. Then go on to consider their transportation, communication, religion, politics, government, sex life, buildings, recreation, etc., etc., etc. Break up into small groups if you wish for various aspects and have the small groups report back. When you are all finished, evaluate this project as a set-breaking experience.

2.6 Describe brainstorming. What are its rules? Why is it so difficult to do?

2.7 Brainstorm a problem of interest to you. Evaluate your group's performance afterwards. How could your brainstorming technique be improved?

2.8 What are the types of analogy described in the chapter?

2.9 What does perceptiveness have to do with inventiveness?

2.10 Pick an object or machine part or something and identify yourself with it, using personal analogy. How does it feel? What do you need to do your job better? What do you want?

2.11 In this chapter it is assumed that one's inventive performance can be improved by a study of the inventive process, by practice, and by use of certain methodologies. Do you agree? Discuss.

2.12 Practice the techniques in this chapter on some problems which are real to you. Evaluate your sessions so you can learn from them. What did you do wrong? What did you do right? How could you improve?

2.13 Get item 7 of the Suggested Reading from your library. Examine the "twenty best designs" given there from the viewpoint of originality, beauty, elegance, and simplicity. Would you recommend substitutions for the list?

2.14 Consider the design of a home clothes dryer and perform a systematic search for new combinations. See item 6 in the Suggested Reading for a good discussion of this method applied to this problem *after* you have tried it for yourself.

2.15 Conduct a systematic search for new combinations in the design of (*a*) a toy rocket, (*b*) an orange-juice squeezer, (*c*) refrigerators, and (*d*) your choice. Note the importance of beginning this approach with a proper list of the functions of the system to be invented.

3
INVENTIVENESS CASE STUDIES

3.1 INTRODUCTION

At the end of the preceding chapter, it was said that there is no such thing as "instant creativity." Increasing one's productivity of ideas requires practice, discipline, and hard work. It is not easy to break set. Brainstorming is difficult to learn. The use of inversion or analogy is a little easier but also requires practice for good results. When these methods are tried the first few times, do not be discouraged about lack of success. Keep trying and practicing and analyzing to see how and why the method failed. There are problems at the end of this chapter which can be used for practice. However, if there is a real problem available, it will also make an excellent trial example for practicing the various techniques outlined in the last chapter.

In this chapter, some case studies of inventiveness are presented. The different techniques or methods previously discussed are illustrated in these cases. In these examples, there is no concern with evaluating or choosing among the various ideas developed as possible solutions to the problems. It is important, however, that real-life idea sessions do have a consummation and that the solution reach fruition. *This is vital.* Too many idea sessions without an ultimate connection with reality will promote sterility. For this reason, using real problems is better than using textbook problems. Textbook problems are fine, however, if some of them are followed through to a real conclusion. Again, let it be repeated: Do not expect quick and easy creativity. With work and practice, however, results can be expected—as the examples in this chapter show.

3.2 SOME INVENTIVE SOLUTIONS AND SOME UNSOLVED PROBLEMS

Most of this chapter will be devoted to reports of actual problem-solving sessions. There is something to be learned also from the observation and study of certain creative solutions already available. Here is a short list of creative products that are familiar to most people today:

Rear-engine jet aircraft
The ball-type head on the new IBM electric typewriter
Drafting machines
Floating vacuum cleaners
Circular slide rules
Ball-point pens

Students should add to the list.

How might these ideas have been originated? Consider them one at a time. What delayed their conception? Which of the various mechanisms described in Chap. 2 could have led to their invention?

There are also many examples of unsolved problems. Here are a few:

Window washing—still very much a hand operation
House building—the same as years ago, and very expensive (Prefabs are new, but are simply the same old design assembled at the factory instead of on the site.)
Noise in cities from autos, trucks, jets, lawn mowers, etc.
Auto safety
Water supplies for growing megalopolises

There is still much for engineering to do!

Students should be able to add to the above list. Also, such a list is a good source of ideas for practice brainstorming sessions.

3.3 THE VIBRATING TURNTABLE

A student brought the following problem to a class problem session. His record turntable in the dorm was not sufficiently isolated from vibration to prevent the needle from skipping. He wanted to know what to do about it. Here is a transcript of the group conversation which followed. On the left are notes which help analyze what is going on.

PROBLEM STATEMENT	BOB:	Well, here's my problem. The table I have my turntable on is kind of wobbly. The needle skips and jumps whenever anything touches the table or moves very hard around the room. I need a good way to fix it.
PROBLEM NOT TOTALLY CLEAR	JIM:	Aren't there any vibration isolators on it?
	BOB:	Yes. There are four springs—one at each corner.
FIRST IDEA, BUT NOTE SET	JOHN:	Maybe you need different springs—stiffer, or maybe springier.
STILL GETTING PROBLEM DEFINED	BILL:	Well, you say the table is wobbly. Does the trouble come from vertical or horizontal bumps?
	BOB:	Mostly from horizontal motion, I guess. The springs seem to do okay for the vertical aspect.
PROBLEM DEFINITION	BILL:	So that's our problem—the springs in there are probably only designed to deal with vertical vibration isolation. What we have to come up with is a way to also isolate it horizontally. Okay, let's see—
SECOND IDEA, BUT NOTE SET	JOHN:	Horizontal springs! The springs work for the vertical. All you have to do is add some springs in the horizontal plane.
NOTE USE OF A KIND OF INVERSION; ALSO BROKE SOME OF JOHN'S SET	JIM:	Springs don't have to be either vertical or horizontal. They can be at any angle. We could put the springs in at an angle and go back to four springs.
	BOB:	The springs are vertical now. This angle idea might do the trick.
	JOHN:	Okay, next problem.
	BILL:	Hold on, now. We've just begun. Let's look for some other ideas. Let's try some mechanisms—empathy. Al, you're good at this. Get in under that turntable and try to hold it still while I shake the table. How is it under there?
	AL:	(Stands out in room with hands up and spread as if he were holding a large weight over his head.) Well, I need swivel hips: As you move the table around, I keep my feet on the table but my upper body still. I need a ball-joint hip—like a girl at the Roxy!
	JIM:	A ball joint—
	AL:	(Still holding table.) One thing I could do is set this box down and see if I could stop the table. The table could be

stabilized better. Bob, you said the table was wobbly. Could it be stiffened?

BOB: Well, we've already done as much as we can. Actually, I have the same trouble, though not as bad, when the turntable is on the floor. The whole silly building shakes!

JIM: This ball-joint thing ought to be good for something. How about just having a ball? Let the turntable roll around up there on some ball bearings.

NOTE ANALOGY

BILL: You know, we need a kind of "tablecloth effect"—like when you pull the tablecloth out from under some dishes. We want to shake the table—horizontally—without moving the turntable.

AL: (Back to his seat.) The tablecloth effect works because you exceed the coefficient of friction. If we had a low coefficient of friction—

BOB: Lubrication—I float the thing in an oil bath!

ANALOGY

JOHN: Or you could use air lubrication. That would be springy in the vertical direction too! An air cushion one way and air-lubricated the other. Like one of those floating vacuum cleaners. But where would the air come from?

AL: Never mind that now.

JIM: I still think this ball thing has something in it we haven't got out of it yet. If we had it sitting on balls, it would stay put pretty well when you shake the table.

BILL: But that would lose the vertical isolation.

AL: How about a bouncy ball! Soft. That would act as a vertical spring and as a horizontal lubricant.

INVERSION AND ANALOGY

BOB: Good idea: Balls ought to be a sort of reverse "silly putty." That stuff deforms when you stress it slowly, but resists rapid deformation. What I need stays stiff for slow motions, but gives when it gets a good whack.

JIM: Okay, let's try some more empathy on a different part. The problem is that the arm holding the needle won't stay in the groove when the box gets bumped horizontally.

JOHN: So, make the arm heavier—then it'll stay down. Oh, no—would ruin the record.

BOB: Yeah, the pressure on the record has to stay fixed.

EMPATHY

JIM: Now wait. Let's say I'm the tone arm. My hand is the needle. (Jim gets down on floor and stretches out one arm, pointing fingers down to simulate needle.)

BOB: That's not quite it, Bob. The tone arm is balanced from a pivot.

JIM: Okay. I'm balanced from a pivot. I can imagine that. Now shake the floor—I mean the table. I've got to try to keep up with the motion—move with it. It's my inertia that keeps me still while the rest of everything moves.

AL: Lighten the tone arm!

BOB: Got to keep the pressure the same.

AL: You can do that by adjusting the counterbalance and pivot. Reducing the mass of the whole thing will help.

BOB: Yeah, I guess you're right. It could be done.

JIM: (Getting up.) Well, that got me a dirty shirt and Bob an idea!

INVERSION

JOHN: Have we given up on springs altogether? Bob, you said it rests on springs. Would it help to *hang* it from springs?

AL: At an angle.

BOB: What's the advantage?

JOHN: I don't know.

JIM: Why not suspend it from the ceiling?

BOB: The whole building shakes.

JIM: Yeah, but with long springs, with maybe a damper in there too, you'd get a real long period—sort of a slow-motion sway. No quick jerks. The needle might stay put.

BOB: Might work at that.

FANTASY

AL: What you need is a magic carpet to set that thing on. A nice calm magic carpet.

JOHN: Like the air cushion.

AL: Why not on a real cushion. A pillow. A whole bunch of feathers!

BOB: I don't think it will work.

JIM: It's an idea—and easy to try!

BOB: I've got a lot of ideas. Let's quit. I'll try some and let you know what happens.

The preceding example was not brainstorming. These problem solvers were deliberately using analogy and inversion to try to get an idea. They kept criticism and evaluation pretty well out of the discussion, however, in the manner of brainstorming. In the following example, brainstorming is used exclusively. Note the differences.

3.4 THE GREEN-TOMATO PROBLEM

The problem is to separate green (unripe) tomatoes from ripe tomatoes quickly and automatically in large numbers. This is a real problem. Tomato farmers in some places are finding it more economical to pick all their plants of all their tomatoes at one time. This can be done by machine. Now they have all the ripe and unripe tomatoes mixed up in a hopper. How do they separate them? A brainstorming session was scheduled.

TOM: We separate them by color. A color meter ought to be practical.

ED: Emissivity or reflectivity—green one ought to have a higher reflectivity.

DAVE: Hardness. We squeeze them—easy—or poke them.

DICK: Electrical conductivity.

TOM: Electrical resistance.

DAVE: Magnetism!

DICK: Size. Won't the green ones be smaller?

ED: Weight. The ripe ones will be heavier.

TOM: Size and weight together ought to correlate.

DAVE: Size and weight is density.

ED: Specific volume.

TOM: They must be mostly water—and have the specific volume of water.

DAVE: Do they float or sink?

DICK: Maybe that's it—separate them by density—by whether they float or sink in water.

ED: Wouldn't have to be water—could be anything.

TOM: Nontoxic.

DAVE: Salt water.

DICK: X ray—size of seeds, or something like that.

TOM: Smell, odor.

ED: Sound—can you hear a tomato?

DICK: Can a tomato hear?

DAVE: Heat—infrared radiation.

ED: Thermal conductivity.

TOM: Specific heat.

DICK: Ability of a juggler to juggle them.

ED: Just let a woman look at them—push a botton.

DAVE: Statistically—check only every other one.

DICK: Just shake the hopper—ripe ones will rise—or fall.

ED: Blow air through as you shake.

TOM: Use random numbers—let 3s and 7s be ripe.

This session went on for a while, but let's stop here. Ideas are thrown out, picked up, followed, in a free-wheeling way. The lack of criticism makes this possible. Afterward, a long list of possible solutions can usually be compiled for evaluation or further investigation. A tape recorder or a good stenographer is a must for effective use of brainstorming.

3.5 A NEW JUICE SQUEEZER

Home orange-juice squeezers are far from satisfactory. Most have a number of parts which must be washed and require a great deal of force from the housewife. A device into which one could toss an orange, and out of which one could obtain fresh orange juice, might be marketable. The results of a problem-solving session (not really brainstorming this time) follow.

JIM: Okay, that's the problem—a home orange-juice squeezer.

HARRY: A can opener! Who wants to be bothered with fresh oranges? Besides, they're too expensive. Frozen is just as good.

ROGER: I can't remember the last time I had fresh orange juice! Does anybody use it anymore?

KEN: We do. It's a lot better than canned. You guys have lost your taste for good food—probably you like TV dinners!

JIM: Yeah, maybe if it wasn't so much trouble and so messy squeezing the juice out more people would use the real thing.

HARRY: Could be. The market boys seem to think so, judging by this assignment. What's wrong with the ones on the market now? My wife uses a tin thing she bought for a dime—when she even bothers. Seems pretty simple. She cuts the orange in half, then this tin thing fits over a cup, and she uses it like a kind of washboard only she twists back and forth. It does make a mess—table—hands—half the juice gets spilled.

KEN: We have a more mechanized affair. It's terrible. Mostly we use a tin job like Harry's describing. With ours, you still have to cut the orange in half, put it in this thing and lean on a lever. My wife's not heavy enough. Then there are three or four parts. Also, it makes a mess.

JIM: Well, it looks like we've got a problem, after all!

ROGER: Yeah, get the juice out of an orange and into a glass—presto—no mess!

RAY: I saw a machine once, in Chicago I think, in a train station or airport or someplace. It took real oranges, sliced them—so you could see—and squeezed 'em out. They sold a lot of the stuff.

KEN: How do you suppose they cleaned it? The housewife only does this once a day—not all day long like a business.

RAY: I don't know how—or if—they cleaned it.

JIM: How'd they do the cutting, Ray?

RAY: With a motorized blade—sort of like a saw.

JIM: And the squeezing?

RAY: I don't remember.

ROGER: Sounds to me like they have just automated and motorized what the housewife does.

RAY: Exactly! They had arms and things moving around. They made a robot housewife squeezing an orange.

JIM: We need a new idea. Something for the home.

HARRY: We usually push the juice out. How about pulling it out?

RAY: With a vacuum. Suck it out.

ROGER: Sure, cut a hole and slurrrp! Out it comes.

JIM: I doubt it! Isn't the juice in little sacks in the orange?

RAY: Yeah, I guess it is. How about a kind of combined blade and sucker? Cuts and sucks.

HARRY: A roving probe—with a hole in it. We poke into the orange and wiggle it around. It cuts and sucks as it goes.

JIM: Not bad. Whole thing could be motorized. Only thing to clean is the probe.

RAY: We'd need a vacuum pump—or something.

JIM: Whatever we do is going to have to be motorized.

ROGER: How else can we get juice out of things?

HARRY: So far we've squeezed it and cut it and sucked it.

ROGER: There is more than one way to squeeze something. (Laughter.)

JIM: How else do we get liquid out of things? Dry them?

RAY: Centrifuge.

HARRY: Good, Ray! We punch holes in the skin and whirl it.

JIM: What about those damned little sacks?

HARRY: We'll whirl it fast enough to break them.

RAY: We twist things to dry them—like you do with a bathing suit.

ROGER: I'm orange juice in a little sack in an orange. It's dark in here and I want out. It's like being in four jails at once. My little cell is in a bigger cell, a a cell block, I guess—you know, the sections of an orange. And the whole thing is inside this wall—the skin. How to get out? Well, I need help! Somebody could cut me out of the cell and then suck me out of the cell block.

JIM: We discussed that. How else? What could *you* do?

ROGER: I don't know. I feel pretty helpless. I need to push out on this cell wall but I can't—I don't have any backbone. If somebody—something—could push on me, I could maybe push on the wall. Like if somebody squeezed the cell one way, then I could push the other—pretty hard, too—I'm incompressible!

RAY: Maybe all we need to do is crush the orange—like Ray says—that would break the sacks too.

HARRY: Or roll it, like through a wringer. Sure! Put the orange through a wringer.

JIM: Won't we get sour skin juice, too?

HARRY: Maybe—we ought to be able to design the shape and tension on the wringer to avoid this.

RAY: How do you clean it?

HARRY: I don't know.

JIM: Why does everything have to be cleaned? You guys are determinod that this is going to make a mess. Let's think clean! Couldn't we avoid getting things dirty in the first place?

ROGER: Only by not letting the juice touch anything.

JIM: Okay, how do we do that?

RAY: Put it in a bag. Put the bag and all through the wringer!

ROGER: Okay, now we've got juice, pulp, skin, and all in a bag. Now what do we do?

HARRY: Pour it through a strainer into a glass.

JIM: Watch out for set!

RAY: The bag would have to be disposable or you'd have *it* to wash. If the bag is disposable, why not have it perforated?

HARRY: The juice would leak out all over.

RAY: Okay, put a spout in it—let the juice run out into the glass. Then throw the whole mess away, bag and all.

ROGER: Yeah! A disposable plastic container with a spout in it. Put the orange in it—maybe even sell the oranges in the bag! Put the bag in the machine. The whole works goes through the wringer, the juice comes out, then throw bag and all away. Beautiful!

JIM: How much will the bags cost?

RAY: That could be a problem. They'll have to be cheap, and yet strong enough not to break.

HARRY: You know, the wringer thing sounds expensive, too. Could we do that cheaper?

JIM: Well, I suppose we could just squeeze it in one of these scrap-metal balers.

RAY: Sure, that would work!

HARRY: We'll have to build this crazy thing—it just might work.

JIM: How else do we get juice—liquid—out of things?

ROGER: We heat them.

RAY: We could boil the juice out.

ROGER: Could we do it chemically?

JIM: The Food and Drug people just tuned in. Just about the only things you can touch food with is water or salt.

HENRY: Salt—

Let's stop the discussion at this point. You can see how they work. Superimposed on some of the brainstorming ideas is the conscious use of inversion, empathy, and analogy. They think freely. They let one idea lead to another. They criticize, or judge, mildly, as is not done in brainstorming —but no one lets his ego get involved in the criticism. Brainstorming is a way of collecting a large number of ideas for future consideration. The session above was more—they wanted a solution now. Either method, or both, is valid and if properly used will often lead to new, good ideas.

3.6 COMPUTER-PRINTER PROBLEM

Most computer printers work by having a hammer, which is activated by a magnet, strike a piece of type which then impacts the paper. If an inked ribbon is interspersed between the type and the printer (just like a typewriter) the desired record is obtained. Several years ago, printer speed became a factor limiting the speed of computers. The printing was a relatively slow mechanical process compared with the speed of the electronics. How to speed up the printing process became a very important problem, and the engineers involved have shown a great deal of inventiveness in solving the problem. Printers today are several orders of magnitude faster.

Without going into detail on this problem here (the details are too many and varied), it is possible to describe two inventive ideas and show how the techniques being discussed could have led to them.

One of the first steps in increasing printer speeds came as a result of inverting the process, in this case by making moving parts stop and stationary parts move. Instead of driving the print into the paper, the engineers drove the paper onto the type! The reduction in inertia of the parts moved, and the time saved in waiting for the parts to settle down after an operation, enabled speeds to be greatly improved. Note how conscious use of inversion could have led to this idea.

The second idea, which we can describe conceptually, came because of a failure of an old part to function when speeds were increased. The type pieces, over a hundred characters, were formerly fastened to a chain, very much like a bicycle chain. This chain was mounted on two sprockets that turned the chain with the type around and around. As the type traveled past, the magnet-operated hammers were activated selectively as desired by the electronic circuit. When the speed of the operation was greatly increased, the chains began to break. A new way had to be found to carry the type on its never-ending circular path.

Suppose you are a piece of type no longer fastened to a chain. How could you do your job? Looked at in this way, your job is to stay in line and keep moving. Solution: Put the pieces of type on a track, like a train, only they don't even need to be fastened together. Drive only the "engine." If the whole closed track is filled, all the type must follow. A whole chain is unnecessary. The idea worked. The chain was simply eliminated.

What next? Back to driving the type? Vertical type chains instead of horizontal? At an angle? Air-powered hammers? Analogy, empathy, inversion—these may well help get the key ideas along the way.

3.7 WATER

As a final example, a group brainstormed the water problem. Water shortage presents a very serious problem in parts of America and in many parts of the world. As nations industrialize, and as populations grow, it will become an even more critical problem. More and more good, fresh water will be needed at more and more places.

GEORGE: Rainwater.
GORDON: Everyone could have a rain barrel.
DON: Or catch it on his roof.
STAN: A big funnel over every house when it rains.
GORDON: A roof that stores rain.
GEORGE: Seeding the clouds to make it rain.
STAN: How about catching the clouds in a big bag and taking them to where we want them?
DON: Could we have a magnetic or electrical "bag"?
GORDON: Big fans on the ground to control air flow.
GEORGE: Big hot or cold spots on the ground to create up or down air currents.
STAN: Build a big wall, like a mountain range, to control air flow.
DON: You could put louvres in it that would open and close to provide some control.

GEORGE: How about saving the winter's snow—we always have too much water in spring—not enough later.

STAN: In a big refrigerator.

DON: Or just in a big pile.

GEORGE: Hell, let it melt—but put it in a big reservoir first.

GORDON: Build more regular dams and reservoirs.

STAN: How could we stop them from losing so much water by evaporation?

DON: Cover them over with a radiation shield.

GORDON: Enclose them.

STAN: Freeze them.

DON: Make them deeper—reduce the ratio of surface area to volume.

GORDON: Put a chemical film over them.

GEORGE: Dig more wells.

STAN: Find a way to collect ground water before it runs off.

DON: Little reservoirs on every street—on every lot.

GEORGE: Regulate lawn sprinkling.

STAN: Make people use automatic lawn sprinklers that won't overdo it.

DON: Find a substitute for water for things like shaving, toilets, washing clothes—

STAN: I like the substitute idea. We've got electric shavers, chemical toilets, dry cleaners—what else?

GEORGE: Alcohol!

DON: Beer!

STAN: Wine!

GORDON: Who needs water? Let's take a break!

DON: Lots of things need water. Industries need lots of it—grape vines, beer breweries—

GEORGE: A toilet that uses less water—a lot less.

DON: A washing machine that reuses water.

STAN: A drinking fountain that reuses water.

DON: A dishwasher that reuses water.

STAN: We need a water purifier.

GORDON: Yeah, for polluted water or for seawater.

DON: Boil it.

STAN: Freeze it.

DON: A membrane.

GORDON: Use chemicals.

GEORGE: I'll bet a lot of water is wasted in drippy faucets. How about a leaky-faucet signal?

STAN: Every house should have one.

GORDON: How about having two kinds of water—one for drinking, one for other things?

GEORGE: Water could be used two or three times before being put in the sewers.

DON: Trees. Trees use a lot of water. What does it do for them?

GEORGE: I don't know. How about other plants?

STAN: Or animals?

GEORGE: Could they live on something else?

STAN: My old school book called water the "universal solvent." Why couldn't we make another "universal solvent"?

GORDON: Water is used in air conditioners.

STAN: Air-cooled condensers.

GEORGE: More efficient cooling towers.

GORDON: A waterless cooling system.

STAN: Cover the ocean to prevent or control evaporation.

GEORGE: No, we want to speed up evaporation from the ocean. There's only so much water in the whole earth system. What we want to do is speed up the rate of circulation.

STAN: Right! How can we speed up evaporation from the ocean?

GEORGE: Heat it.

DON: Stir it. Make it bubble up—like a whole bunch of fountains.

GORDON: Put something in it to increase its vapor pressure.

STAN: Keep the air about it turbulent—move the moist air away quickly.

DON: Pipe air in from the desert.

This, of course, could go on forever. An hour or so of brainstorming can result in a long list of ideas, some of which can themselves be brainstormed later, such as the seawater-conversion plant mentioned in the above session. Remember the rules: do not criticize; think freely—the farther out the better; let the ideas flow. These rules aren't easy to follow, so remember to practice!

3.8 SUMMARY

This will conclude the discussion of inventiveness. In the next chapter begins a more detailed study of engineering analysis. The first step in the solution of most engineering problems is to get an idea for how the objective might be accomplished. The whole rest of the problem-solving process can be only as good as the idea or concept conceived for a solution. This is engineering and it is creativity. Engineering creativity is inventiveness.

In Chap. 2 the discussion of inventiveness began with a statement that heredity, environment, and past training have limited the creativity of us all but that few people are working up to the natural limit. Thus all people can try to improve their creative performance. Chapter 2 discussed the problem from a psychological viewpoint, and it presented some specific techniques for increasing creative performance: brainstorm, break the set, use inversion, analogy, empathy, take advantage of set-breaking experiences if possible, etc. Chapter 2 also told about some general things that can be done: practice being perceptive, avoid judgment until necessary, practice forming new relationships, etc. In Chap. 3, some examples were outlined to show these techniques in action. None of the techniques are panaceas, but they should show that there is no need to despair when in need of an idea. There are definite things to be done.

A final repetitious word: Practice and hard work are required to make these methods work. The examples presented in the chapter were all pretty successful problem-solving sessions. They all "clicked." It is not always so, especially with inexperienced participants. Sometimes nothing happens. Sometimes the sessions are fatiguing and boring. Improvement comes with practice, however, and as the examples show, the various techniques do work and often lead to fruitful results.

PROBLEMS

3.1 Point out places where the various inventive mechanisms are used in the home juice-squeezer example in the chapter.

3.2 Brainstorm the problem of air pollution.

3.3 Brainstorm the problem of automobile accidents and casualties.

3.4 Consider the problem of making an automatic ice-cube maker.

3.5 Consider the problem of designing a way to measure surface tension of liquids. Viscosity of liquids.

3.6 Consider the problem of getting people from floor to floor in a department store.

3.7 Consider the problem of window cleaning.

3.8 Make up some problems of your own and solve them.

3.9 Brainstorm the problem of training engineers.

3.10 Brainstorm the world's food problem. (Despite the temporary abundance in the United States much of the world is undernourished.)

3.11 Consider the problem of a landing gear for a rocket to land on the moon. Remember, the nature of the surface is unknown.

3.12 Brainstorm the problem of absorbing energy in a collision.

3.13 Brainstorm the problem of storing energy.

3.14 Toy rockets can be designed to use water and air as a fuel. They are filled about one-third to two-thirds with water and pumped to, say, 50 psia. They work very well, going 200 or 300 ft high! Consider the problem of making a toy two-stage rocket of this type.

3.15 Brainstorm the problem of helping engineering students to like new and poorly defined problem situations better.

3.16 Brainstorm the problem of teaching engineering students to be more creative.

II
ENGINEERING ANALYSIS

4
INTRODUCTION TO ENGINEERING ANALYSIS

4.1 INTRODUCTION

Engineering analysis, as described briefly in Chap. 1, is the process of applying basic principles to problems in order to arrive at meaningful answers in a reasonable time. The key phrases here are *basic principles, meaningful answers,* and *reasonable time.* In performing engineering analysis, the engineer has to be aware of the constraints on the problem-solving situation. He has to know, for example, whether "meaningful" means within 100 percent, 10 percent, or 1 percent of the correct (that is, nature's) answer. He also has to know whether "reasonable time" means a day, a week, or a year. He has to be aware of his own limitations—and strengths—and of the facilities for computation and experiment available to him. It is a basic viewpoint of this book that no engineering analysis problem has a "right" solution or a "right" method of solution except in the light of the constraints placed on the problem solver. Good problem solvers choose methods which are consistent with their goals and problem-solving constraints. From this chapter through Chap. 10, this book will be concerned with developing a good analysis problem-solving methodology which stresses the use of basic laws (as distinguished from specialized knowledge), the techniques of making assumptions, and the need for awareness of problem-solving constraints.

4.2 THE ENGINEERING ANALYSIS METHOD

No single methodology can handle all situations perfectly. To be useful, a methodology must be used flexibly as a guide, not dogmatically as a ritual. A methodology can call attention to the high spots and to the rough spots which give trouble to many problem solvers. It can provide a checklist,

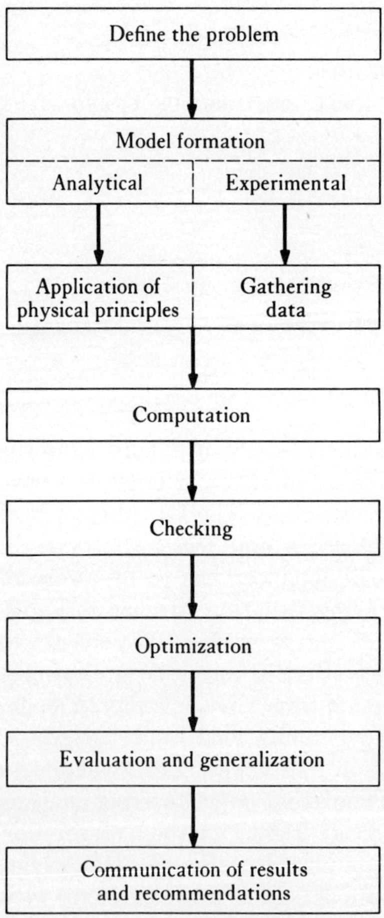

Fig. 4.1 Block diagram of the analysis process.

a port in a storm, and a starting place for students to grow from. But do not expect a "plug and chug" problem-solving equation. There is no such thing. Thinking—good, clear, hard, creative thinking—is required at every step. Knowledge and understanding are required. A problem-solving methodology is a mobilizer of existing skills, not a substitute for them. In Fig. 4.1 a step-by-step outline of a good engineering problem-solving approach is diagrammed. How does it agree with your own ideas on the subject?

Defining the Problem Operationally and Specifically Though this sounds like an obvious first step, it is often the step upon which much failure rests. An operational question is one which can be answered in terms of parameters that can be calculated or measured. In other words, a question must be asked that can be answered quantitatively. This is not always easy. It involves translation of a real physical situation into a paper-and-pencil question. For example, it is not enough to ask of a system, "Will it work?" Such a question is neither operational nor specific. Instead, questions must be asked such as "Given an input of 5 Btu/sec, what will be the power output?" Defining the problem specifically and operationally requires skill, understanding, and judgment. Usually anybody can define problems generally (e.g., "Will it work?" "Will it get too hot?" "Will the vibration settle down quickly?"). But good engineering method calls for first determining objectively what such generalizations really mean. What temperature is "too" hot? What does "quickly" mean in seconds? An engineering analyst must begin by defining a *quantitatively answerable question.*

Model Formation and Making Assumptions Model formation is another step in the problem-solving process which causes much trouble. A model is an idealized approximation to the real situation. Forming a good analytical model involves making assumptions that take into account the relative importance of the various elements of the problem. How many times students say to professors, and young engineers to senior engineers: "But how did you know to make that assumption?" Sometimes it is simply lack of experience that causes poor assumption making. But many times the cause is the failure to consciously make an analytical model of the situation, with the result that the possibility of the assumption does not occur to the student or young engineer. When asked, most young engineers will make fairly good assumptions most of the time. But when working alone, one has no teacher or senior engineer to bring up the question. If, however, a clear model of the real situation is first made, it will call attention to the critical questions about assumptions. Figure 4.2 shows a schematic view of the role of model formation in engineering analysis.

Analytical models are not the only models used to solve problems.

Many problems are more quickly and easily solved by making an experimental model. The experiment may not exactly duplicate the real physical situation—hence it is a model—but it may still yield pertinent results.

Of course, many problems require combined analytical and experimental models. Often special experimental data are needed for use in theoretical expressions—or theoretical work may suggest the kinds of experiments most useful.

Model formation is a process of abstraction. The model is not real. It is a figment of the engineer's imagination. The trick is to form models

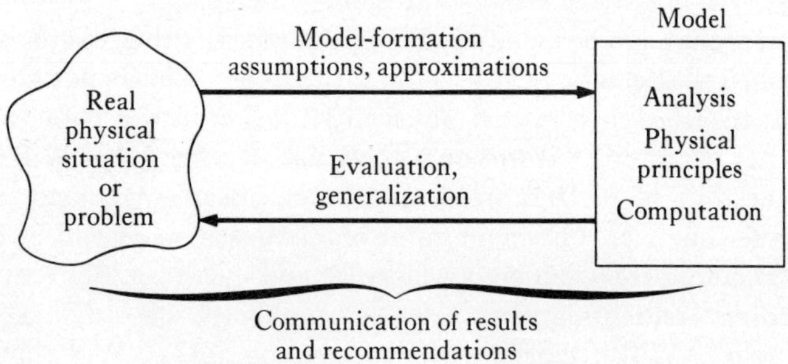

Fig. 4.2 Schematic of the analysis process.

which are simple enough to solve but which still contain the essence of the problem so that the results are meaningful.

Application of Principles or Gathering Data Once an analytical model is formed, then use can be made of science and engineering science knowledge and understanding. Newton's law or the first law or the momentum equation or another appropriate principle is applied to the model. Students get much practice at this in their specialized course work. A good rule, not always stressed in course work, is: Always use the *most basic principle* available for use in any given problem. Use special and complex techniques or equations only when absolutely certain they apply and when something simpler will not work.

When an experimental model is used, the next step is data gathering. Laboratory courses are intended to teach the necessary skills for this. They should also teach how to do it quickly and cheaply. Remember, there are always problem-solving constraints—time, money, facilities, etc.

Computation With the theories applied and the equations written, the problem is now one of getting a numerical answer. When this can be done analytically, fine. Do so. But when it can't—quite often the case in new, real problems—one still *must* get a number in reasonable time at reasonable cost. Graphical solutions are often very helpful in such cases. And, of course, the computers now available make numerical analysis quite common. Modern engineers must develop facility with numerical methods of getting answers. If high-powered math fails (sometimes it will), one can still go on. Remember, engineering analysis usually has *got* to get a number by *some* means.

Finally, if the model has been experimental, at this stage the data must be analyzed. This will often require statistical or dimensional analysis.

Checking Checking should occur at every step of the way, not all at once at the end. In general, there are two kinds of checks which can be made: mathematical checks and engineering-sense checks. Certain mathematical checks, such as checking arithmetic, ensuring correct equations, etc., are common, and everyone is used to making these. Dimensional checks are also mathematical in nature but most students do not use them nearly enough. *Every equation must be dimensionally consistent.* Boundary conditions also must be satisfied. Checking them is a good way to catch both careless and fundamental errors.

Engineering-sense checks are also very helpful—and very seldom used by students. Do derived equations look right? If one factor is increased, do the other factors behave as expected? Limit checks are very good engineering-sense checks. If one factor approaches some limit (zero, infinity, some other factor), does the equation behave as it should? Are there other factors which should be important but which do not appear in the equation? Has everything important been considered? Finally, do the numbers make common sense? One student reported that a *toy* rocket would go 7,200 ft high!

Checks are important. They can save a great deal of time, money, and embarrassment. At one company, a saying among engineers is: "There is never time to do it right, but always time to do it over." Leave a little time for checking. It's usually cheaper to do it right instead of over.

In this regard, another point should be made. Paper is cheap. Mistakes are expensive. Use lots of paper, one side only. Keep things neat.

Evaluation and Generalization The steps called problem definition and model formation are seldom discussed in engineering science courses. These courses quite properly concentrate on the application of principles and on the computation steps. Following computation come two more steps also seldom discussed in ordinary subject-matter courses: evaluation and generalization. Now that a number has been obtained for an answer, what does it mean? Also, can generalizations be made which give more than just the answer to the specific problem? These are important steps because, in a way, without them properly done the whole solution has been done in vain. In the examples which follow in this chapter and the next, evaluation and generalization will be elaborated upon.

Optimization The evaluation may call for an optimization, or optimization may have been a part of the analysis and computation. Optimization will be treated in a later chapter more completely. It is a factor which, like checking, permeates the entire design, analysis, and decision process.

The steps of engineering analysis have been presented in linear form. Actually it seldom happens that way. There is continual iteration. Plans are changed. It may take several iterations before an answer is obtained which is satisfying. Let it be emphasized one more time: The whole problem-solving process is subject to constraints of time, money, facilities, and ability. Optimizing the problem-solving process, subject to these constraints and for the purpose of getting an answer of desired accuracy, is the subject of this book. There is no "right" answer to a problem except in the light of the objective and the constraints. Good problem solving requires practice.

Presentation and Communication of Results and Recommendations One's work is not done when the results are known. They must be communicated to others. Recommendations often must be formed. These, too, must be communicated to others. Communication is another process to be optimized. Efficient communication must take into account the circumstances (constraints) as well as the information to be delivered.

Spelling and grammar are a part of the communication process. Poor spelling or incorrect English usage will tend to destroy the confidence of the reader or listener in the ability of the writer or speaker. Furthermore, effective and appropriate use of words can greatly enhance both a person and his ideas. This is not phony. Just as an idea is not creative until it is put into practice, neither is a recommendation any good until it is accepted. The ability to communicate is vital to the successful engineer.

A 100-page detailed report is often necessary for the engineering files. Someone, someday, may wish to redesign or make other use of work already done. But a much briefer report is usually more appropriate for communicating at times of decision making. (The author once worked for a man who insisted on reports of one page only. He refused to read even two-page reports! This executive wanted engineers to give him the *essence* of the results and their recommendations. If he wanted details, he would ask. Another employer would not accept written reports at all. He wanted brief verbal summaries of recommendations. Then he would have the engineer explain away every objection he could think of.) The point is this: An engineer must communicate the essence of his results and recommendations in a form which can be quickly and clearly understood by the *intended* reader or listener. These short reports are hard to write or prepare—much harder than just putting everything down.

In engineering analysis, finding out what the problem is, is the first step, and communicating the solution is the last. Both steps are engineering, as much as is the use of physical principles in between. This chapter now concludes with a case study intended to illustrate the engineering analysis process in action.

4.3 A CASE STUDY IN ENGINEERING ANALYSIS: VISCOUS LUBRICATION IN WIRE DRAWING

The process of wire drawing is an important one in the metal-working industry. Drawn wire is used to produce nails, coat hangers, electrical wire, and a thousand other products. The process itself is simply that of making a longer, thinner wire from a shorter, fatter one. Usually the lengths involved are so long that the process can be considered essentially continuous. Speeds vary about 100 fps for small, soft wire to about 1 fps for large, hard wire.

Lubrication is very important to the wire-drawing process, as it is to all metal-working processes. In wire drawing, lubrication conventionally is done by allowing the wire to pass through a "soapbox" just prior to its passing through the die. The soapbox contains a dry, powdered mixture of various hydrocarbons and often contains the equivalent of ordinary laundry soap. Companies, and even individual machine operators, have their own mixtures. Lubrication in wire drawing is very much an art today.

Die wear is a major expense in producing drawn wire, not only because of the cost of the dies, but also because of the production lost during frequent die changes.

The kind of lubrication which results when a soapbox is used is called *boundary lubrication.* It is characterized by the presence of "slippery" stuff (e.g., soap) between the moving surfaces. The slipperiness is caused by chain molecules in the lubricant which orient themselves normal to the surfaces. Since such molecules are weak in shear, little resistance is offered to the sliding motion. Unfortunately, it is difficult to get a very thick film of such lubricants between the surfaces, and hence the separation of the surfaces is not large. A close-up of the situation might look schematically as shown in Fig. 4.3a. In boundary lubrication, surface-to-surface contact is not totally prevented, because the separation of the surfaces is of the same magnitude as the surface roughnesses.

A newly proposed scheme (Christopherson and Naylor, Promotion of Fluid Lubrication in Wire Drawing, *Proc. Inst. Mech. Engrs.*, vol. 169, p. 643, 1955) would provide for *full-film* or hydrodynamic lubrication of the wire in the die (Fig. 4.3*b*). Separation of the surfaces is complete in this case, and the lubrication results from fluid mechanical effects, just as it does in ordinary oil-lubricated journal or thrust bearings. Students unfamiliar with this kind of lubrication should take a few minutes to look at a standard text on the subject (e.g., "Lubrication of Bearings," by Radzimovsky, or "Engineering Applications of Fluid Mechanics," by Hunsaker and Rightmire).

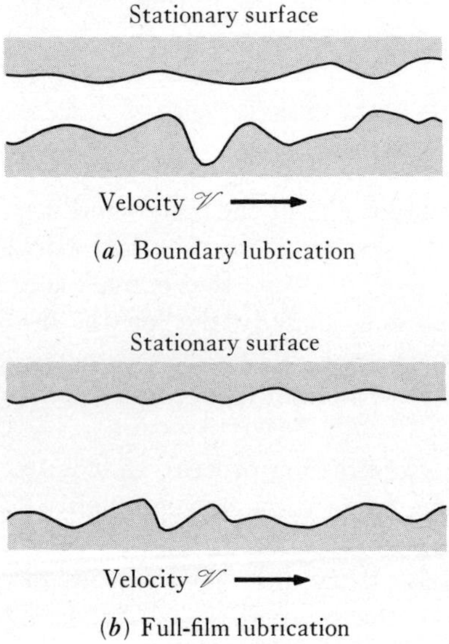

Fig. 4.3 The difference between boundary lubrication and full-film lubrication.

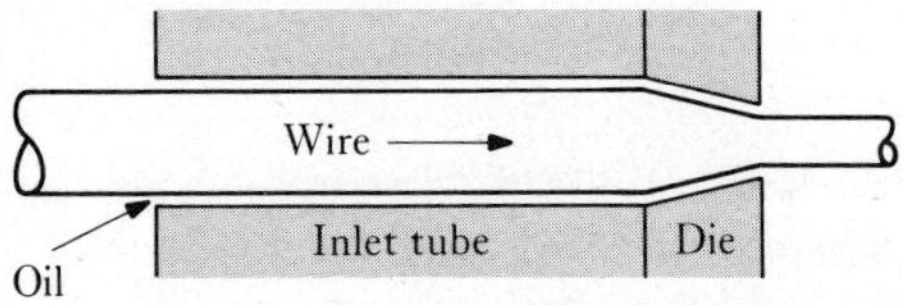

Fig. 4.4 A proposed scheme for providing viscous lubrication in wire drawing.

Two ideas for providing lubricant under high pressure to the die inlet have been proposed. One is to use a pump. The other is to install an *inlet tube*, as shown in Fig. 4.4. The inlet tube would be a close-tolerance tube and would, if practical, have the advantage of eliminating the pump, piping, and other complications associated with the pump idea.

Let us assume that a group of engineers has been assigned to this problem area. One engineer is to look into the pumping possibility. Another is to study the wire and lubricant in the die. And one, the one to be followed in detail here, is to investigate the inlet-tube idea. Let us say that they have agreed to meet with the section head not later than one week from now to discuss their findings and decide on what, if anything, should be the next step.

That which follows now is the engineer's attack on the inlet-tube portion of the problem. In the pages which follow, the right page contains his notebook writings. The left page is a kind of train of consciousness showing you his possible thought process. Of course, it is condensed considerably. However, the purpose is to illustrate the engineering analysis process, and a condensed version can do that more effectively. There are two ways to go about reading the case study. You may read it through, omitting a detailed study and comprehension of each technical step and equation. Alternatively, you may go through it step by step, thereby reviewing or learning some fluid mechanics. In either case, however, be sure that you understand how the engineering analysis process is being used by this engineer as he proceeds through his problem.

At certain points in the case, you will be asked to make the next step or answer a question pertaining to the problem on your own. (This will be true in later examples, too.) The steps or answers are always given to you on the next page in the text, but you can benefit a great deal from honestly and conscientiously attempting to answer the questions before looking at the author's suggested answer.

"Well now, let me see, I've got to look at the inlet tube. Better draw a good picture of this thing, What's supposed to happen? Oil enters at o, gets dragged into the tube, and supposedly the pressure will go up like this." (Pressure sketch.) *"Don't suppose that's a straight-line—linear—increase. Why should the pressure go up? I'll have to look at the fluid flow in that annular space between wire and tube. Find out when—under what conditions—speed, viscosity, oil thickness, etc.—the pressure will rise if at all."*

"Is that my problem—to find out what makes the pressure go up and when it will? So what if it goes up—I think it will under the right conditions—again, so what? I've got to find out if this screwball idea is practical. Practical. What does "practical" mean in this case? Just having the pressure go up doesn't make it practical. Go up to what? How high does the pressure have to get? Got to see John (the engineer assigned to the die portion of the problem). *He's got to say what pressure he needs at the die inlet—because my old inlet tube has to deliver it to him. Hey! He also has to say what oil flow rate we need. I can't look just at the inlet tube—John and I are going to have to work together a little here."*

(Note how the engineer is struggling here with *problem definition*. He has to translate the vague situation into a quantitative question which, when answered, will provide pertinent information about the feasibility of the scheme.)

(A conversation with John at this point results in the conclusion that the oil pressure at the die inlet will have to be sufficient to yield the wire material. Since there will probably be little tension in the wire at this point, the pressure will have to be approximately equal to the compressive yield stress of the material, σ_y. σ_y is about 60,000 psi for steels, about 20,000 psi for aluminum, and about 6,000 psi for copper. These numbers, it is agreed, will serve as first approximations. John cannot yet say anything about the necessary flow rate. He has quite a lot more analysis to do before this can be determined.)

"Okay, the pressure I need to deliver is σ_y, as the case may be. I'll just have to let the flow rate be a parameter until John can come up with some numbers. Might as well keep it all general and just use σ_y until time for getting some numbers. Now let's see—where was I? Oh yes, what does "practical" mean in this problem?"

"Well, first of all it means that the tube must not be too long. If the tube is too long, the whole idea is no good. That's one thing I need to determine: How long must the tube be in order to get σ_y at the die inlet?"

(At this point the reader should attempt to define the problem specifically before going on to the next page.)

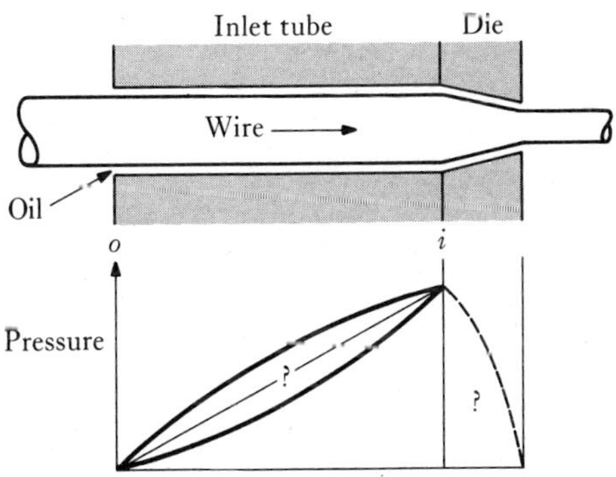

Pressure needed at die inlet $i \approx \sigma_y$

For steel	$\sigma_y = 60{,}000$ psi
For aluminum	$\sigma_y = 20{,}000$ psi
For copper	$\sigma_y = 6{,}000$ psi

Flow rate unknown? John working on this.

"This problem statement isn't really specific enough yet, however. I need to specify more conditions which affect the length: oil viscosity, speed of the wire, clearance between wire and tube—maybe even some more things."

"Maybe instead of trying to solve a specific case, I should look at the general situation." (See new sketch)

"What I need is a relationship among all these parameters. Let's be specific. I need to find the relationship among μ_0, Q, U_0, h_0, D_0, L, *and* σ_y. *Then if I pick values of* U_0, D_0, h_0, μ_0, σ_y, *and John gives me* Q, *I can compute* L."

(Note how the engineer now has his problem defined quantitatively.)

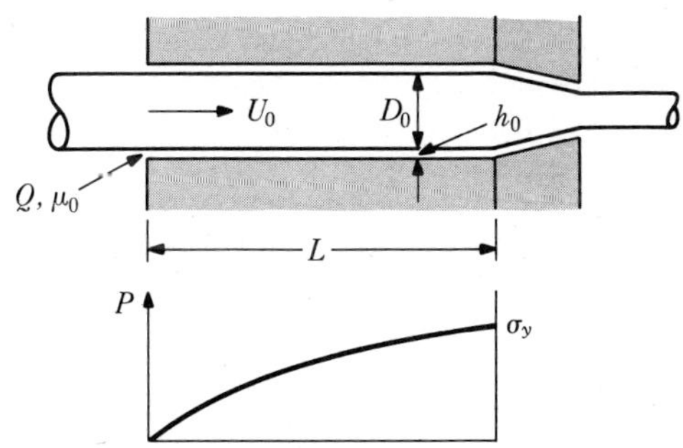

Problem: Find the relationship among μ_0, Q, D_0, h_0, L, U_0, and σ_Y.
Plan: Analyze fluid flow using momentum and continuity.

"Find the relationship among μ_0, Q, U_0, h_0, D_0, L, *and* σ_y. *That's my problem. How shall I begin? This is a fluid-flow problem. I had better begin by analyzing the fluid flow of the oil in the inlet tube, using the momentum and continuity equations. I'll do that and see where it leads."*

"I need a new picture to do this fluid analysis. I'll have to make some additional assumptions. In fact I'd better make a clear model of the problem and list all the assumptions and approximations.

(The engineer, having defined his problem, is now forming a model.)
(What assumptions would you make? List yours before going on.)

"My sketch assumes that the wire is in the center of the tube."

(Note how easy it is to make an assumption without realizing you've made it: Watch out for this.)

"Why should it be in the middle? The die looks like a pretty good centering device. I'll assume it's in the center—concentric—but I'd better discuss this with John. What if it's not in the center? Would pressure rise faster or slower? I don't know! The center assumption ought to be okay for a start, but I've got to look at this some more later."

"Viscosity will rise with pressure and fall with temperature. There'll probably be some viscous heating. At these pressures and high speeds, both pressure and temperature effects may well be important. But these are opposite effects. I'll let it go for now but must come back to this assumption, too."

"Laminar flow looks good to me. Let's see, I can estimate the Reynolds number—sort of—even now. Oh, no I can't. What is a good value for h_0? μ_0 *for oils is about* 10^{-3} ft^2/sec. *Wire-drawing speeds are maybe* 1 *to* 100 *fps, say* 10. *That makes* $Re \approx 10^4 h_0$. *If* h_0 *is even* 0.1 *ft—oh, it'll be more like* 0.01 *ft, so* $Re \approx 10^2$. *Laminar flow is good."*

(Note that for flow between flat plates, the length parameter in the Reynolds number is to be taken as the distance between the plates. Note also how the engineer has made an order-of-magnitude analysis, very quickly, to justify his assumption.)

"Let's see. The curvature can be ignored if h_0 *is small compared with* D_0. *That will certainly be the case."*

"If the tube or wire deforms elastically, as they certainly will as the pressure increases down the tube, h_0 *will get larger. This will tend to increase the length required to get a given pressure. But is it significant? This doesn't seem as important as the possible pressure and temperature effects on the viscosity. I'll come back to it, though, later. It seems good as a first approximation."*

"Okay, is that it? I guess so."

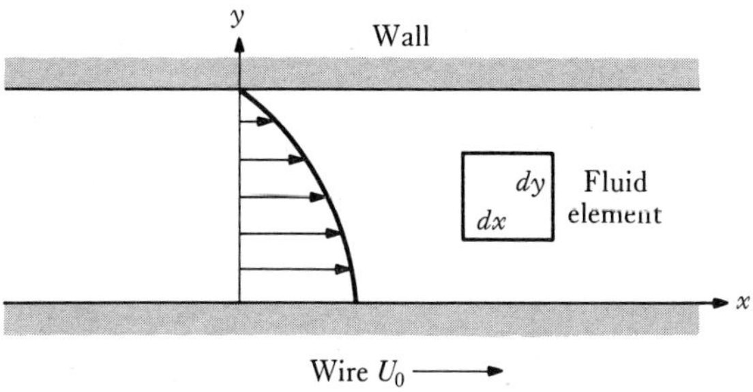

Assumptions:

1. Wire is concentric in tube. (See John about this.)
2. Viscosity is constant—neglect p and t effects.
3. Flow is laminar. ($Re = \rho VD/\mu = Vh_0/\mathcal{V} = 10^2$)
4. Curvature may be neglected. ($h_0/D_0 \ll 1$)
5. Elastic deformations of wire and tube are negligible.

"Now on with the plan: apply momentum and continuity to fluid flow to see if I can get a relationship among Q, h_0, μ_0, D_0, L, σ_y, *and* U_0.*"*

"Let's make a nice big sketch of that fluid element—and put in forces."

Application of momentum to fluid element:

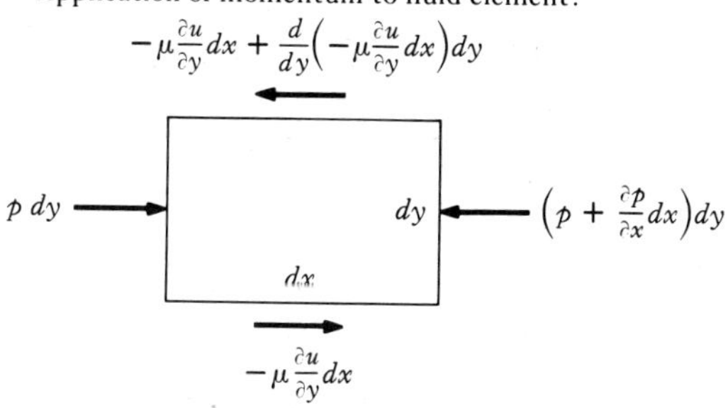

"Now neglect y component of velocity. This is consistent with laminar-flow assumption. I'm assuming fully developed flow, too. No inertia. But what about entrance region? Must neglect now, but better check on it later."

(With the model now formed, the engineer is applying physical principles.)

"Better check the dimensions of this equation—they check."

$$p\,dy - \left(p + \frac{\partial p}{\partial x}\,dx\right)dy - \mu\,\frac{\partial u}{\partial y}\,dx - \left[-\mu\,\frac{\partial u}{\partial y}\,dx + \frac{\partial}{\partial y}\left(-\mu\,\frac{\partial u}{\partial y}\,dy\right)\right]dx = 0$$

$$-\frac{\partial p}{\partial x} + \mu\,\frac{\partial^2 u}{\partial y^2} = 0$$

$$\frac{\partial p}{\partial x} = \mu\,\frac{\partial^2 u}{\partial y^2}$$

$$\frac{\mathrm{lb}_f/\mathrm{ft}^2}{\mathrm{ft}} = \frac{(\mathrm{lb}_f)\,(\mathrm{sec})}{\mathrm{ft}^2}\;\frac{\mathrm{ft/sec}}{\mathrm{ft}^2}$$

$$\frac{\mathrm{lb}_f}{\mathrm{ft}^3} = \frac{\mathrm{lb}_f}{\mathrm{ft}^3}$$

"Now does the thing make sense? It says that the pressure gradient is proportional to the second derivative of the velocity. If $\partial p/\partial x = 0$*, then* du/dy $=$ *constant—that would be a linear velocity distribution, like between two plates. That's as it should be. Also, the proportionality constant is the viscosity. The bigger* μ*, the bigger* $\partial p/\partial x$*. Okay. How about the sign? A positive* $\partial p/\partial x$ *means a positive* $\partial^2 u/\partial y^2$*. That gives a velocity like this."* (See sketch, line A.) *"Looks okay. Say, it could even go negative."* (See sketch, line B.)

"Wait, this can't be integrated unless $\partial p/\partial x$ *is independent of* y*. Of course, it will be in such a thin film. Also,* p *is a function of* x *only."*

"Differentiate to check integration. Checks. Integrate again. Differentiation checks. Now what are the boundary conditions?"

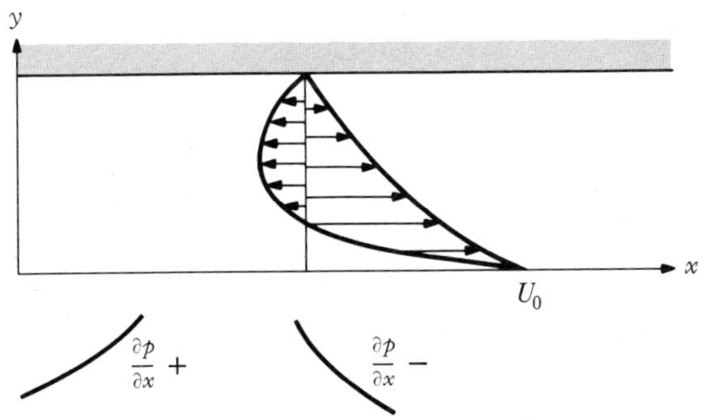

Additional assumption:

6. p and $\dfrac{\partial p}{\partial x}$ are $f(x)$ only.

$$\mu_0 \frac{\partial^2 u}{\partial y^2} = \frac{dp}{dx}$$

$$\mu_0 \frac{\partial u}{\partial y} = \frac{dp}{dx} y + f_1(x)$$

$$\mu_0 u = \frac{dp}{dx}\frac{y^2}{2} + yf_1(x) + f_2(x)$$

Boundary conditions:

(1) At $y = 0$, $\quad u = U_0$

(2) At $y = h_0$, $\quad u = 0$

"The first boundary condition gives

"Then the second gives

"Now the whole equation is

"Solving for u

(Now, what should the problem solver do next?)

Boundary conditions:

(1) $\mu_0 U_0 = f_2(x)$

(2) $$0 = \frac{dp}{dx}\frac{h_0^2}{2} + h_0 f_1(x) + \mu_0 U_0$$

$$f_1(x) = -\frac{dp}{dx}\frac{h_0}{2} - \frac{\mu_0 U_0}{h_0}$$

$$\mu_0 u = \frac{dp}{dx}\frac{y^2}{2} - \frac{dp}{dx}\frac{h_0 y}{2} - \frac{\mu_0 U_0 y}{h_0} + \mu_0 U_0$$

$$u = \frac{1}{2\mu_0}\frac{dp}{dx}(y_2 - h_0 y) + U_0\left(1 - \frac{y}{h_0}\right)$$

"Now let's check this result. First, dimensions."

(Notice how the engineer continually checks, using dimensions, boundary conditions, engineering sense, etc.)

"Next, check limits."

"Okay, now does it make sense? The second term, $U_0\,(1 - y/h_0)$, is a linear distribution." (See sketch, line A.) *"The other term $(y^2 - h_0 y)$ is parabolic and always negative and symmetrical."* (See sketch, line B.) *"Oh, the linear term is what we get with no pressure gradient and moving plates, and the parabola term is what we get with a pressure gradient, but no motion. Seems okay. They add up to give total results."*

"Okay, so now I have the velocity distribution. I ought to be able to get flow rate now."

"Wait, I have assumed $h_0 \ll D_0$. Let's just stick with that."

$$\frac{\text{ft}}{\text{sec}} = \frac{1}{(\text{lb}_f)(\text{sec})/\text{ft}^2} \frac{\text{lb}_f/\text{ft}^2}{\text{ft}} \text{ft}^2 = \frac{\text{ft}}{\text{sec}}$$

at $y = 0$, equation gives $u = U_0$

at $y = h$, equation gives $u = 0$

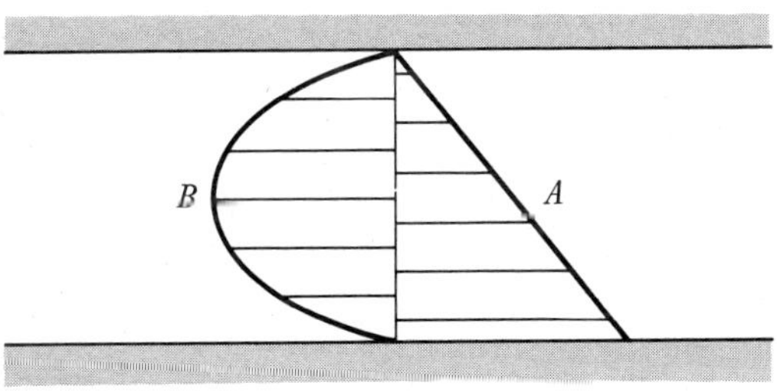

$$u = \frac{1}{2\mu_0} \frac{dp}{dx} (y^2 - h_0 y) + U_0\left(1 - \frac{y}{h_0}\right)$$

$$Q = \int_{D_0/2}^{D_0/2+h_0} \pi D u \, dy \qquad \text{for flow rate}$$

"There."

"Now let's check this. First, dimensions."

"Let's see now, does it make sense? The $\mu_0 u_0 D_0 h_0/2$ *term is the flow with no pressure gradient and is as it should be. The other term is the flow due to pressure gradient. It gets longer as* dp/dx *gets larger, which seems reasonable. The* D_0, h_0, *and* μ_0 *parts also seem correct. It sure is sensitive to* h_0*!"*

(Does the equation for Q seem correct to you? What other checks could be made?)

$$Q = \pi D_0 \int_0^{h_0} u\,dy$$

$$= \pi D_0 \int_0^{h_0} \frac{1}{2\mu_0}\frac{dp}{dx}(y^2 - hy)\,dy + U_0\left(1 + \frac{y}{h_0}\right)dy$$

$$= \pi D_0 \left[\frac{1}{2\mu_0}\frac{dp}{dx}\left(\frac{h_0^3}{3} - \frac{h_0^2}{2}\right) + U_0\left(h_0 - \frac{h_0^2}{2}\right)\right]$$

$$= \frac{\pi D_0 U_0 h_0}{2} + \frac{\pi D_0 h_0^3}{12\mu_0}\frac{dp}{dx}$$

$$\frac{\text{ft}^3}{\text{sec}} = \text{ft}\ \frac{\text{ft}}{\text{sec}}\ \text{ft} = \frac{\text{ft ft}^3\,\text{lb}_f}{\text{ft}^3}\ \frac{\text{ft}^2}{(\text{lb}_f)(\text{sec})}$$

$$\frac{\text{ft}^3}{\text{sec}} = \frac{\text{ft}^3}{\text{sec}} = \frac{\text{ft}^3}{\text{sec}}$$

Note: Very sensitive to h_0. Must get more information on practical values for h_0.

"Now how about the sign? If dp/dx is positive, the pressure should be pushing the flow back. That plus sign looks wrong. Seems like it ought to be minus. Better go back and look."

"Oh, yes, made an error. $\frac{1}{3} - \frac{1}{2} = -\frac{1}{6}$*. Okay."*

"This equation defines the limiting condition $Q = 0$*, which might be interesting later."*

"Check units. Okay."

"Well, now, let's not get all bogged down in these details unless they're taking me somewhere. What am I trying to do? Oh, yes, get a relationship among all the parameters. Well, we've got Q as a function of D_0, U_0, h_0, μ_0, *and* dp/dx*. What we really want is L and y—so we integrate! And it's real easy, too. You can't lose them all."*

Correction:

$$Q = \frac{\pi D_0 U_0 h_0}{2} - \frac{\pi D_0 h_0^3}{12\mu_0} \frac{dp}{dx}$$

If $Q = 0$: $\quad \dfrac{dp}{dx} = \dfrac{6\mu_0 U_0}{h_0^2}$

If Q positive: $\quad \dfrac{dp}{dx} < \dfrac{6\mu_0 U_0}{h_0^2}$

$$\frac{\text{lb}_f}{\text{ft}^3} = \frac{(\text{lb}_f)(\text{sec})}{\text{ft}^2} \frac{\text{ft}}{\text{sec}} \frac{1}{\text{ft}^2} = \frac{\text{lb}_f}{\text{ft}^3}$$

"Well, that looks like my relationship! Let's check some things. Units—okay. As σ_y *increases, L increases. Check. As* μ_0 *increases, L decreases. Okay. As* U_0 *increases, L decreases. Can't tell about* h_0*, but it seems okay."*

$$\frac{dp}{dx} = \frac{6\mu_0 U_0}{h_0^2} - \frac{12\mu_0 Q}{\pi D_0 h_0^3}$$

$$p = \left(\frac{6\mu_0 U_0}{h_0^2} - \frac{12\mu_0 Q}{\pi D_0 h_0^3}\right) x + C_1$$

At $x = 0, \quad p = 0 \quad \therefore C_1 = 0$

At $x = L, \quad p = \sigma_y$

$$\boxed{\sigma_y = \left(\frac{6\mu_0 U_0}{h_0^2} - \frac{12\mu_0 Q}{\pi D_0 h_0^3}\right) L} \qquad \text{Final result}$$

$$\frac{\text{lb}_f}{\text{ft}^2} = \frac{(\text{lb}_f)(\text{sec})}{(\text{ft}^2)(\text{ft}^2)}\frac{\text{ft ft}}{\text{sec}} = \frac{(\text{lb}_f)(\text{sec})}{\text{ft}^2}\frac{\text{ft}^3}{\text{sec}}\frac{\text{ft}}{\text{ft}^4}$$

$$\frac{\text{lb}_f}{\text{ft}^2} = \frac{\text{lb}_f}{\text{ft}^2} = \frac{\text{lb}_f}{\text{ft}^2}$$

"What else can I tell from this? What other limits? Let's see, dp/dx must be positive."

"Oh, of course!"

"Well, now, if John can tell me the Q he needs, I can find the length—if I know μ_0, U_0, D_0, and h_0. All of these will be easy to specify except h_0. What determines h_0? We want it to be small. The smaller the better. But, if it gets too small, there'll be all kinds of practical problems with alignment and crooked wire. I suppose h_0/D_0 of maybe 0.05 or 0.10 or 0.15 would be reasonable but there is no way to tell without an experiment. The whole process is pretty sensitive to h_0. We'll probably have to do that experiment eventually, but we can determine the range of values from my equation, so we shouldn't spend the time and money just yet."

$$\frac{dp}{dx} = \frac{6\mu_0 U_0}{h_0^2} - \frac{12\mu_0 Q}{\pi D_0 h_0^3} > 0$$

$$\frac{6\mu_0 U_0}{h_0^2} > \frac{12\mu_0 Q}{\pi D_0 h_0^3}$$

$$\boxed{Q < \frac{\pi D_0 U_0 h_0}{2}}$$

Restriction on Q for positive pressure gradient

Let us break into the plot of this engineering story at this point. Much work remains to be done. In particular, the engineer hasn't considered the possible optimization of his inlet tube, nor has he done a real job of evaluating and generalizing his results. Before he can do much of this he really needs some numbers from his colleagues. We are going to skip over some of his work and some of his colleagues' and interject you back into his mind and notebook at the point where the following determinations and decisions have been made.

1. Oil flow rates of approximately 10^{-4} ft³/sec will probably be satisfactory in the die.
2. An oil viscosity of 2×10^{-3} $(\text{lb}_f)(\text{sec})/\text{ft}^2$ will probably be satisfactory in the die.
3. An h_0 equal to approximately one-tenth of D_0 will probably be practical.
4. The following speed, yield stress, and diameter numbers are typical:

Case	*Wire material*	*Yield stress, psi*	*Diameter, in.*	*Speed, fps*
1	Steel	60,000	$\frac{1}{32}$	100
2	Aluminum	20,000	$\frac{1}{16}$	100
3	Aluminum	20,000	$\frac{1}{8}$	10
4	Copper	6,000	$\frac{1}{8}$	100
5	Copper	6,000	$\frac{1}{4}$	10

Actually, the engineer investigated more possible cases than shown above—the list is abbreviated here for purposes of clarity and to save space. You can well imagine the discussions, thinking, phoning, and reading which could go into making the above determinations.

Now our engineer has some numbers to use in his equation

$$L = \frac{\sigma_y}{6\mu_0 U_0/h_0^2 - 12Q\mu_0/D_0h_0^3}$$

He plugs and chugs and finds:

Case	*L, ft*
1	7.85
2	0.85
3	62.7
4	0.82
5	36.0

Earlier in his analysis he took the concrete, real situation and translated it into a model—a fictitious approximation. This model has now given him some real, concrete numbers. The question now is: What do these numbers mean with respect to the original real physical situation? Answering this question is the stage of analysis we have called evaluation. A related question is: What do these numbers mean or imply with respect to general physical situations of this type or of analogous types? This is the stage we call generalization. This is an important stage, the stage where learning takes place. Before rejoining the engineer now as he considers the numbers he has just computed, consider for yourself their implications. Then go on to the next page.

"Wow! Sixty feet for case 3. *No good at all! But let's look at these one at a time."*

"Case 1 *is definitely too long. Eight feet is an order of magnitude too large. Let's see, what could help?"* (He studies his figures.) *"A larger viscosity—well, that could be arranged within limits. A higher speed—not too much to be expected here, though reduced friction might make higher speeds possible. A lower flow rate—no, a higher flow rate. Would have to satisfy the restriction just noted, but this could be arranged. John certainly wouldn't object to more flow. Hey, with enough flow, I can make L as small as I want! That's a key to this thing!* h_0 *is another key, though. It appears squared and cubed. Boy, small changes in it are really critical. If* h_0 *could be just a little smaller, the length would be a lot shorter. Even though* 8 *ft is too long, I think this case has possibilities. And look, if* D_0 *(and hence* h_0*) is made smaller, the length would go way down."*

"Cases 2 *and* 4 *look good just the way they are. Less than a foot! Very good."*

"Cases 3 *and* 5 *look out of it. Bad news."*

"Now let's see—what's the reason? They both have slow speed, 10 *fps. And both are fairly large sizes."*

"Now let's see—what does all this mean? It means that this thing might just be feasible under certain conditions and that we ought to look at it some more, at least so far as the engineering aspects are concerned. It also means that probably only the small-sized, high-speed cases will work—but that those cases will work with any wire material. And all this is dependent on the practical problem of keeping h_0 *small—very small. A high-viscosity oil will help too."*

CALCULATION SUMMARY

Case no.	y, lb/ft^2	μ_0, $(lb)(sec)/ft^2$	U_0, fps	D_0, ft	h_0, ft	h_0^2, ft^2	h_0^3, ft^3	Q, ft^3/sec	$A=\frac{6\mu_0 U_0}{D_0 h_0^2}$, lb/ft^3	$B=\frac{12\mu_0 Q_0}{D_0 h_0^3}$, lb/ft^3	$L=\frac{y}{A-B}$, ft
1	8.64×10^6	2×10^{-3}	100	2.6×10^{-4}	2.6×10^{-4}	6.7×10^{-8}	1.8×10^{-11}	10^{-4}	1.8×10^7	1.69×10^7	7.85
2	2.88×10^6	2×10^{-3}	100	5.2×10^{-3}	5.2×10^{-4}	2.7×10^{-7}	1.4×10^{-10}	10^{-4}	0.45×10^7	0.11×10^4	0.85
3	2.88×10^6	2×10^{-3}	10	1.04×10^{-2}	1.04×10^{-3}	1.1×10^{-6}	1.1×10^{-9}	10^{-4}	11.3×10^4	6.65×10^4	62.7
4	8.64×10^5	2×10^{-3}	100	1.04×10^{-2}	1.04×10^{-3}	1.1×10^{-6}	1.1×10^{-9}	10^{-4}	11.3×10^5	6.65×10^4	0.815
5	8.64×10^5	2×10^{-3}	10	2.08×10^{-2}	2.08×10^{-3}	4.3×10^{-6}	9×10^{-9}	10^{-4}	2.82×10^4	0.42×10^4	36.0

Let's interrupt the engineer again. Notice the evaluation and generalization taking place. He started with a real physical situation and made a model of it, a model simple enough to be tractable but complex enough to be meaningful. His model gave him some numbers. Strictly speaking, the numbers apply only to the model. *By evaluation and generalization he now goes back to the real physical situation.* Schematically, it's just as shown in Fig. 4.2. A great deal of judgment is required in crossing this gap between real situations and models, regardless of which direction you are crossing. A too complex model will prohibit your getting results in time to be useful. A too simple model will give erroneous results which can lead to poor action decisions. Of course, some situations do not have both simple and meaningful models. The engineer must be able to recognize these, too, for they require longer-range research and development.

Here is the point to be emphasized: Most academic work starts with the model and ends with the computation. In this book the model-formation, evaluation, and generalization stages are also included, because they are an integral part of most engineering activities.

Now, this wire-drawing lubrication problem that has been used for an example is not finished yet, by any means. One more thing must be done with it, however, before leaving it. Figure 4.2 shows another step: communication of results and recommendations. Back at the beginning of this case study, the section head said there would be a meeting to discuss the results of their preliminary study and to plan what, if anything, to do next. Well, that meeting is coming up! How shall our engineer communicate what he has learned?

Since no special format has been prescribed for him, he is free to choose his own method. He could go in prepared with a verbal report of his findings. He could supplement his verbal report with paper work—graphs, tables, etc. He might choose to have a written report, to be supplemented verbally as needed. Such a written report could be of any length. He might even prepare slides or other visual aids.

In this case, he considers carefully the circumstances and purposes of the meeting. His results are crucial to the decision to be made: Should the company pursue this new idea? If so, in what way? Both his engineering colleagues and his administrative superiors will be interested in what he found and in how accurate his results are. Time, however, will be limited. He therefore decides on a brief, summarized, written presentation, which he can distribute to those present, and which he will supplement very briefly verbally. In this way he feels he can effectively communicate his findings and his recommendations in no more than 10 min. (All this work reduced to 10 min? Right!) His report is presented on the next page. Before looking at it, though, you may wish to prepare one of your own as a point of comparison.

Assignment: Investigate the feasibility of providing viscous lubrication in wire drawing using the inlet-tube concept.

Problem: Can inlet tubes of reasonable lengths provide the necessary lubricant flow and pressure?

Subproblem: Determine the relationship among tube length L, oil viscosity μ_0, wire speed U_0, yield stress σ_y, film thickness h_0, flow rate Q, and wire diameter D_0.

Model assumptions	*Comments*
1. Pressure required is yield stress.	**1.** Back tension will reduce this and shorten required length.
2. Viscosity not affected by pressure.	**2.** Pressure increase will increase viscosity; reduce length.
3. Viscosity not affected by heating.	**3.** Viscosity heating will raise temperature and reduce viscosity; increase length.
4. $h_0 = \frac{1}{10}D_0$; h_0 constant.	**4.** Results very sensitive to h_0.
5. $Q = 10^{-4}$ ft³/sec; $\mu_0 = 2 \times 10^{-3}$	**5.** Larger Q's might be desirable.

Model results: $L = \dfrac{\sigma_y}{6\mu_0 U_0/h_0^2 - 12\mu_0 Q/D_0 h_0^3} \qquad Q = \dfrac{\pi D_0 U_0 h_0}{2}$

Example case	*Material*	*D_0, in.*	*U_0, fps*	*L, ft*
1	Steel	$\frac{1}{32}$	100	7.85
2	Aluminum	$\frac{1}{16}$	100	0.85
3	Aluminum	$\frac{1}{8}$	10	62.7
4	Copper	$\frac{1}{8}$	100	0.82
5	Copper	$\frac{1}{4}$	10	36.0

Conclusions: Concept appears feasible for small-sized, high-speed cases, for all materials. Larger, slow-speed cases appear not feasible.

Recommendations:

1. An experimental program should be undertaken to get better data on the practical values of h_0 as a function D_0, U_0, and other possible factors.
2. Analytical work should be continued to relax assumption of constant viscosity.
3. With (1) and (2) completed, a development program should be reevaluated in the light of the new results.

If the engineer's recommendations are accepted—and that depends on a lot of things out of his control (e.g., financial matters, other technical crises which may be developing, his colleagues' findings, etc.)—but *if* his recommendations are accepted, he will still have some more work to do on this problem.

4.4 SUMMARY AND PREVIEW

This chapter has presented an overview of the engineering analysis process, first in words and then with an example. Again the major points are:

1. Define the problem in specific and answerable terms.
2. Form a model which is simple enough to solve, but complex enough to give meaningful results.
3. Use physical fundamentals, not specialized formulas, wherever possible in your analysis.
4. Check your work. Use limit tests, common sense, unit checks, etc.
5. Evaluate your results and learn or generalize from them. What do your model answers imply about the real, physical problem?
6. Communicate your results in an appropriate and efficient way. Present the essence, not the details.

The next few chapters contain a more detailed discussion of each of the above points individually. Case studies are used for examples. Chapter 5 begins with the problem of defining problems.

SUGGESTED READING

Johnson, W. C.: "Mathematical and Physical Principles of Engineering Analysis," McGraw-Hill Book Company, New York, 1944. This is another good book on engineering problem solving. It stresses mathematics more than Ver Planck and Teare. Another good source of problems.

Ryder, Frederick L.: "Creative Engineering Analysis," Prentice-Hall, Inc., Englewood Cliffs, N.J., 1961. Many engineering analysis problems are solved.

Ver Planck, D. W., and B. R. Teare: "Engineering Analysis," John Wiley & Sons, New York, 1954. In a sense, this book was a forerunner of the one you are reading. It introduces the concept of the "professional method" of problem solving. The book is well written and treats many of the topics discussed in the middle third of this book. *You are strongly advised to read Ver Planck and Teare's treatment of this subject.* Also, in the back of the book there is a long list of excellent engineering problems.

5
PROBLEM FORMULATION

5.1 INTRODUCTION

In the preceding chapter the overall engineering analysis problem-solving approach was introduced and illustrated with a case study. In this chapter, and in the next three chapters, the individual steps of engineering analysis will be considered in more detail, starting here with defining the problem and forming models. These two steps are conveniently combined under the single heading of *problem formulation*. Because of the time required to teach the basic engineering science principles, most problems given to students of engineering are already formed, with models drawn and assumptions given or clearly implied. With the problem formulation done by teachers and textbook authors, students can spend more time learning and applying the physical principles. The neglect of the problem-formulation stage has its disadvantages, however. In real life, the problems encountered seldom are well defined and the models never preformed. Difficulty with these initial steps often is the major cause of inability to solve engineering problems. The objective of this chapter is to familiarize the reader with problem formulation through discussion and example. Three problems are worked in detail, again in such a way that the reader can try his skill at various points in the solution.

5.2 DEFINING THE PROBLEM

The objective in the problem-definition step is to translate vague, general problem statements into specific, answerable questions. No amount of mathematical analysis will ever answer such questions as: Will it work? Is it feasible? Will it deflect too much? Will the process function at low temperatures? Etc. It is the task of the problem solver faced with such qualitative questions to translate them first into specific questions which can be answered by engineering-type analysis. He must determine which of the many parameters determine or limit function or feasibility. He must be able to comprehend the physical situation in such a way that the significant and essential features are clear. He must translate the above general questions into such questions as: Given an input of 5 Btu/sec, at what temperature will the manifold operate? How much space is required for the lever arm to clear the area? What is the relationship among the variables p, v, and T? Etc.

Experience and judgment are very helpful in the problem-definition stage, but simply realizing that a question amenable to analysis is needed will encourage progress. Also it is useful to bear in mind that the question asked should be meaningful so that an answer will definitely contribute information relevant to the general question, and it should be reasonable so that it is possible to obtain an answer within the resources available.

5.3 FORMING MODELS

Once the problem is defined or the question asked, the next step is to form a model. Figure 5.1 is repeated from Chap. 4, where the concept of a model was first introduced. The figure is intended to illustrate the fact that engineering models are *not* exact replicas of real physical situations. Most nonengineers feel or believe that engineering is an exact and precise activity. By comparison with the activities of, say, psychologists, this is quite true. In an absolute sense, however, it is not true at all. The real physical situations with which engineers work are very complex and are never analyzed exactly or completely. No one even tries to analyze them exactly or completely. Assumptions, idealizations, and approximations are expected.

The formation of a fictitious model of the real physical situation is standard procedure for an engineer. It is important that he be aware of the distinctions between his model and the real world. Those distinctions are his assumptions, whether he made them consciously or not.

Forming a model is always a compromise. The model must include enough of the details and complexities of reality so that it is *meaningful.* At the same time, it must be *simple* enough to allow for solution within the problem-solving constraints of time, money, facilities, ability, etc. The great success of engineering has been its ability to find such models.

Models which can be solved within reasonable constraints, but which also give meaningful results, are not always available. This is the problem of the social scientist at present. If he gets a model of social man simple enough to use, he gets results that do not check with his observations. If he tries a more complex model, he cannot keep track of all the variables. Some engineering problems work in the same way. In such cases, experimental or empirical methods are generally used to get the answers. In other

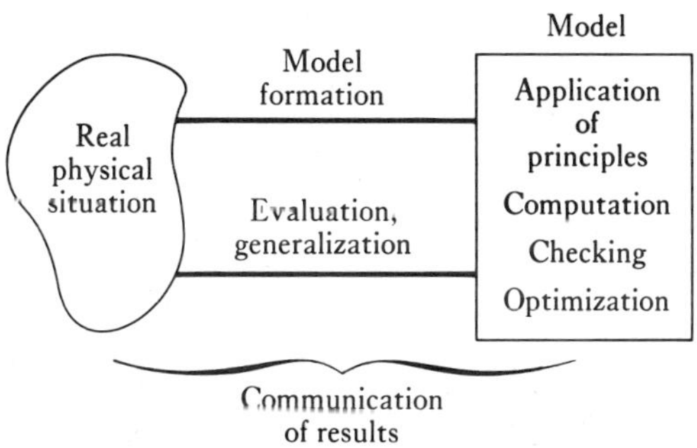

Fig. 5.1 Schematic of the analysis process.

words, research is done to gather more knowledge. Empirical data are used to do the designing until someone develops a theory (a generalized model) to handle the process or problem area. Hence, models can be either experimental or analytical.

This book, simply because it is a book, will deal mostly with analytical models. This is an unfortunate necessity, because experimental solutions are too often slighted in many engineering curricula. They are much used in practice, however. Nevertheless, the major responsibility for the design of simple but meaningful experiments must be left for your laboratory courses.

5.4 SOME COMMENTS ON MAKING ASSUMPTIONS

There cannot be any hard and fast set of rules for making assumptions in engineering problems. Engineers try to construct (in their minds) a model of the real physical situation which is simultaneously meaningful and simple. Which assumptions are consistent with this purpose depends a great deal upon the problem-solving constraints. If an order of magnitude is all that is needed, one set of assumptions is appropriate. If $\pm$ 1 percent precision is needed, another set of assumptions will be "correct." If there is only an hour to get an answer, the assumptions made will be quite different from the ones made if ten days are allowed. Judgment is required. Experience helps a lot. However, a few more specific comments are possible.

There is grave danger involved when an assumption has been made unknowingly. Errors of omission can come back to haunt an engineer. Careful and complete model formation and assumption stating can help in avoiding this pitfall. But it's always good to check at the beginning and end of an analysis for hidden assumptions. In other words, *be aware of assumptions.*

Now, let us turn attention to some of the common idealizations (assumptions) used by engineers.

The Adiabatic Wall When is a surface *adiabatic?* None *really* is, of course. However, generally speaking a surface is adiabatic if the amount of heat flow through it is small compared with other heat-flux terms significant in the problem. A wall can be considered adiabatic if it is well insulated and the problem is a steady-state one. Or it can be adiabatic if there is only a very small temperature difference across it.

The problem of the insulated wall and unsteady state is worthy of mention. The insulation may have considerable amount of heat capacity. Thus, heat may flow into (or out of) the insulation itself during a transient situation. In such cases, the insulated wall may be far from adiabatic from the viewpoint of the material adjacent to the wall.

Heat transfer is a rate process. It requires time. Hence surfaces can be considered adiabatic for all practical purposes in some problems, even though the heat flux is quite high. If the time is short, the total quantity of energy transferred may still be quite small compared with other energy changes in the problem.

Thus there can be insulated walls that aren't adiabatic and uninsulated walls that are. What counts is the effect of the heat transferred on the solution. If the effect is small, it is neglected, and the surface through which it passes is adiabatic.

What does "small" mean? Well, that depends on how accurate an answer is needed and how much time is available to work the problem. Get the idea?

The Ideal Gas There is, of course, no such thing as an ideal gas. It is a model. However, under certain conditions, assuming that real gases are ideal (that is, behave according to $Pv = RT$) causes very little error. The usual criterion for near ideal gas behavior is that the pressure be low or the temperature high compared with the critical pressure or critical temperature for the substance. Liquids or liquid-vapor *mixtures* can *never* be assumed to be an ideal gas without large errors. (Sometimes the vapor phase of a two-phase mixture can be considered ideal, however, if the pressure is low enough.) Again, how much accuracy is desired determines whether the gas is ideal or not. The gas itself has nothing to do with it. It is what it is.

The Plane Beam When are the assumptions of simple elastic beam theory satisfied? To answer "when plane cross sections remain plane" is simply to repeat the question. Under what circumstances do plane cross sections remain plane? Of course, first of all, only *elastic* deformations are being considered. (When is a deformation elastic?) But more than this is required. The beam height must be small compared with its length. How small? A ratio of 1:100 is usually ample. But what about 1:50 or 1:25 or 1:10? That, of course, depends upon the needs of the problem.

Constant Properties Engineers are forever dealing with the properties of substances and materials. Let's list a few: thermal conductivity (k), specific heat (C_p, C_v), viscosity (μ), modules of elasticity (E), emissivity (ϵ), density (ρ), electrical resistance, capacitance, and inductance (R, C, L), coefficient of thermal expansion (α or β), compressibility (K), etc., etc., etc. In addition, engineers deal with such things as spring constants, heat-transfer coefficients, nozzle or orifice coefficients, etc., etc., etc. It is often very helpful, in terms of reduced work, to consider these properties constant and these coefficients invariant. But this is almost never exactly correct. Thermal conductivity is a strong function of temperature for some materials. So is specific heat. Viscosity varies markedly with temperature and to some extent with pressure. Emissivity is nearly proportional to absolute temperature for metals. Electrical properties also show temperature dependence. And so on.

Space doesn't permit even a summary classification of all these variables and their dependencies. Our purpose here is to remind you that these common parameters, so often unconsciously assumed to be constant, are in fact variables. Engineers should be aware of their assumptions. In many problems, assuming certain factors constant is perfectly proper and leads to very little error. In other cases, careful study must be given to assumptions about constants. A little experience soon provides a young engineer with the necessary background to make such decisions easily. There are a lot of easily accessible data in the literature, however, when doubts arise.

The Newtonian Fluid The major assumption producing the Newtonian fluid is that stress is proportional to strain rate. (Note that the "elastic" solid is based on the assumption that stress is proportional to strain.) The proportionality factor is called the viscosity, μ. As has been said, viscosity is a function of temperature and, to a lesser extent, pressure. Almost everything in the engineering literature on fluid mechanics and heat transfer is based on the assumption of the Newtonian fluid, and, indeed, most normal fluids satisfy the Newtonian assumption quite well in engineering applications.

However, it must be remembered that the fluids themselves are not Newtonian, just as no gas is ideal. These substances simply are what they are. People invent the abstractions—Newtonian, ideal, etc.—and under some conditions, some real fluids approximate the abstractions quite well, or at least well enough for practical purposes. Thus, some common liquids (such as water) exhibit strong non-Newtonian properties under certain conditions (such as at very high strain rates). And some liquids (such as glue) are quite non-Newtonian in their usual state. Engineers have to handle *all* substances, so be aware of the Newtonian-fluid assumption when it is made, and be sure that it is valid.

The Lumped Parameter The most common lumped-parameter assumptions are made in electric circuit theory and in dynamics. It is so common that it is seldom mentioned. The resistance of a long wire is really not all lumped into that one spot marked *R*, however. And no mass is lumped together at a point. There are circumstances when such assumptions are perfectly valid, and others when they render the resulting computation meaningless. In most cases it is not difficult to determine the effect of the lumped-parameter assumption. But keep *aware* of these things, because they aren't *always* valid!

Listed above are only a few of the common engineering assumptions used in forming models. Others are brought out in problems at the end of the chapter. Others are left for you to recall. New ones come along as new developments occur. *Be aware.*

5.5 ORDERS OF MAGNITUDE

Making assumptions need not always be based on subjective feelings or on experience. More often than not, assumptions can be justified with quick, approximate calculations known as *order-of-magnitude checks*. Students are probably familiar with this type of thing in mathematics work where so-called higher-order terms are neglected. Thus, if δ is a small number less than unity, it is permissible to neglect δ^2 *compared with* δ, or δ^3 compared with δ^2. In such things as Taylor-series expansions, one often drops the higher-order terms. In physical problems, therefore, if it can be shown that one term is very much smaller (i.e., an order-of-magnitude smaller) than another, then much work can be saved. Note that it is not necessary to compute the exact values of the terms involved, but only to estimate the orders of magnitude. It is, however, often necessary to use numerical values in these calculations, and so some experience with the physical world helps one with this procedure. An order of magnitude is usually considered to be roughly a factor of ten.

To consider one wall of an enclosure to be adiabatic, it must be shown that the heat flow through it will be very small compared with the heat flow through some other wall. In such a case, one can compare $kA/\Delta x$ for two walls. Suppose by putting in some numbers which are realistic but approximate it is found that

$$\frac{(kA/\Delta x)_{section\ 1}}{(kA/\Delta x)_{section\ 2}} = O\,\tfrac{1}{10}$$

The O stands for order of magnitude. Now can the heat transferred through section 1 be neglected? The answer, of course, depends upon the desired accuracy of the result and the time available for getting a solution. But if the ratio were 1:100 or 1:1000, the justification for neglecting section 1 would be clearly established, even though the number is only approximate. It doesn't matter whether the ratio is 1:8 or 1:10 or 1:12. It only matters whether it is *about* 1:10 or *about* 1:1000 or *about,* say, 1:100.

In making order-of-magnitude analyses, it is often necessary to expand certain functions in series. Most students have done this in mathematics, of course, and it can be very helpful to remember such things for application to real problems. For example, if, in a heat-flow problem, the term $1/(T - \Delta T)$ appears, it can be expanded as:

$$\frac{T}{T - \Delta T} = \frac{1}{1 - (\Delta T/T)} = 1 + \frac{\Delta T}{T} + \left(\frac{\Delta T}{T}\right)^2 + \cdots$$

If $\Delta T/T$ is small compared with unity, the approximate result is

$$\frac{T}{T - \Delta T} \cong 1 + \frac{\Delta T}{T}$$

There are other series which should be known, or easily found, such as those for the trigonometric functions, exponentials, etc. The teaching of these is left to math courses, but it should be remembered that such things are very useful in engineering analysis.

Problem formulation—defining the problem and making a model—is best described by example, and hence the remainder of this chapter consists of three problem solutions. In presenting these solutions in this chapter, the problem-formulation step is emphasized, though the complete solution is included.

5.6 THE HEATING-DUCT INSULATION PROBLEM

Statement of the Problem In a rapidly growing city, a new building code was recently instituted. In the section on warm-air heating, the code requires that contractors cover all warm-air ducts with $\frac{1}{32}$-in. asbestos insulating paper. One contractor, however, refuses to comply with the code in this respect because he claims adding the insulation increases, rather than decreases, the heat loss from these ducts. A special hearing has been called to consider the problem, and an engineer has been hired by the city to comment on the contractor's contention.

Now before going on, the student, imagining himself to be the consulting engineer, should define the problem for himself.

Did you really do a good serious job of defining the problem? If so, go on. If not . . . .

Definition of the Problem The engineer in this situation is faced with a rather vague problem situation, though it is not an extreme case. He decides quite readily that what he must do is determine the heat loss from the duct under a set of typical conditions, with and without the insulation. But what are "typical" conditions, if indeed they exist at all? Perhaps this question can be avoided if he derives a *general* expression for the heat loss from a duct first. Then it will be a simple matter to insert whatever numbers the city and/or the contractor feel would be typical. In any event, once the general expression is obtained, the numerical aspect will not be difficult.

The engineer still must prescribe some specific things. For example, he decides to work with a horizontal round duct, but notes that he could

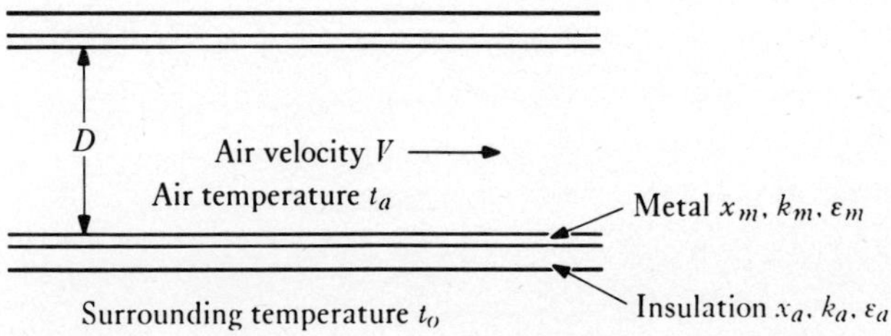

Fig. 5.2 Schematic for the heating-duct insulation problem.

easily also repeat his work for a rectangular duct if necessary. It seems doubtful that the results will be qualitatively different.

The engineer states his problem as follows (see Fig. 5.2): Determine an expression for the heat loss to the surroundings per foot of duct in terms of V, D, t_a, t_0, x_a, x_m, ϵ_a, and ϵ_m, with and without insulation.

Now notice what the engineer has accomplished. He has gone from a general assignment ("comment" on the problem) to a specific problem, which can be solved by engineering principles. Later, of course, when he has his equations derived and his numerical examples worked, he will have to go back and make his evaluations and generalizations. In the present example, it is a fairly straightforward job to see what problem should be defined. But the student reader should not be deceived by the apparent simplicity. Hindsight often makes things seem simple.

Making a Guess For both fun and experience, try now to make a guess about the outcome. Doing so is a good way to gain engineering judgment. Guessing can also be an aid to understanding the problem because it forces one to consider only the most essential factors. In this case, for example, the engineer realized that adding the insulation will increase the thermal resistance of the duct wall and will hence decrease the convection loss because the outer surface temperature will be lower. However, most ducts are made of galvanized sheet metal or aluminum, both of which have a low emissivity. The asbestos paper, on the other hand, has a relatively high emissivity, which will tend to increase the radiation loss. Which effect will dominate? Make a guess before going on.

The engineer being observed here guesses that adding the insulation has practically *no* effect on the heat loss, either positive or negative.

Incidentally, many students when given this problem recall a discussion of "critical radius" in their heat-transfer course and try to apply that idea here. Why is that not an important consideration in this problem?

Model Formation The engineer has done a little model formation already, because he has decided to consider a round duct. This illustrates the fact that it is impossible to keep these stages of problem solving from overlapping, reversing order, etc., depending on the problem. However, most of the time most of the assumption making takes place in the model-formation stage.

In this problem there are several simplifying assumptions which do not change the numerical results significantly but which reduce the amount of work necessary by a great deal. First of all, since the duct wall and the

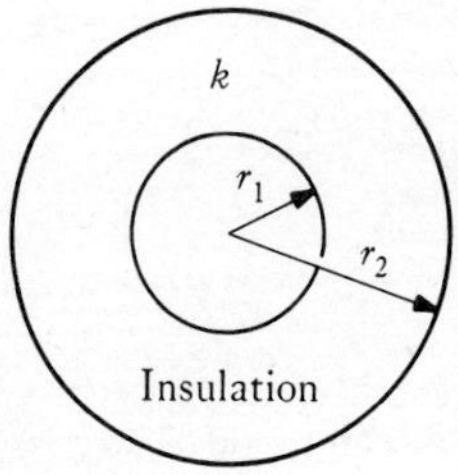

Fig. 5.3

insulating paper are both quite thin compared with the radius of any heating duct, it is appropriate to neglect the curvature of the pipe. The engineer will, of course, compute his heat-transfer coefficients based on pipe flow, but he will use the values obtained as if he were dealing with a flat surface. To show that this is a valid assumption, the engineer can use an order-of-magnitude calculation. Writing the exact expression for the resistance of the curved insulating section (refer to Fig. 5.3), he gets

$$R = \frac{\ln (r_2/r_1)}{2\pi k l}$$

The logarithm can be expanded as

$$\ln x = (x - 1) - \tfrac{1}{2}(x - 1)^2 + \tfrac{1}{3}(x - 1)^3 \cdots$$

If $x = r_2/r_1$ and $r_2 = r_1 + x_a$, then

$$x - 1 = \frac{r_2}{r_1} - 1 = \frac{r_1 + x_a}{r_1} - 1 = 1 + \frac{x_a}{r_1} - 1 = \frac{x_a}{r_1}$$

In our problem $x_a/r_1 = (1/32)/6 \cong 1/200$, so $(x_a/r_1)^2$ is negligible compared with x_a/r_1. Thus

$$\ln \frac{r_2}{r_1} \cong \frac{r_2}{r_1} - 1 = \frac{x_a}{r_1}$$

The resistance is therefore

$$R = \frac{\ln (r_2/r_1)}{2\pi kl} \cong \frac{x_a}{2\pi k r_1 l}$$

which is the same as the expression for a flat plate with area $2\pi r_1 l$.

Certainly one does not need to go through this kind of calculation every time one is confronted with such a case. The criterion is the magnitude of x_a/r_1, as would be expected.

The second assumption the engineer must make has to do with the flow conditions inside the pipe. Is it turbulent or laminar? The engineer can quickly decide from an order-of-magnitude check of the Reynolds number. He gets

$$Re = \frac{\rho VD}{\mu} = \frac{VD}{\nu} = \frac{(10)(1/2)}{2.0 \times 10^{-4}} = 2.5 \times 10^4$$

Checking units,

$$\frac{(\text{fps})(\text{ft})}{\text{ft}^2/\text{sec}}$$

This is well into the turbulent range, and the assumption of turbulent flow is clearly justified. Note how the engineer needed *numbers* to get this check. He used a 6-in.-diameter pipe and a flow rate of 600 fpm, both reasonable but on the low side for the situation. The kinematic viscosity ν he took from a table for air at about 150°F.

The engineer must make some other assumptions before he has a final and complete model. On the outside of the duct (which he has assumed horizontal), he assumes "still air" so that natural convection heat-transfer correlations will apply. Also, he wonders whether the thermal resistance of the metal wall is significant compared with the others involved. He decides to compare them with an order-of-magnitude analysis, which the reader should do for himself before going on.

$$R_{\text{inside film}} = \frac{1}{h_i} \cong \tfrac{1}{5} = 0.20 \text{ (hr)(ft}^2\text{)(°F)/Btu}$$

$$R_{\text{metal}} = \frac{x_m}{k_m} = \frac{1/32}{12 \times 20} \cong \tfrac{1}{700} = 0.0014 \text{ (hr)(ft}^2\text{)(°F)/Btu}$$

$$R_{\text{asbestos}} = \frac{x_a}{k_a} \cong \frac{1/32}{12 \times .03} = \tfrac{1}{10} = 0.10 \text{ (hr)(ft}^2\text{)(°F)/Btu}$$

$$R_{\text{outside film}} = \frac{1}{h_0} \cong \tfrac{1}{2} = 0.50 \text{ (hr)(ft}^2\text{)(°F)/Btu}$$

Comparing these, he can see quickly that neglecting the thermal resistance of the metal duct will not change his results, so he decides to neglect it in his calculations.

Finally, he has to consider radiation. Probably this is negligible at these temperatures, so he decides to omit it. Then, however, he remembers that if it is negligible, the contrary contractor is obviously wrong in his belief that adding the insulating paper might increase the heat transfer. In fact, radiation is the *key* to this problem. Nevertheless, remembering that radiation is usually quite small at these relatively low temperatures, the engineer wonders how this could possibly be so important. Instead of just speculating, however, he goes ahead to do an order-of-magnitude check. For this purpose he assumes he has a surface at 120°F in an ambient at 70°F, with an emissivity 0.9 and a convection film coefficient of 1 Btu/(hr)(ft²)(°F). The convection transfer from this surface would then be

$$q_{\text{conv}} = 1(120 - 70) = 50 \text{ Btu/(hr)(ft}^2\text{)}$$

The radiation, to a black surroundings, would be about

$$q_{\text{rad}} = \sigma\epsilon\,\Delta(T^4) = (0.173)(1)(5.8^4 - 5.3^4) = 58 \text{ Btu/(hr)(ft}^2\text{)}$$

It appears from this very rough check that radiation is just about as important as convection under these circumstances. Certainly there is *no* justification for neglecting it! So he must leave it in. Because the pipe is a rather small thing in large surroundings, he will assume that the shape factor is unity and that the surroundings are black at the ambient temperature.

Now he has his model: a flat surface, turbulent flow inside the pipe, natural convection outside, radiation to black surroundings with a shape factor of unity. The next step is the application of physical principles.

Application of Physical Principles The problem is defined and the model formed. The problem is now amenable to solution by the use of engineering principles. This is the chapter on problem formulation and that step has now been done. The remainder of the solution is included for completeness but it will also serve as a preview to the stages which follow. However, in the later discussions, such as on applying physical principles or on checking, it will be assumed that the reader has read through these earlier problems completely.

The engineer considers first the case where the insulation is present. He may, then, be able to solve the other case just by letting the insulation

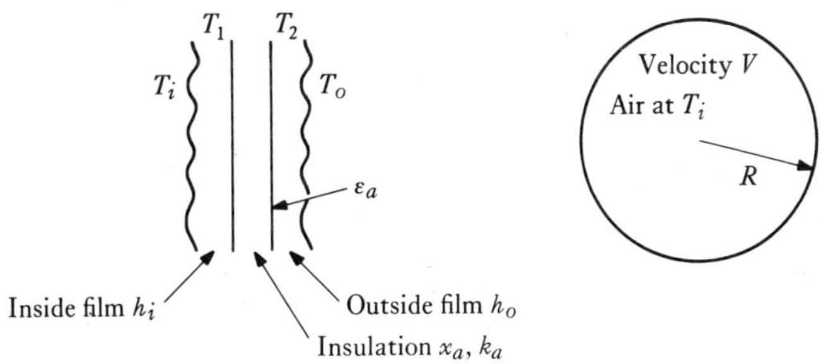

Fig. 5.4

thickness go to zero. Figure 5.4 shows the essence of the situation and also contains some symbol definitions.

Because of the combined convection and radiation transfer from the outer surface, solution for the heat transfer does not yield simply to computation of an overall thermal resistance. An analogous electrical circuit for the situation looks like the following:

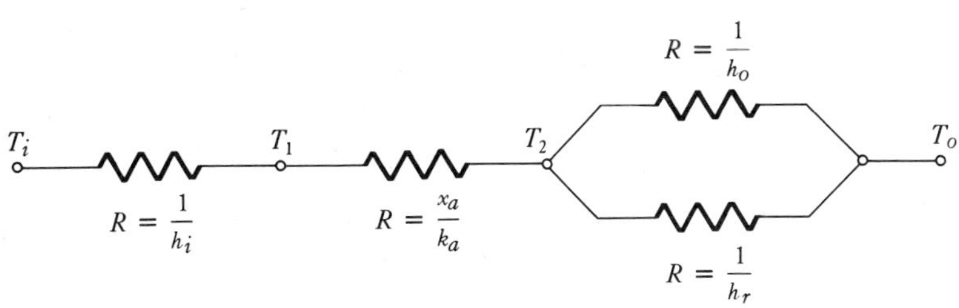

A radiation coefficient in this scheme can be defined by

$$h_r = \frac{\mathfrak{F}_{23}\sigma(T_2^4 - T_0^4)}{T_2 - T_0}$$

In our case $\mathfrak{F}_{23} = \epsilon_a$ since the shape factor is unity and the surroundings are black. (Note that evaluation of h_r requires knowledge of T_2 and that evaluation of T_2 requires knowledge of h_r!) From the electrical analogy, the total resistance is given by

$$R = \frac{1}{h_i} + \frac{x_a}{k_a} + \frac{1}{h_0 + h_r}$$

Hence $$q = \frac{\Delta T}{\sum R} = \frac{T_i - T_0}{(1/h_i) + (x_a/k_a) + [1/(h_0 + h_r)]}$$

Unfortunately, the engineer does not know the value of h_r, because he does not know T_2. How can he find T_2? The reader should think of a method before going on to the next page.

In the steady state, which is the case being considered, the heat transferred through the inside film and insulation is equal to the sum of the heat transferred by convection and radiation from the surface. In symbolic notation

$$q_{\text{total}} = q_{\substack{\text{inside film}\\ \text{and metal}}} = q_{\substack{\text{conv}\\ \text{surface}}} + q_{\substack{\text{rad}\\ \text{surface}}}$$

The motivation for writing the equation in this way is that each of the above heat fluxes can now be expressed in terms of T_2. The resulting equation gives us an equation which can be solved for T_2. Notice that the concept here is that *all* the heat flux passes through the inside film and insulation, but that the total flux is divided into two modes for transfer to the surroundings from the surface. This is an example of a problem situation which cannot be solved by application of some formal equation such as $\nabla^2 T = 0$.

Expanding the terms above gives

$$\frac{1}{(1/h_i) + (x_a/k_a)}(T_i - T_2) = h_0(T_2 - T_0) + \epsilon_a\sigma(T_2^4 - T_0^4)$$

This is a single equation in T_2 which must be solved. Once T_2 is obtained, one can get the heat flow by evaluating either side of the above equation.

All the terms in the above expression are known except ϵ_a, h_0, and h_i. For the emissivity, the engineer chooses $\epsilon_a = 0.90$ for the insulation and 0.3 for the plain duct, because these are rather typical values for these things as given in the literature. For h_i, he remembers (or looks up) the dimensionless correlation for turbulent flow in pipes:

$$Nu = 0.023(Re)^{0.8}(Pr)^{0.4}$$

or

$$\frac{hD}{k} = 0.023\left(\frac{\rho VD}{\mu}\right)^{0.8}\left(\frac{C_p\mu}{k}\right)^{0.4}$$

For h_0, he finds the empirical correlation for heat transfer by free convection from single-heated horizontal cylinders:[1]

$$Nu_D = 0.53(Gr_D Pr)^{1/4} \qquad Pr > 0.5$$

$$10^3 < Gr < 10^9$$

$$\frac{hD}{k} = 0.53\left[\frac{C_p\rho g\beta(\Delta T)D^3}{\mu k}\right]^{1/4}$$

[1] See, for example, W. H. McAdams, "Heat Transmission," 3d ed., McGraw-Hill Book Company, New York, 1965.

A simplified correlation is also available

$$h = 0.27\left(\frac{\Delta T}{D}\right)^{1/4}$$

where h is Btu/(hr) (ft²) (°F), T is °F, and D is feet. Using these correlations he can find all the unknowns and solve for the heat transfer. It is also quite easy to repeat the calculations for the case of no insulation. Comparison of the results will give an answer to the question asked.

Computation As noted above, the first step in the solution requires solution for T_2 (see Fig. 5.5).

$$\frac{1}{(1/h_i) + (x_a/k_a)}(T_i - T_2) = h_0(T_2 - T_0) + \epsilon_a\sigma(T_2^4 - T_0^4)$$

Solving for the case with the insulation applied, the engineer takes the

T_2, °F	$\frac{1}{1/h_i + x_a/k}(T_i - T_2)$	h_0	$h_0(T_2 - T_0) + \epsilon_a(T_2^4 - T_0^4)$
150	0	0.532	149.0
60	225.0	—	0
115	87.5	0.470	82.3
116	85.0	0.473	84.2
117	82.5	0.476	86.0

Answer: $T_2 = 116.2$°F
$q = 84.5$ Btu/(hr) (ft²)

Fig. 5.5 Results for the heating-duct insulation problem.

following numbers as typical:

$$\begin{aligned} x_a &= \tfrac{1}{32}\text{in.} \\ k_a &= 0.096 \text{ Btu/(hr)(ft}^2\text{)(°F)} \\ \epsilon_a &= 0.90 \\ t_i &= 150\text{°F} \\ T_0 &= 60\text{°F} \\ V &= 800 \text{ fpm} \\ R &= 3 \text{ in. } (D = 6 \text{ in.}) \end{aligned}$$

For the outside film coefficient, he will use the simplified equation

$$h_0 = 0.027\left(\frac{\Delta t}{D}\right)^{0.25} = 0.027\left(\frac{T_2 - T_0}{D}\right)^{0.25}$$

Since T_2 is unknown, h_0 cannot be obtained directly. For the inside film coefficient h_i:

$$h_i = \frac{k}{D} 0.023 \left(\frac{\rho V D}{\mu}\right)^{0.8} \left(\frac{C_p \mu}{k}\right)^{0.4}$$

$$= \frac{0.016}{0.5} 0.023 \left(\frac{800 \times 0.5}{60 \times 2 \times 10^{-4}}\right)^{0.8} (0.75)^{0.4}$$

$$= \frac{0.016}{0.50} 0.023(.41 \times 10^4)(0.981) = 2.69 \text{ Btu/(hr)(ft}^2\text{)(°F)}$$

Check units: $$\frac{\text{Btu}}{\text{(hr)(ft)(°F)}} \frac{1}{\text{ft}} \rightarrow \frac{\text{Btu}}{\text{(hr)(ft}^2\text{)(°F)}}$$

Now the quantity

$$\frac{1}{(1/h_i) + (x_a/k_a)} = \frac{1}{(1/2.69) + [1/(32 \times 12 \times 0.096)]}$$
$$= 2.50 \text{ Btu/(hr)(ft}^2\text{)(°F)}$$

The equation above which defines T_2 is a single equation in one unknown but it cannot be solved explicitly for T_2. Such equations are called *transcendental* and must be solved by numerical methods. In this case, a simple trial-and-error procedure will do. In working trial-and-error problems it is almost always best to get the correct answer *bracketed* as quickly as possible; that is, to make a guess that is too big and one that is too small right at the outset. Then linear interpolation can be used to make the third and subsequent guesses. In the example under consideration, it is clear that T_2 lies between $T_i = 150$°F and $T_2 = 60$°F. Thus these are ready made brackets. They are used to establish limits, and the first two rows of Fig. 5.5 are easily completed. These data are now plotted as shown in Fig. 5.6, and by linear interpolation a good third trial is found to be $T_2 = 115$°F. Plotting the results of this calculation on the graph (see the dashed lines) suggests that a fourth trial be made at 116°F. The additional results are shown in the table, together with the result:

$$T_2 = 116.2\text{°F} \qquad q = 84.5 \text{ Btu/(hr)(ft}^2\text{)}$$

A similar computation for the uninsulated case (left as an exercise for the interested and/or skeptical student) gives

$$T_2 = 128.5\text{°F} \qquad q = 57.5 \text{ Btu/(hr)(ft}^2\text{)}$$

The surprising result is that addition of the insulation *increases* the total

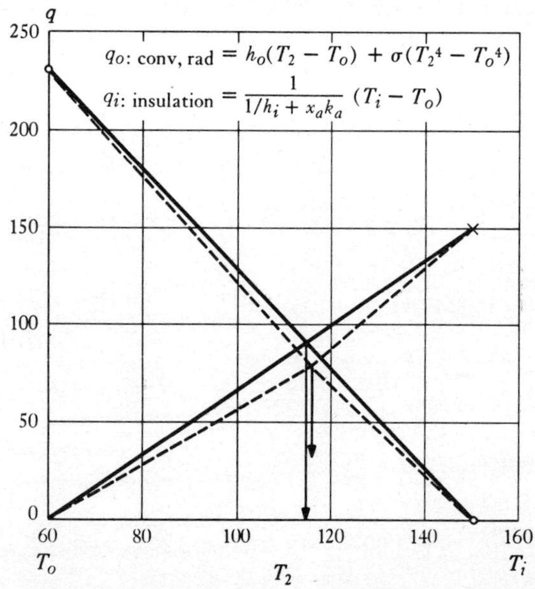

Fig. 5.6 Graphical aid for solving equation in heating-duct problem.

heat loss. Note that this is due to the radiation effect, which definitely is not negligible.

Evaluation and Generalization and Communication For his own personal future use, the engineer learns from this project that in natural-convection situations, radiation is not to be neglected without careful consideration. He also concludes that the contrary contractor is correct in his feeling that the insulation increases the heat transfer in this typical situation. Different numbers could be used, of course, but the difference [84.5 Btu/(hr) (ft^2) vs. 57.5 Btu/(hr) (ft^2)] is so great that the conclusion seems certain. The engineer prepares a report for entrance into the court records for the case. Before going on the student should prepare a report of his own for this purpose.

5.7 THE TOY BAZOOKA

Statement of the Problem Examine the sketch of the toy bazooka shown in Fig. 5.7. The barrel of this vicious weapon is made of a hollow cardboard tube. The "shell" is a thin, somewhat deformable plastic cap that clips into place as rim Ⓐ hooks into position Ⓑ. When the handle or "trigger" is moved rapidly forward, the cap is forced off and "fires." The cap will travel about 15 or 20 ft and makes a satisfying (to a child!) pop as it leaves the bazooka. Toys of this gentle ilk were popular a few years ago before more realistic, sturdy, and deadly toys became popular. See Prob. 5.

The bazooka shown is to be redesigned so that it can be made shorter. The present maximum length of the piston stroke is 9 in. A number of trials indicate that, starting from the 9-in. position, the shortest possible stroke which will pop the cap is 4 in. That is, a rapid motion of the piston from 9 in. to 5 in. will pop the cap. After obtaining this data, the engineer realizes that what he wants to do is not start at the 9-in. position but rather end the stroke at the 0-in. position. Therefore he wants to find the

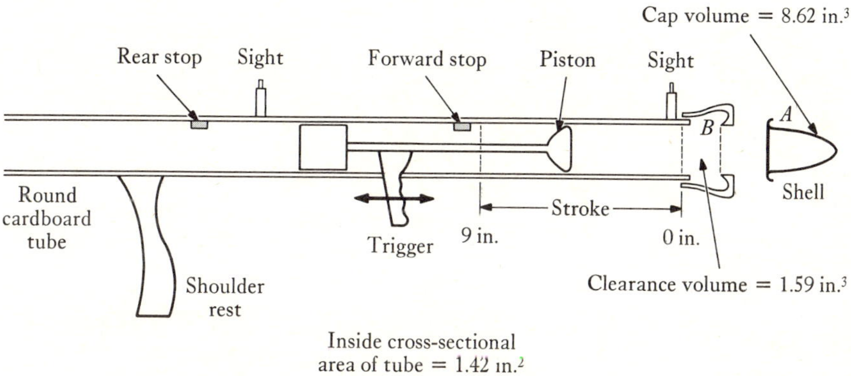

Fig. 5.7 Schematic of the toy bazooka.

location from which he can pop the cap if he ends at the 0-in. position. This will enable him to shorten the bazooka by shortening the length of the available piston stroke.

Also, the designer of the cap itself is interested in knowing what pressure is required to pop the cap.

Define the problem before going on. You will first have to study Fig. 5.7 to ascertain how the bazooka works.

Definition of the Problem In order to define the problem, the engineer must decide what it is that causes the cap to fire off. He quickly decides that it is primarily pressure which does it (and not some kind of compressive shock or impulse). He reasons that a hand-operated device such as this is not going to produce sonic speeds. (Some student engineers do not see this, however, and hence define a completely meaningless problem.) If it is pressure which pops the cap, then the engineer must find the pressure which does the job. That becomes his first problem. Then he must find the position which will produce that pressure with the 0-in. position as an end point.

Making a Guess As a crude approximation, the pressure will vary inversely as the volume. The volume will be roughly proportional to the length to the piston position plus the clearance, which by eyeball appears to be about 2 in. Hence the volume reduction from the data is to the ratio of about $(5 + 2)/(9 + 2) = 7/11$. Hence the pressure ought to be about $(\frac{11}{7})(15) = 24$ psia. This seems pretty reasonable. To get this pressure from the shorter position requires $l + 2 = \frac{11}{7}(2 + 0) = 3$ or $l = 1$ in. This seems short to the engineer, but the general idea and order of magnitude appear okay. He decides to leave it as a guess and go on to a more refined analysis.

Notice how these guesses require one to get to the essence of the problem. Force yourself to do them quickly and to get a number. It will also help sharpen your engineering judgment.

Model Formation The problem defined is to determine the shortest initial piston position which will pop off the cap and also to find the pressure required to pop the cap. See the sketch on page 117. What assumptions are needed in order to model this situation? State them below and then go on to the next page.

First of all, of course, the engineer must assume something about the properties of the air compressed in the bazooka. Air is a very good ideal gas under normal atmospheric conditions and, though compressed slightly in this problem, it will remain so. Its critical pressure is about 37 atm and its critical temperature is about 240°R (−220°F). Hence even at, say, 2 atm pressure, it can be expected to behave very much as an ideal gas.

The second assumption needed has to do with the process of compression. Is it isothermal, adiabatic, or something else? Here is a case where the most reasonable assumption seems to be that it is an adiabatic and reversible compression. It will be nearly adiabatic, the engineer reasons, not because there is insulation, but because of the speed of the compression process. There simply isn't enough time for a significant amount of heat transfer to take place. As a check to justify this assumption, the engineer notes that if *no* heat is transferred, the air will change temperature according to

$$T_1 P_1^{(1-k)/k} = T_2 P_2^{(1-k)/k}$$

Using an approximate pressure ratio $P_2/P_1 = 2$, and $T_1 = 70°\text{F}$, he finds $T_2 = 647°\text{R} = 187°\text{F}$. Then the order of magnitude of the amount of heat removal required to keep the air isothermal is

$$Q - W = \Delta E = 0$$

$$Q = W = p\,dV = mRT_1 \ln \frac{P_2}{P_1}$$

From the volume of the cylinder, clearance, and cap, he estimates the mass of air and finds $Q = 0.0211$ Btu. Next he must consider the heat-transfer rate. He decides to choose a convection coefficient of about 5 Btu/(hr) (ft²) (°F) since the motion of the piston will create some turbulence. Furthermore, he notes that the compression process takes place in something less than a second so he decides to use 1 sec as a time, though this is probably too long. Finally, he computes an average area for heat transfer from the bazooka to the confined air during the compression. He can then write

$$Q = 0.0211 = hA(\Delta T)_{\text{av}}(\text{time})$$

and solving for ΔT_{av} gives

$$T_{\text{av}} = 50°\text{F}$$

This means that in order to transfer the necessary amount of heat, an average temperature difference of at least about 50°F would be required. This is hardly an isothermal situation. On the other hand, since an average ΔT of about 50°F is just about what the air would experience adiabatically (from 70° to 187°F), the heat transfer is not conclusively negligible. The engineer notes that some polytropic expansion Pv^n with $1 < n < k$ is probably "correct," and he decides to use the adiabatic assumption as a limiting case. Perhaps he will also compute the isothermal case as the other limit.

The model is an ideal gas being compressed reversibly and adiabatically. What is required for the reversible assumption? It must be frictionless (internally—so far as the *air* itself is concerned, that is) and also slow. How slow is "slow"? Slow enough to avoid pressure waves or shock fronts. So "slow" means less than the velocity of sound. Now a man can create a nice compressible shock with a whip, but he can hardly do so by direct one-to-one action on the piston of a toy bazooka. (How fast would he have to move the piston in, say, 6 in.?) There will be turbulence and friction in the air, of course, but he will neglect it as being small compared with the other effects in the problem. The air compression, then, can be considered internally reversible.

Thus the engineer has his model.

Computation In this problem, the computation is quite simple and straightforward. For the reversible adiabatic compression

$$P_1 \nu_1^k = P_2 \nu_2^k$$

where $P_1 = 14.7$ psia

$\nu_1 = 8.62 + 1.59 + (9)(1.42) = 21 \text{ in.}^3$

$\nu_2 = 8.62 + 1.59 + (5)(1.42) = 17.3 \text{ in.}^3$

$k = 1.4$

and P_2 is to be found. The result is

$$P_2 = P_1 \left(\frac{\nu_1}{\nu_2}\right)^k = (14.7)\left(\frac{21.0}{17.3}\right)^{1.4} = 19.3 \text{ psia}$$

The pressure required to pop the cap is thus 19.3 psia. A stroke which ends at the 0 position must have this pressure at 0 and atmospheric pressure at its beginning. Since $P\nu^k$ is again a constant, one can find the new position two ways. By using

$$\left(\frac{\nu_1}{\nu_2}\right)_{\text{old}} = \left(\frac{\nu_1}{\nu_2}\right)_{\text{new}}$$

the result is

$$(\nu_1)_{\text{new}} = (\nu_2)_{\text{new}} \left(\frac{\nu_1}{\nu_2}\right)_{\text{old}} = (8.62 + 1.59) \frac{21}{17.3} = 12.4 \text{ in.}^3$$

The new starting position is thus given by x:

$$8.62 + 1.59 + 1.42x = 12.4 \qquad x = 1.55 \text{ in.}$$

Alternatively one can calculate $(\nu_1)_{\text{new}}$ from

$$P_1\nu_1^k = P_2\nu_2^k$$

where $P_1 = 14.7$ psia
$P_2 = 19.3$ psia
$\nu_2 = 10.21$ in.^{3}

The result is the same.

If the compression is isothermal, the result with respect to the required new position is unchanged. However, the pressure required will now be computed from

$$P_1\nu_1 = P_2\nu_2$$

giving $P_2 = 17.8$ psia. In either case, therefore, it should be possible to shorten the length in such a way that the stroke is from the 2-in. position to the 0-in. position. The engineer notes also that the stroke can be shortened by other means. If the clearance volume or cap volume is re-

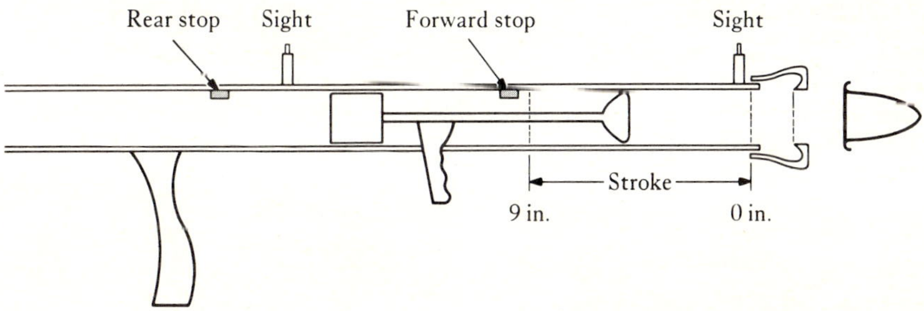

Problem: 1. Investigate the possibility of shortening the bazooka by providing a shorter stroke.
2. Determine the pressure required to pop the cap.

Results: 1. Theoretical analysis shows that the stroke can be shortened to 2 in. if the stroke is from the 2-in. position to the 0-in. position shown in the sketch. Alternatively, redesign to reduce the cap volume or clearance volume, or to increase the piston area, would contribute to a shorter stroke.
2. If the compression is isothermal, the pressure required is 17.8 psia. If adiabatic, the pressure is 19.3 psia. The true answer will be between these limits. Experimental work will be necessary if a more accurate answer is required.

Fig. 5.8 Communication of toy bazooka results.

duced, or the piston area increased, required volume change can be effected with a shorter stroke.

Communication In a simple problem of this kind, probably a verbal communication is most appropriate. At most a very brief report would be required. See Fig. 5.8, for example.

5.8 A HIGH-SPEED BELT DRIVE

Statement of the Problem A company that builds conventional drills and drill presses has designed a new, very high-speed, low-power drill.

BELT 4—XR2 HIGH-SPEED DRILL PRESS

Specifications: Flat leather belt $\rho = 0.035$ lb/in³, $\sigma_{\text{working}} = 275$ psi
Pully diameter $d = 6$ in.
Angle of contact $\theta = 180°$
Speed = 3,000 rpm
Power = 0.26 hp

Assumptions: Coefficient of friction $\mu = .20$

Derivation of equations: $$\text{Power} = \frac{(T_1 - T_2)}{550} V \qquad T_1, T_2 \text{ in lb}_f, \quad V = \text{fps} \tag{1}$$

$$\frac{T_1}{T_2} = e^{\mu\theta}$$

Solution for T_1 and T_2 gives: $T_1 = 3.91$ lb$_f$, $T_2 = 2.08$ lb$_f$

For the minimum belt area: $$A = \frac{3.91}{275} = 0.0142 \text{ in}^2$$

Thus a belt 1/4 by 1/16 in. will be sufficient ($A = 0.0156$ in.).

Fig. 5.9 Communication of old belt design results.

A flat leather belt and pulley drive in the prototype of the new drill has repeatedly failed. The belt was designed on the same basis used successfully by the company on past machines, and the report on this design is shown in Fig. 5.9. An engineer has been assigned the job of recommending corrective action and of finding out why the regular company design procedure has failed in this case.

Before going on, the student should try for himself to define the problem, form a model, and solve the problem.

Definition of the Problem The engineer in this project must convert his "find the trouble" assignment into an analyzable problem. What *could* be the trouble? To avoid some obvious error, he checks the figures given in the design report and finds that the arithmetic there is okay. He also checks the specifications given and finds them to be standard and conservative. No previous difficulties have been encountered with this design procedure.

Because the high speed involved in the new drill is a new feature, he suspects that it may affect the belt design in some way not accounted for in the analysis. But how? This he doesn't know (Do you? Think about it!), but he decides to start at the beginning and rederive the basic equations, making sure not to make any unwarranted assumptions. He is especially interested in the origin of the equation $T_1/T_2 = e^{\mu\theta}$ which appears in the report. His problem definition then is this: Analyze the mechanics of a high-speed belt on a pulley to determine the relationship among T_1, T_2, μ, θ, r, w, and horsepower.

Model Formation After sketching Fig. 5.10, the engineer very nearly has his model. He assumes that the angle of contact is 180° and that

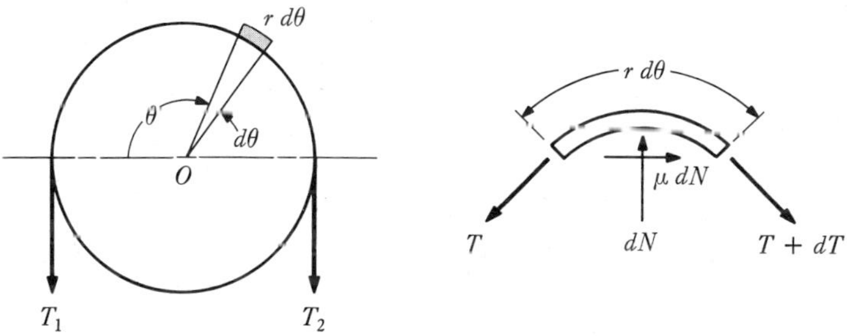

Radial force balance: $$dN = T \sin\frac{d\theta}{2} + (T + dT)\sin\frac{d\theta}{2} = T\,d\theta$$

Tangential force balance: $$\mu\,dN + (T + dT)\cos\frac{d\theta}{2} = T\cos\frac{d\theta}{2}$$

Fig. 5.10

μ is constant. He neglects bending stresses in the belt and any area changes in the belting due to the varying tension. He also decides to neglect the effects of centrifugal force on the belt.

Application of Physical Principles If one neglects the energy used in bending the belt, the power equation given in the report is correct:

$$\text{Power (hp)} = \frac{(T_1 - T_2)V}{550} \qquad \begin{array}{l} T \text{ in lb}_f \\ V \text{ in fps} \end{array}$$

A radial-force balance on the little element of belting gives

$$dN = T \sin \frac{d\theta}{2} + (T + dT) \sin \frac{d\theta}{2}$$

Since for small angles $\sin \alpha = \alpha$, this equation becomes

$$dN = T\, d\theta$$

A force balance in the tangential direction gives

$$dN + (T + dT) \cos \frac{d\theta}{2} = T \cos \frac{d\theta}{2}$$

Since for small angle $\cos \alpha \cong 1$, this becomes

$$\mu\, dN = -dT$$

Combining $dN = T\, d\theta$ with $\mu\, dN = -dT$ to eliminate dN and integrating gives

$$\frac{T_1}{T_2} = e^{\mu\theta}$$

Evaluation and Generalization This is the equation used in the design that has failed, and the engineer now feels certain that either the bending energy or the centrifugal force, both neglected here, becomes important at the new, higher speeds. A little additional thought reveals that the bending energy is not very likely to become *relatively* more important at higher speeds, probably being nearly proportional to V. The centrifugal-force component, however, is proportional to V^2. His decision, therefore, is to relax the assumption about centrifugal force and include it in the analysis.

Model Formation Same as before only do not neglect centrifugal effects.

Application of Physical Principles The power equation and the tangential force balance remain the same. The radial balance now becomes

$$dN + \frac{wV^2}{gr} r\, d\theta = T \sin\frac{d\theta}{2} + (T + dT) \sin\frac{d\theta}{2}$$

$$dN = \left(T - \frac{wV^2}{g}\right) d\theta$$

Combining and integrating as before gives

$$\frac{T_1 - (wV^2/g)}{T_2 - (wV^2/g)} = e^{\mu\theta}$$

This result shows clearly that unless the term wV^2/g is small compared with the T's, the centrifugal effects are not negligible.

Computation Checking the old design, the engineer finds $wV^2/g = 1.26\ \text{lb}_f$ which is clearly not negligible compared with $T_1 = 3.91\ \text{lb}_f$ and $T_2 = 2.08\ \text{lb}_f$. Redesign with the above correct equation gives $T_1 = 5.20\ \text{lb}_f$ and $T_2 = 3.37\ \text{lb}_f$. The required belt area would then be

$$A = \frac{5.20}{275} = 0.00189 \text{ in.}^2$$

(Note that the stress in the previous belt would be $5.20/0.00156 = 333$ psi, well in excess of the recommended working stress. This explains the failures.) A new belt $\frac{3}{32}$ by $\frac{1}{4}$ in. should do the job.

Evaluation, Generalization, and Communication The engineer has learned that at "high" speeds, centrifugal force is important in belt design, and he also knows how to determine what "high" means: If wV^2/g is not small compared with T_2, the speed is high. A slightly thicker belt should correct the failures, and future designs of high-speed belting in the company will have to use the procedure outlined in the revised derivation above. Composition of a one-page report on the results of this investigation is left for the student.

5.9 SLOPPINESS

Before this chapter ends, a word about sloppiness is in order. There are two kinds of sloppy work. One kind of sloppiness is the paper-and-pencil kind. It consists of illegible writing and figuring, use of dirty, irregular paper, disorganized bookkeeping, etc. The other kind of sloppiness is mental. The engineering analysis process attempts to eliminate *both* types because there is often a connection. Of course, neatness is no substitute for competence, but it can contribute to competence by reducing errors of commission and omission.

There is one other point to be made about sloppiness. Students sometimes show their personal and professional immaturity by carrying the ideas of order-of-magnitude checks, assumption making, and model formation to a ridiculous extreme. They want to take crude estimates used to justify assumptions as exact or final figures. Or, instead of taking the trouble to integrate $\int_{x2}^{x1} f(x)\,dx$, they just take some average $\bar{x}$ and compute $f(\bar{x})(x_2 - x_1)$. There is no excuse for this kind of sloppiness unless it is definitely dictated by the constraints on the problem solver. The point is this: Do not confuse the valid process of model formation with the immature, unprofessional, and lazy process of just being sloppy. The way to avoid this error in judgment is to be consciously aware of what is being done and why. There is no place for sloppiness in engineering. Models and approximations must be meaningful and must lead to results which will check with nature when translated back to reality.

5.10 SUMMARY AND PREVIEW

This chapter has been primarily concerned with the definition of problems and the formation of models for engineering analysis. It was pointed out that problem definitions must be quantitatively answerable questions and that the models formed must be simultaneously simple enough to solve within the given constraints and complex enough to yield meaningful answers. Some ways of making good assumptions were also discussed, especially through the use of order-of-magnitude analysis. Finally, three problem examples were worked through, showing the processes in action.

It is now time to consider in more detail the application of physical principles to engineering analysis problems. As noted in the first chapter, it is best to learn to apply principles in their most fundamental and simple form. The discussion in the next chapter is directed to this end.

PROBLEMS

5.1 When is a process thermodynamically reversible? Why do engineers use the model of the reversible process? How do engineers often adjust for the fact that we know certain processes are not reversible?

5.2 What is a frictionless fluid? When are these assumptions meaningful or useful?

5.3 Metals are generally tested in simple tensile, compression, or torsion tests. What are the common models used to predict when failure due to combined stresses will occur?

5.4 Can you think of some idealizations not discussed in the text that are made in forming engineering models? List them and discuss.

5.5 A cork stopper is used in an ordinary glass bottle to contain a cold liquid. Would you say the cork is adiabatic? The same cork is used in a thermos bottle under similar conditions. Would you say it is adiabatic there? Justify your assumptions.

5.6 Examine some of the problems you have worked or are working in your engineering science courses (heat transfer, dynamics, fluid dynamics, strength of materials, etc.). What sort of assumptions are you required to make? How do you know how to make them? What "built in" assumptions concerning the problem statements make them different from real problems? Why is it legitimate to skip the problem-defined and model-forming stages in these courses?

5.7 An open-top box is made of balsa wood 1 in. thick. The internal dimensions of the box are 12 in. wide by 12 in. long by 3 in. deep. This box is filled $2\frac{1}{2}$ in. deep with water at 36°F and placed on a rack in a large walk-in type cooler kept at 0°F. Suppose you are asked to estimate the time required to freeze a layer of ice $\frac{1}{2}$ in. thick. What model would you form? Guess an answer for the time. Compute an answer. Actually perform the experiment to see how well you have done.

5.8 An aluminum billet 5 in. in diameter and 24 in. long has been heated to 1000°F in a furnace. If this billet is allowed to cool on an open rack in the factory, how much time should be allowed for the billet to cool to 900°F? Form a model. Make a guess. Calculate an answer.

5.9 How long after a 60-watt light bulb is turned off can a person hold onto it with comfort? Define the problem, form a model, guess, and then solve.

5.10 For practice in making order-of-magnitude analyses, estimate the following:

a The number of gasoline service stations in the United States

b The average daily domestic water consumption per person in the United States

c The circulation of *Time*, *Playboy*, *The Atlantic Monthly*, and *Mechanical Engineering*

d The water lost by evaporation daily per square foot of freshwater reservoirs

e The magnitude of the thermal expansion and contraction in the Golden Gate Bridge.

f The volume of a bathtub

g The rate of flow of water from a typical spigot

h The temperature of a parked car's roof on a sunny day in August in Chicago

i The efficiency of a good golfer on a practice driving range

j The efficiency of a window air-conditioning unit

k The amount of water used to flush a typical toilet

l The heat loss from a typical insulated 30-gal domestic water heater

m The magnitude of the thermal expansion in a steam line 100 ft long

n The force exerted on a civil-rights demonstrator hit with a stream from a fire hose

o The stress in a cross-country high-voltage power line

p The deflection if a standard 2-in. I beam 10 ft long is loaded with 400 lb at a free end and cantilevered at the other

q The air-drag force on a 1965 Chevrolet traveling at 60 mph; at 30 mph

Make up and answer some questions of your own of this nature.

5.11 When one makes estimates and assumptions, it is very helpful to have a quantitative "feel" for the nature of things. For example:

a What is the density of steel, aluminum, balsa wood, oak, plastic, air, etc.?

b What is the melting point of steel, aluminum, mercury, carbon dioxide, etc.?

c What is the yield stress of steel, aluminum, balsa wood, oak, plastic, glass, etc.?

d What is the thermal conductivity of steel, aluminum, balsa wood, water, ice, air, etc.?

e How does thermal conductivity change (magnitude and direction) with temperature for steel, aluminum, wood, air, etc.?

5.12 Suppose that you are John, the engineer in the case study in Chap. 4 who is assigned to the die portion of the problem. Define the problem. Form a model stating the appropriate assumptions for a first try at an analysis in the die. (For a complete solution to this problem, see "Viscous Lubrication in Wire Drawing," by J. F. Osterle and J. R. Dixon, *Trans. Am. Soc. Lubricating Engrs.*, vol. 5, no. 1, April, 1962.)

5.13 A 20-ft-wide room is spanned by an I beam from which a crane is suspended. The crane is rated at 1,000 lb. What model would you form to design the I beam? Would you account for shock loads or other nonstatic conditions?

5.14 A cylindrical pressure vessel is to be designed for holding air at 400 psig. What model would you form to design the rivet system for this vessel?

5.15 In the study of traffic problems, an explanation is sought for the common accident where, in a long line of traffic, when the first car slows down or stops quickly, it is usually the third or fourth car that collides with the car in front of it. What model would you form of the factors that influence the car's motion under these circumstances?

5.16 In considering the need for future highways, how would you go about predicting the traffic flow between two cities, or between two given areas within one city?

5.17 What assumptions are inherent in most lubrication studies?

5.18 In designing a table fan for an office, what model would you form of the air to be handled by the fan? What model would you use for the air when designing an air compressor?

5.19 A recent graduate of an engineering school went to work as an engineer for a company that used a number of large cranes. Since this young man was alert and perceptive, he noted that the crane hooks in operation were excessively heavy. With the permission of his boss, therefore, he redesigned the hooks and had one made for trial purposes. On the first trial the boss instructed the crane operator to hoist the heavy trial load rapidly and with an initial jerk. The new hook broke. What was wrong with the young engineer's model? How can this sort of error be avoided?

5.20 Find out how each of the following vary with temperature, pressure, and humidity:

a Emissivity of aluminum

b Thermal conductivity of copper

c Coefficient of thermal expansion of steel

d Modulus of elasticity of water

e Specific heat of air

f Length of human hair

g Electrical conductivity of copper

5.21 It is common to neglect the mass of springs in dynamic calculations, though of course the spring itself is in fact a distributed mass. Find the mass M which when attached to a weightless spring with constant k will have the same natural frequency as a distributed spring of mass m with no weight attached. Neglect gravity.

6
PHYSICAL PRINCIPLES

6.1 INTRODUCTION

Most engineering-science courses are concerned with having students learn the principles, laws, rules, and empirical knowledge in the given subject-matter area. A design course, on the other hand, is concerned with having students learn to use purposefully what they already know. This is the reason engineering-science courses properly omit much emphasis on the problem-definition and model-formation steps of engineering problem solving. By omitting these steps, such courses can concentrate on the principles themselves. Design courses have the mission of bringing the total problem-solving process into perspective. In this text, for example, prior to this chapter, the problem-definition and model-formation steps have been emphasized. Later, the stages of evaluation, generalization, and communication will be covered. In this chapter, however, the material presented is a summary and overview of the physical principles used by engineers in solving problems. Obviously, space does not permit more than a brief discussion of the various principles, laws, and rules. The purpose here is not to teach the principles but simply to review their high points and to outline a kind of philosophy for their use. It is always best to use principles, laws, and rules in their purest and most simple and fundamental form. Derived, specialized, and sophisticated equations are special-purpose devices, and in the proper special circumstances they are very helpful. However, in general, and especially in new and unusual circumstances, it is only the pure and simple principle itself which is useful.

This point about using principles in their most elementary form is so important that some more elaboration is appropriate. Consider these two groups of words:

symbols	reality
equations	principles
knowledge	understanding

The symbol is not the thing. Many meaningless symbols (e.g., doleveıt, abalotics) can be devised. Words or other symbols don't have to stand for anything. They are just so much ink on paper, chalk on blackboard, or pressure waves on the air. Furthermore, equations, which are relations among symbols, need not be either true or meaningful.

All this is by way of saying that knowledge is not understanding. Writing down $d'Q - d'W = dE$ implies next to nothing about one's understanding of the first law of thermodynamics. Knowledge of any equation and its symbols implies next to nothing about one's understanding of the principle and the physical realities for which the symbols stand.

How does one know when he understands a principle? By *using* it to get an answer which checks reality, of course! There is no other way to prove understanding except by demonstrating it in action. This statement is true in a larger sense, but an example close to home will suffice. Students often say to their professors: "I really understand the material but I just can't work the problems." How utterly ridiculous! Such students are only kidding themselves and not their teachers. Professors know that these students have the *knowledge* of equations and symbols but not an *under-*

standing of the principles and the reality which the symbols symbolize. In this book, therefore, the use of principles in their most fundamental form will be stressed.

Before we go on, a word must be said about vectors. This book is intended both for students who have had and those who have not had vectors. All equations written in vector notation are repeated or explained in nonvectorial form. In either case, of course, what is important is the reader's ability to translate reality into the symbolic notion, whatever it is, and back again.

6.2 THE FORMAL AND THE CONCEPTUAL APPROACHES

It was said above that engineers should apply the *basic* laws of nature to their problems, rather than special or particular equations derived or developed from the more basic laws. Thus the *direct* application of Newton's second law, the first and second laws of thermodynamics, and conservation of mass to problem situations is encouraged. At this point a question always arises about the meaning of "basic." Just what is "basic"? The following example will illustrate the point.

In a heat-conduction problem it is always possible to start with the formal form of the Fourier heat-conduction equation

$$\alpha\,\nabla^2 T + \frac{G}{\rho C} = \frac{\partial T}{\partial t}$$

Working a simple steady-state heat flow problem with no internal heat generation and in one dimension, one can formally strike out terms of the above equation and obtain

$$\frac{d^2 T}{dx^2} = 0$$

This deduction of specific equations from the general governing equation, which has presumably been previously derived and memorized and is now "understood," will be called the *formal approach* in this book. It somewhat satisfies the desire to work with basic principles. No one would want to have to remember the particular result that in one-dimensional steady-state heat-flow problems with no sources, the equation is $d^2T/dx^2 = 0$, because there would be thousands of equally trivial things to remember. It is preferable to remember the much more general $\alpha\,\nabla^2 T + (G/\rho C) = \partial T/\partial t$. However, as will be seen, this general equation is not as basic as it seems. Formal equations of this nature are themselves derived from more basic principles. Thus, whenever possible, students should derive the differential equation from direct application of the principle involved in their problem. In the heat-flow problem, the more basic *concepts* of Fourier's law and the first law of thermodynamics are used to develop the formal equation

$$\alpha \nabla^2 T + \frac{G}{\rho C} = \frac{\partial T}{\partial t}$$

and they can be used directly in the one-dimensional problem as follows by starting with a differential element.

$$dq_{\text{in}} = -k\,dA\,\frac{\partial T}{\partial x} \rightarrow \boxed{dx} \rightarrow dq_{\text{out}} = -k\,dA\,\frac{\partial T}{\partial x} - \frac{\partial}{\partial x}\left(-k\,dA\,\frac{\partial T}{\partial x}\right)dx$$

An energy balance (note use of first law) gives

$$-k\,dA\,\frac{\partial T}{\partial x} = -k\,dA\,\frac{\partial T}{\partial x} + \frac{\partial}{\partial x}\left(-k\,dA\,\frac{\partial T}{\partial x}\right)dx$$

Now if k and A are constants (will the equation plugger remember that the formal equation above, thought to be so general or basic, assumed constant properties?), this reduces to

$$\frac{d^2T}{dx^2} = 0$$

This second approach is called the *conceptual approach*, as distinguished from the formal approach of applying a general but derived equation.[1] It is preferred because it brings the problem solver closer to his problem, closer to the physical laws, and it forces him to be aware of assumptions. Furthermore, and this is perhaps the most significant reason for preferring the conceptual technique, it puts the problem solver in a much better position to work new problems, deriving equations as needed, and to work other problems for which the principle applies but for which the formal equation does not. [Try, for example, to solve for the temperature distribution in a thin copper wire *with convection* from the surface using the formal $\alpha \nabla^2 T + (G/\rho C) = \partial T/\partial t$.] In terms of the example, the point is this: The truly basic idea is that heat conduction is proportional to the temperature gradient, the proportionality being dependent on the material properties. This idea can be applied to a differential element in *any* problem. This idea plus the first law, *together with some assumptions* (e.g., constant properties), are used to derive $\alpha \nabla^2 T + (G/\rho C) = \partial T/\partial t$, an equation which is now seen to be not so basic as it seemed at first glance.

Much of the same kind of discussion can be built around fluid mechanics (use of the formal Navier-Stokes equations vs. conceptual application of Newton's second law to a differential fluid element), thermodynamics (use of a formal vector general energy equation vs. a conceptual

[1] I am indebted to my colleague Professor Alden Emery, Jr., of Purdue University, for suggesting the words "formal" and "conceptual."

energy balance), dynamics (use of formal Lagrange equations vs. conceptual application of $F = ma$), and others.

It would not be proper to give the impression now that students should not know and use, when appropriate, certain widely applicable formal-type general equations. The heat-conduction equation, the energy equation, the Navier-Stokes equations, Lagrange's equations, etc., are powerful analytical tools. They should be learned well in engineering science courses. Students should learn in these courses the formal approach of applying these equations to particular situations. Those who go on to graduate school and those who specialize in one engineering area will remain familiar with one or more of these formal equations and will know how to use it. The more basic conceptual approach should be placed on at least an equal footing, however, so that those who forget Lagrange's equations can still apply $F = ma$ to a dynamic system. Those who forget the heat-conduction equation can still apply Fourier's law and the first law to a differential element. Also, many new problem situations *require* the conceptual method. In other words, students must learn to use *both* the formal and conceptual techniques. Most real engineering analysis problems growing out of design situations are best solved by the conceptual approach. This approach, now being deemphasized in most schools because of the time and discipline necessary to teach formal equations, should be reemphasized. In this chapter, however, both the formal and conceptual use of principles will be discussed.

6.3 TWO COORDINATE CONCEPTS: THE LAGRANGIAN AND THE EULERIAN

The following sections will discuss in brief and summary fashion the basic physical laws used by engineers, and some of the more common rules, definitions, and techniques. Before this, however, it is helpful to point out that the principles can be expressed in two fundamentally different coordinate systems. One is known as the lagrangian system and fixes a coordinate framework in a particular particle of material. As that material moves, so does the coordinate system. In the lagrangian system, therefore, attention is focused on a particular chunk of material, and it is followed wherever it goes.

The other system is known as the eulerian system. In this system a coordinate system is fixed in space. Instead of following a particular particle of material, attention is now focused on a particular point (or region) in space. This is the coordinate viewpoint which leads to the concept of the "control volume."[2]

[2] Here the word "system" will be used to denote a prescribed collection of matter, and the words "control volume" to denote a prescribed region in space.

There are no hard and fast rules about which viewpoint is best in any given problem situation. If properly used, both will lead to the same correct answer. Often, of course, one will be easier to use than the other. There is, however, regardless of which viewpoint is selected, a general failure of beginning engineering students to carefully define and describe the system or control volume to which they are applying principles. Literally, then, people who fail to make this definition do not know what they are talking about! In statics (since nothing is moving, the eulerian–lagrangian distinction does not exist), students' most common failure is to isolate a free body. The same is true in dynamics. In thermodynamics, failure to define the system (or control volume) is a common source of inability to solve problems. Thus some people who "understand" the principles involved never get them properly applied, because they have failed to define an appropriate, clearly identifiable system or region to which to apply them. It can be said that this failure really shows a lack of understanding of the principle, but, regardless, the point is this: Start the application of principles by defining specifically, in system or control-volume terms, what it is to which the principles are to be applied.

6.4 AN EXAMPLE: CONSERVATION OF MASS

Thus far in this chapter several new ideas have been introduced: formal and conceptual approaches to principles; lagrangian and eulerian coordinate systems; and some thoughts about symbols in general. Since conservation of mass is an easily understood principle, it is convenient to use it to illustrate the meanings of all these new terms.

Conceptual Approach—Lagrangian View In using the lagrangian (or system) viewpoint, attention is focused on a particular piece of matter, which is followed wherever it goes. Conceptually, one can imagine riding along with the system as it proceeds along its path with the coordinate axes also attached. The conservation-of-mass concept is simply that the mass of the defined system is constant. If the symbol used to denote the mass of the piece of matter in question is M, then the conceptual view of the principle of conservation of mass is simply that $M =$ constant or $dM = 0$ or $dM/dt = 0$ or $(\partial/\partial t)\int_{\mathcal{V}} \rho \, d\mathcal{V} = 0$.

Conceptual Approach—Eulerian View The eulerian viewpoint focuses attention, not on a system of matter, but on a region of space. The region may move. (It may even change size or shape, but this is complex to handle and is seldom done.) Conceptually, one imagines standing back to watch the specified region (control volume), while material or work and heat enter, leave, or pass through. The conservation-of-mass concept applied to such a region means simply that the mass which enters minus the mass which leaves equals the increase of mass inside the region. Symbolically, $M_{\text{in}} - M_{\text{out}} = \Delta M_{\text{inside}}$. Since it is often a convenience, the concept can be expressed in terms of rates also: The rate of mass flow into the region minus the rate of mass flow out of the region equals the rate of change of mass inside the region. Symbolically,

$$\left(\frac{dm}{dt}\right)_{\text{in}} - \left(\frac{dm}{dt}\right)_{\text{out}} = \left(\frac{dM}{dt}\right)_{\text{inside}}$$

where m stands for mass entering or leaving and M stands for mass inside. In vector notation, this same equation is more elegantly written as

$$-\oint_A \rho(\mathbf{V} \cdot \mathbf{n})\, dA = \frac{\partial}{\partial t}\int_{\mathcal{V}} \rho\, d\mathcal{V}$$

where $\mathbf{n}$ is the outward-drawn unit normal vector, ρ is the local density, $\mathbf{V}$ is the velocity, A is the surface area of the region, $\mathcal{V}$ is the volume of the region, and t is time.

When the conceptual view is used in a problem, the equations derived above are actually rederived for the special case of the problem, starting simply with the concept itself. In the formal approach, however, the derived general equations are the starting points, and the equations are applied deductively to the specific problems.

Formal Approach—Lagrangian View The derived equations are simply M = constant or $dM = 0$ or $dM/dt = 0$ or $\partial/dt \int \rho\, d\mathcal{V} = 0$. Applied to a problem, normally the appropriate form is merely treated mathematically with the other equations governing the situation. No further derivation or engineering thought is required unless there are simplifications which can be made in the particular problem. For example, if a system is incompressible, then ρ = constant, and the equation can be reduced to $\mathcal{V}$ = constant or $d\mathcal{V} = 0$ or $d\mathcal{V}/dt = 0$. This simply means that the volume of the incompressible system is constant, which is conceptually obvious too.

Formal Approach—Eulerian View The derived equations in terms of rates of flow for the control volume are $(dm/dt)_{\text{in}} - (dm/dt)_{\text{out}} = (dM/dt)_{\text{inside}}$ or vectorially $-\oint_A \rho(\mathbf{V} \cdot \mathbf{n})\, dA = (\partial/dt)\int_{\mathcal{V}} \rho\, d\mathcal{V}$. In a formal approach to a problem these equations would be applied to the problem as they are, terms being struck out or simplified as appropriate. A steady-flow problem means no time variation inside, so $(\partial/\partial t)\int_{\mathcal{V}} \rho\, d\mathcal{V} = 0$ and

$(dM/dt)_{\text{inside}} = 0$. The equations then become $(dm/dt)_{\text{in}} = (dm/dt)_{\text{out}}$ and $\oint_A \rho\mathbf{V}\cdot\mathbf{n}\,dA = 0$. Conceptually, in a steady-flow situation the engineer would begin with the concept and note that the mass which enters must equal the mass which leaves. Or in rate terms $(dm/dt)_{\text{in}} = (dm/dt)_{\text{out}}$. Expressed vectorially, this is also $\oint_A \rho\mathbf{V}\cdot\mathbf{n}\,dA = 0$.

In this simple case of conservation of mass, the formal and conceptual approaches are equally simple and productive. There are problems, however, where one or the other is vastly superior. Students should be able to handle both methods comfortably and with facility.

6.5 NEWTON'S LAWS OF MOTION—LAGRANGIAN VIEW

Newton's laws of motion are often subdivided into the first, second, and third laws. The distinction is not really necessary, however, since the first and third are only special cases of the second, which is the familiar $F = ma$. The first law says that if there are no external forces, there will be no acceleration, and hence the velocity of a body will not change. The third law says that if there is no acceleration, the external forces on the body must balance. This is often stated as action equals reaction.

The third law is the basis for the area of analysis called statics. Thus the principle underlying statics is simple enough. Students' difficulties with statics, when there are some, do not involve inability to write down $\sum F_x = \sum F_y = \sum F_z = \sum \tau = 0$. The difficulties that arise are caused by an unwillingness to isolate free bodies to which to apply the above equations. Such unwillingness is a demonstration of lack of understanding, because the principle is meaningless except as applied to an isolated body. It is a contradiction for anyone who fails to isolate free bodies to say he understands statics.

Newton's third law is also part of the basis for the engineering science of elasticity, together with the concepts of stress and strain, the constraints imposed by geometric compatibility, and constitutive equations pertaining to the material itself (e.g., Hooke's law). When certain assumptions are permissible, such as that plane cross sections of beams remain plane, elasticity simplifies to the area called strength of materials. The concept of static balance of forces underlies all these topics, however.

Newton's second law involves much greater conceptual difficulty than the first and third laws. Relativity effects will be neglected in this book. It is not possible, however, to neglect all the complicating factors which are implied by $\mathbf{F} = m\mathbf{a}$. First of all, the equation must be applied to isolated free bodies with constant mass. As in statics, many difficulties which underlie the use of $\mathbf{F} = m\mathbf{a}$ result, not from lack of knowing the equation, but from lack of understanding the principle. When one writes $\mathbf{F} = m\mathbf{a}$, there are a lot of unwritten words which define the total meaning of the principle. It is to be applied to isolated systems with constant mass. $\mathbf{F}$ stands for the sum of the external forces. What else is implied?

It is also important to bear in mind that the principle is directional. In vector notation, this is obvious because vectors are themselves directional:

$$\mathbf{F} = m\mathbf{a}$$

In nonvector notation, there is

$$F_x = ma_x$$
$$F_y = ma_y$$
$$F_z = ma_z$$

In the above equations, it is taken for granted that all parts of the constant mass m have the same velocity. In the event that this is not the case, one can make a lumped-parameter type of assumption and work with the center of mass. This will introduce increasing amounts of error, however, as the motions involved and the size of the system approach the same dimensions. Also, the lumped-parameter assumption is unsuitable for fluid systems. In these cases, of course, the distributed nature of the system must be handled by integration, starting with

$$d\mathbf{F} = dm \cdot \mathbf{a}$$

or

$$d\mathbf{F} = dm \cdot \frac{d\mathbf{V}}{dt}$$

By chain-rule differentiation

$$\mathbf{a} = \frac{d\mathbf{V}}{dt} = V_x \frac{d\mathbf{V}}{dx} + V_y \frac{d\mathbf{V}}{dy} + V_z \frac{d\mathbf{V}}{dz} + \frac{\partial V}{\partial t} \equiv \frac{D\mathbf{V}}{Dt}$$

where D/Dt is called the substantive derivative. Thus

$$d\mathbf{F} = \frac{D}{Dt} (dm \cdot \mathbf{V})$$

Integration gives

$$\mathbf{F} = \frac{D}{Dt} \int_{\mathcal{V}} \rho \mathbf{V} \, d\mathcal{V}$$

where $\mathcal{V}$ is volume and ρ is density. In nonvectorial notation the above equations in the x direction are

$$dF_x = dma_x$$

$$dF_x = dm\frac{dV_x}{dt}$$

$$a_x = V_x\frac{dV_x}{dx} + V_y\frac{dV_x}{dy} + V_z\frac{dV_x}{dz} + \frac{\partial V}{\partial t} = \frac{DV_x}{Dt}$$

$$dF_x = \frac{D}{Dt}dmV_x$$

$$F_x = \frac{D}{Dt}\int_{\mathcal{V}} \rho V_x \, d\mathcal{V}$$

Finally, and most important, it must be remembered that the principle applies only in what is called an inertial or Newtonian frame of reference. This means that the acceleration **a** must be expressed in terms of a reference frame which is either stationary or moving at constant velocity. For most engineering systems, the earth serves as a suitable inertial reference frame. When does it not?

When the times or motions are such that the motion of the earth during the time of the analysis is not negligible, then the earth is not a suitable reference. Rocket and satellite problems are examples. In such cases, either the sun or the so-called fixed stars must be used as a coordinate frame, and accelerations must be expressed with reference to them.

Often it is necessary to deal with bodies in terms of a noninertia reference frame xyz which is in motion with respect to a proper inertial frame XYZ. Kinematic considerations lead to the following equation, which relates the acceleration relative to the noninertial XYZ frame to the acceleration in the inertial xyz frame:

$$\mathbf{a}_{XYZ} = \mathbf{a}_{xyz} + (\ddot{\mathbf{R}}) + (2\boldsymbol{\omega} \times \mathbf{V}) + (\dot{\boldsymbol{\omega}} \times \mathbf{r}) + (\boldsymbol{\omega} \times (\boldsymbol{\omega} \times \mathbf{r}))$$

where $\mathbf{R}$ is the position vector from the XYZ origin to the xyz origin, $\mathbf{V}$ is velocity relative to the xyz frame, $\mathbf{r}$ is the position vector from the xyz origin, and $\boldsymbol{\omega}$ is rotation relative to the xyz frame. In nonvector notation, the above equation is

$$(a_x)_{XYZ} = (a_x)_{xyz} + \ddot{R}_x + 2(\omega_y V_z - \omega_z V_y) + (\omega_y V_z - \omega_z V_y) + \omega_x \omega r \cos \theta_{\omega r} - \omega^2 r_x$$

where the subscripts refer to components of the various quantities as defined above in the x, y, or z directions, ω is the magnitude of the rotation, and $\theta_{\omega r}$ is the angle between ω and r. Students should recognize the $2\boldsymbol{\omega} \times \mathbf{V}$ or $2(\omega_y V_z - \omega_z V_y)$ terms as the so-called Coriolis effect.

The above general equations may be applied formally, but a great deal of care must be used to ensure that the symbols are used properly and the implied rules obeyed. Conceptually, the principle $F = ma$ can be used, the necessary expressions for acceleration being developed in terms of an inertial frame appropriate to the particular situation.

One type of student error is so common that it deserves special mention. For a variable-mass system, such as a rocket, it is wrong to say $F = (d/dt)mV = m(dV/dt) + V(dm/dt)$. The mathematics of this may be all right, but the physics is incorrect. The principle applies to a constant-mass system. Variable-mass systems can be best handled using an eulerian view, which will be discussed below. The point is this: One cannot go around indiscriminately applying $F = ma$ to everything that moves. You must understand the principle behind the equation.

Newton's second law is often conveniently applied as a moment equation. The familiar result $\tau = I\alpha$ is analogous to $F = ma$. There are, of course, unwritten implications which go along with $\tau = I\alpha$, just as there are with $F = ma$, the major one being the necessity for an inertial reference frame. The rotational form of the second law can be obtained formally in the following manner:

$$d\boldsymbol{\tau} = \mathbf{r} \times d\mathbf{F} = \mathbf{r} \times \frac{D}{Dt}(dm \cdot \mathbf{V}) = \frac{D}{Dt}(\mathbf{r} \times dm \cdot \mathbf{V}) = \frac{d\mathbf{H}_{xyz}}{dt}$$

where $\mathbf{H}_{xyz}$ is the moment of momentum in an inertia frame. Integration for a finite body gives

$$\boldsymbol{\tau} = \frac{D}{Dt}\int_{\mathcal{V}} (\mathbf{r} \times \mathbf{V})\rho \, d\mathcal{V}$$

For a rigid body rotating in an inertial frame, these reduce to three equations of the form

$$\sum \tau_x = I_{xx}\dot{\omega}_x - I_{xy}\dot{\omega}_y - I_{xz}\dot{\omega}_z$$

where $$I_{xx} = \int_{\mathcal{V}} \rho(y^2 + z^2) \, d\mathcal{V}$$

$$I_{xy} = \int_{\mathcal{V}} \rho xy \, d\mathcal{V}$$

Thus for a simple system rotating about a single axis, these simply give

$$\sum \tau = I_{xx}\dot{\omega}_x = I_{xx}\alpha_x$$

This is a result which can be derived by straightforward *conceptual* application of $\mathbf{F} = m\mathbf{a}$ to the simple rotating system. As can be seen, arriving at it by reduction of the complete formal vector equation is a treacherous process. On the other hand, students will also observe that for more complex situations, direct application of $\mathbf{F} = m\mathbf{a}$ conceptually, though always possible, will be a gruesome experience. This is the reason that it is contended here that students should be aware of learning *both* the formal and conceptual approaches, so they will be at home with both and can select the one which is best for the problem at hand.

6.6 MASS, MOMENTUM, ENERGY, AND ENTROPY—LAGRANGIAN AND EULERIAN VIEW—FORMAL APPROACH

Figures 6.1 and 6.2 show a summary of the equations which symbolize the principles of conservation of mass, Newton's second law, and the first and second laws of thermodynamics. As always, formal application of these equations requires an understanding of the underlying principle and the ability to connect the symbols with physical reality. The equations themselves are merely shorthand notation. Many implied conditions are left unwritten, as was seen in the preceding discussion of Newton's second law.

Principle	*System or lagrangian equation*	*Eulerian or control-volume equation*
Conservation of mass	$0 = \frac{d}{dt}\int_{\mathcal{V}} \rho \, d\mathcal{V}$	$0 = \frac{d}{dt}\int_{\mathcal{V}} \rho \, d\mathcal{V} + \oint_A \rho \mathbf{V} \cdot \mathbf{n} \, dA$
First law of thermodynamics	$d'Q - d'W = dE$	
	$\dot{Q} - \dot{W} = \frac{d}{dt}\int_{\mathcal{V}} \rho e \, d\mathcal{V}$	$\dot{Q} - \dot{W} = \frac{d}{dt}\int_{\mathcal{V}} \rho e \, d\mathcal{V} + \oint_A \rho(e + pv) \, \mathbf{V} \cdot \mathbf{n} \, dA$
Newton's second law	$F = \frac{D}{Dt}\int_{\mathcal{V}} \rho V \, d\mathcal{V}$	$F = \frac{dt}{dt}\int_{\mathcal{V}} \rho \mathbf{V} \, d\mathcal{V} + \oint \rho \mathbf{V} \, (\mathbf{V} \cdot \mathbf{n}) \, dA$
Second law of thermodynamics	$\frac{d'Q}{T} \leqq ds$	
	$\oint_A \frac{\dot{Q}}{T} \, dA \leqq \int_{\mathcal{V}} \rho s \, d\mathcal{V}$	$\oint_A \frac{\dot{Q}}{T} \, dA \leqq \frac{d}{dt}\int_{\mathcal{V}} \rho s \, d\mathcal{V} + \oint \rho s \mathbf{V} \cdot \mathbf{n} \, dA$

Fig. 6.1

Principle	*System or lagrangian equation*	*Eulerian or control-volume equation*
Conservation of mass	$0 = dM$	$0 = \frac{d}{dt}\int_{\mathcal{V}} \rho \, d\mathcal{V} + \oint_{A_{out}} \dot{m} \, dA_{out} - \oint_{A_{in}} \dot{m} \, dA_{in}$
	$0 = \frac{d}{dt}\int_{\mathcal{V}} \rho \, d\mathcal{V}$	
First law of thermodynamics	$d'Q - d'W = dE$	$\dot{Q} - \dot{W} = \frac{d}{dt}\int_{\mathcal{V}} \rho e \, d\mathcal{V} + \oint_{A_{out}} (e + pv)\dot{m} \, dA_{out}$
	$Q - W = \frac{d}{dt}\int_{\mathcal{V}} \rho e \, d\mathcal{V}$	$- \oint_{A_{in}} (e + pv)\dot{m} \, dA_{in}$
Newton's second law	$F_x = \frac{D}{Dt}\int_{\mathcal{V}} \rho V_x \, d\mathcal{V}$	$F_x = \frac{d}{dt}\int_{\mathcal{V}} \rho V_x \, d\mathcal{V} + \oint_{A_{out}} \rho V_x \dot{m} \, dA_{out} - \oint_{A_{in}} \rho V_x \dot{m} \, dA_{in}$
Second law of thermodynamics	$\frac{d'Q}{T} \leqq ds$	
	$\frac{Q}{T} dA \leqq \int_{\mathcal{V}} \rho s \, d\mathcal{V}$	$\frac{\dot{Q}}{T} dA = \int_{\mathcal{V}} \rho s \, d\mathcal{V} + \oint_{A_{out}} s\dot{m} \, dA_{out} - \oint_{A_{in}} s\dot{m} \, dA_{in}$

Fig. 6.2

The system equations are relatively straightforward. All that needs to be added here is that it is important to begin by carefully defining the system to which the principle is to be applied.

The control-volume equations are equally easy to interpret when the control volume is stationary. As has been mentioned, however, the principles can be applied to well-defined moving regions as well. It is usually assumed that the region does not change size or shape, but merely moves through space. Always, then, the **V** in $\mathbf{V}\cdot\mathbf{n}$ and $\dot{m}$ refer to the velocity or mass flow *relative* to the control volume. If the motion is constant with respect to an inertial frame, then the control volume itself defines an inertia frame. In this case the **V** in $\mathbf{V}\cdot\mathbf{n}$ and the $\dot{m}$ are still relative to the control volume, but the other **V**'s may be either relative to the control volume or relative to any other convenient inertia frame. If the control volume is accelerating and hence is itself a noninertial frame, then the separate **V**'s and V_x's *must* be relative to some noninertial frame in order to satisfy the implications of Newton's Law. As always, though, the **V** in $\mathbf{V}\cdot\mathbf{n}$ and $\dot{m}$ are relative to the control volume, because their function is to describe the mass flow into or out of the control volume itself.

6.7 CONCEPTUAL APPROACHES TO MASS, MOMENTUM, ENERGY, AND ENTROPY PRINCIPLES

There are formal mathematical techniques[3] converting the system (lagrangian) equations into the control-volume (eulerian) equations shown in Figs. 6.1 and 6.2. It is important for engineering students to be able to do this kind of formal manipulation. When the time comes to use the principle, however, the student must understand the physical significance of each symbol in the equation. Alternatively, the principles can often be applied conceptually without special recourse to the formally derived equation.

It was pointed out in Sec. 6.4 that conservation of mass conceptually applied to a region of space means that the rate of mass flow in minus the rate of mass flow out equals the rate of change of mass inside. That, in words, is the same as

$$0 = \frac{d}{dt}\int_{\mathcal{V}} \rho \, d\mathcal{V} + \oint \rho(\mathbf{V}\cdot\mathbf{n})\, dA$$

or

$$0 = \frac{d}{dt}\int_{\mathcal{V}} \rho \, d\mathcal{V} + \oint_{A_{\text{out}}} \dot{m}\, dA_{\text{out}} - \oint_{A_{\text{in}}} \dot{m}\, dA_{\text{in}}$$

Conservation of energy is conceived conceptually in the same way. The

[3] See, for example, the books by Shames or Shapiro listed in the Suggested Reading.

rate of energy flow into the region minus the rate of energy flow out of the region equals the rate of change of energy inside. One has to account here for energy in all its forms: heat, work, internal, kinetic, potential, etc. Using usual thermodynamic notation and sign convention,

$$\dot{Q} + \int_{A_{\text{in}}} \dot{m}e\, dA_{\text{in}} - \dot{W}_{\text{total}} - \oint_{A_{\text{out}}} \dot{m}e\, dA_{\text{out}} = \frac{d}{dt}\int_{\mathcal{V}} \rho e\, d\mathcal{V}$$

where e is the total energy $(u + \frac{1}{2}V^2 + gz)$ per unit mass of material. $\dot{W}_{\text{total}}$ represents the total rate of flow of work out of the region and this is commonly divided into two parts:

$$\dot{W}_{\text{total}} = \dot{W}_x + \text{flow work}$$

$$= \dot{W}_x + \oint_{A_{\text{out}}} pv\dot{m}\, dA_{\text{out}} - \oint_{A_{\text{in}}} pv\dot{m}\, dA_{\text{in}}$$

For the composition of $\dot{W}_x$ students are referred to Sec. 6.8. Combining these equations and rearranging into the familiar form gives

$$\dot{Q} - \dot{W}_x = \frac{d}{dt}\int_{\mathcal{V}} \rho e\, d\mathcal{V} + \oint_{A_{\text{out}}} (e + pv)\dot{m}\, dA_{\text{out}} - \oint_{A_{\text{in}}} (e + pv)\dot{m}\, dA_{\text{in}}$$

Using vector notation,

$$\dot{Q} - \dot{W}_x = \frac{d}{dt}\int_{\mathcal{V}} \rho e\, d\mathcal{V} + \oint_{A} \rho\, (e + pv)(\mathbf{V}\cdot\mathbf{n})\, dA$$

$(e + pv)$, of course, is the familiar $(h + \frac{1}{2}V^2 + gz)$, where h is the enthalpy defined as $u + pv$. Thus conceptual application of the motion of conservation of energy leads to the same resulting equation as the formal transformation. Either the formal equation or the direct concept may be used in particular problems, of course.

Conceptual application of Newton's second law to a region of space is also not difficult. The idea is this: External forces acting at the boundary tend to increase or decrease the momentum inside the region in accordance with the $F = (d/dt)mV$ idea. In addition, of course, there is the flow of momentum into and out of the region, and there may be a net change of momentum inside the region. Since the principle is Newton's law, all the usual implications of that law are present: the law is directional and applies only in an inertial frame of reference. With this in mind, the concept in words is this: The external forces plus the rate of flow of momentum into the region minus the rate of flow of momentum out equals the rate of change of momentum inside. Symbolically in the x direction,

$$F_x + \int_{A_{\text{in}}} \dot{m}V_x\, dA_{\text{in}} - \int_{A_{\text{out}}} \dot{m}V_x\, dA_{\text{out}} = \frac{d}{dt}\int_{\mathcal{V}} \rho V_x\, d\mathcal{V}$$

In vector notation,

$$\mathbf{F} = \frac{d}{dt}\int_{\mathcal{V}} \rho \mathbf{V}\, d\mathcal{V} + \oint_A \rho(\mathbf{V}\cdot\mathbf{n})\mathbf{V}\, dA$$

The meaning of the **V**'s and V_x's is the same as discussed in the preceding section. Students should be certain they know which V's are relative to the control volume and which are relative to an inertial frame, and why.

Conceptual application of the second law of thermodynamics is equally possible if it is remembered that the law is not a conservation law. The system expression indicates that the entropy of an adiabatic system increases and that for nonadiabatic systems

$$\frac{d'Q}{T} \leqq ds$$

Similarly, the entropy of an adiabatic region must increase or be greater than $d'q/T$ evaluated at its boundary. The situation is a little more complex, however, because of the mass flow which carries entropy with it. It is convenient to think of $d'q/T$ as a kind of "flow." Then keeping the inequality in mind, the word equation is: $d'q/T$ plus the rate of flow of entropy into the region minus the rate of entropy out *is less than* the rate of change of entropy inside. Symbolically,

$$\oint \frac{\dot{Q}}{T}\, dA + \oint_{A\,\text{in}} s\dot{m}\, dA_{\text{in}} - \oint_{A_{\text{out}}} s\dot{m}\, dA_{\text{out}} < \frac{d}{dt}\int_{\mathcal{V}} \rho s\, d\mathcal{V}$$

Vectorially, $$\oint \frac{\dot{Q}}{T}\, dA < \frac{d}{dt}\int_{\mathcal{V}} \rho s\, d\mathcal{V} + \oint_A \rho s(\mathbf{V}\cdot\mathbf{n})\, dA$$

Figure 6.3 is a summary in vector notation of equations which represent the concepts being discussed in this section. Students who will take the

Input terms		*Storage terms*		*Flux terms*
0	$=$	$\frac{d}{dt}\int_{\mathcal{V}} \rho\, d\mathcal{V}$	$+$	$\oint_A \rho\, \mathbf{V}\cdot\mathbf{n}\, dA$
$\dot{Q} - \dot{W}_x$	$=$	$\frac{d}{dt}\int_{\mathcal{V}} \rho\, e\, d\mathcal{V}$	$+$	$\oint_A (e + pv)\, \rho\, \mathbf{V}\cdot\mathbf{n}\, dA$
$\mathbf{F}$	$=$	$\frac{d}{dt}\int_{\mathcal{V}} \rho\mathbf{V}\, d\mathcal{V}$	$+$	$\oint_A (\mathbf{V}\cdot\mathbf{n})\, dA$
$\frac{\dot{Q}}{T}$	$<$	$\frac{d}{dt}\int_{\mathcal{V}} \rho s\, d\mathcal{V}$	$+$	$\oint_A s\rho(\mathbf{V}\cdot\mathbf{n})\, dA$

Fig. 6.3

time and effort to work some problems using these principles both formally and conceptually will be greatly rewarded with increased understanding.

6.8 WORK

Mechanical work is a familiar phenomenon and one that is used early and often in explaining the general concept of work. The result is a general neglect and omission of other very important forms of work. Electrical and magnetic work effects are very significant in many fields. Effects such as surface tension (which becomes important in low-gravity fields), and the areas of thermoelectrics and magnetodynamics, involve work in unfamiliar ways. It should be emphasized that W is the system expression for the first law, and $\dot{W}_x$ in the control-volume expression includes *all* forms of work and not just mechanical. It is not the purpose here to do more than call the student's attention to these work terms. There are thermodynamic books described in the Suggested Reading at the end of the chapter which go into more detail, with examples and explanations.

6.9 HEAT TRANSFER

Conduction As noted in this example at the beginning of this chapter, the fundamental concept of conduction-heat transfer is that the heat flow is proportional to the temperature gradient and area, the proportionality factor being a function of the material itself and often its temperature. This concept is usually written symbolically as

$$d'q_x = -k\,dA_x \frac{\partial T}{\partial x}$$

Also, as noted previously, this concept together with the first law is used to arrive at a formal equation of heat conduction:

$$\nabla^2 T + \frac{G}{\rho C} = \frac{\partial T}{\partial t}$$

As always, it is valuable to be able to use the principle with either the formal or conceptual techniques.

Convection The basic concept of convection-heat transfer is that the heat flow is proportional to the temperature difference between the wall and the bulk or average temperature of the fluid, the proportionality factor being called the heat-transfer coefficient. Symbolically,

$$d'q = h\,dA(T_{\text{wall}} - T_{\text{fluid}})$$

Determination of convection-heat-transfer rates is largely, therefore, a problem in evaluating the coefficient h. It turns out that h is a function of nearly everything, but the fluid properties, the flow regime (laminar or turbulent), and the geometry are the major factors. Some analysis can be used, especially in laminar-flow cases, to compute h, but in general the complications are so great that empirical methods are necessary. Without the use of dimensional analysis, correlation of the experimental results would be nearly impossible. Consider, for example, the familiar equation for the heat-transfer coefficient for turbulent flow inside pipes:

$$Nu_D = 0.023\,(Re_D)^{0.8}(Pr)^{0.4}$$

or

$$h = \frac{k}{D}\,0.023\left(\frac{\rho V D}{\mu}\right)^{0.8}\left(\frac{C_p \mu}{k}\right)^{0.4}$$

This equation is an empirical approximation involving three variables: Nu_D, Re_D, and Pr. Without dimensional analysis it will consist of seven variables, and finding the correlation in the first place would be extremely difficult. Other correlations in terms of dimensionless groups are available in the heat-transfer literature for a wide variety of flow situations. Most of these call for evaluation of fluid properties at a temperature halfway between the wall temperature and the bulk temperature of the fluid.

Radiation Radiation-heat transfer is governed by the Stefan-Boltzmann fourth-power law:

$$d'q_{12} = \mathfrak{F}_{12}\, dA_1 \sigma (T_1^4 - T_2^4)$$

where σ is the universal constant 0.173×10^{-8} Btu/(hr)(ft^2)(°R^4) and $\mathfrak{F}_{12}$ is a factor accounting for geometry and the material property of emissivity. Just as most convection problems become problems in determining h, most radiation problems reduce to determining $\mathfrak{F}$. Details of these determinations, both experimentally and analytically, are the subject matter of courses in heat transfer.

6.10 ELECTRIC CIRCUIT PRINCIPLES

The principles which apply to electric circuit analysis are called Kirchhoff's voltage and current laws. The voltage-law concept is this: The sum of the voltage drops around a closed circuit is zero. The current-law concept is this: The sum of the currents at a junction is zero. Both of these laws require careful consideration of sign conventions.

The voltage law requires use of the voltage drops through such electrical components as resistors, capacitors, and inductors. The voltage drop through a resistor in the direction of current flow is, of course, iR, where

R is the resistance in ohms. The voltage drop in a capacitor in the direction of current flow is

$$\Delta V = \frac{q}{c} = \frac{1}{c}\int i\,dt$$

where c is the capacitance and q is the charge. For an inductance, the voltage drop is

$$\Delta V = L\frac{di}{dt}$$

where L is the inductance. Thus for the circuit shown in Fig. 6.4, application

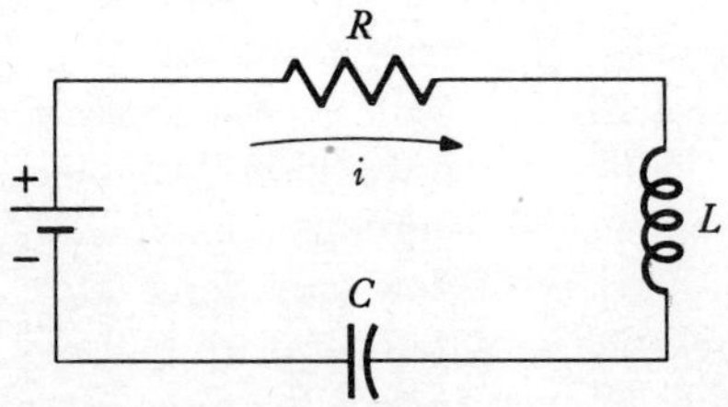

Fig. 6.4

of the voltage law gives

$$-E + Ri + L\frac{di}{dt} + \frac{q}{c} = 0$$

Differentiating and noting that $dq/dt = i$, this can be written as

$$L\frac{d^2i}{dt^2} + R\frac{di}{dt} + \frac{i}{c} = 0$$

6.11 REMARKS ON THE DERIVATION OF DIFFERENTIAL EQUATIONS

In the problem examples which follow, several differential equations are derived—in mechanics, heat transfer, and electric circuits. Several others have previously been derived in Chap. 4 and in this chapter. Students should pay particular attention to these derivations, which are all done from the conceptual approach. Using the formal approach, textbook authors or professors derive the differential equations once and for all. For the most part students watch this derivation, and though they get lots of practice in using the result, they have had very little to do themselves with the derivation. Since people learn mostly from what they do themselves, the result is a certain lack of confidence and ability in deriving differential equations.

The derivation of a differential equation requires the conceptualization of a differential change. This seems to be a hard step until one has successfully done it a few times. "Take a differential element" is and should be a hackneyed phrase in many areas such as conduction-heat transfer and fluid mechanics. "Consider a differential time change" is the frequent starting place for derivations in mechanics. There are no pat rules, however, and it is impossible to generalize the procedure. The example problems will have to serve the purpose. Or see also item 1 in the Suggested Reading. Derivation of differential equations, using principles conceptually, is a valuable skill for the engineer. Students generally need more practice at it. Don't let your textbook authors and instructors do all the work and hence all the learning!

Obtaining the differential equation for a problem is only half the problem, of course. Before the equation can be solved in any meaningful way, the boundary and/or initial conditions must also be specified. This is not always as trivial as it seems, and students quite often make errors of both omission and commission on boundary and initial conditions. In the examples which follow, opportunity will be given for students to derive the equations and specify boundary conditions. Wise students will take advantage of these opportunities to get some practice in an important skill.

6.12 EXAMPLE: AN ELECTRIC CIRCUIT PROBLEM

Statement of the Problem For use in the self-correcting control ("automation") of a complex manufacturing process, a circuit is needed

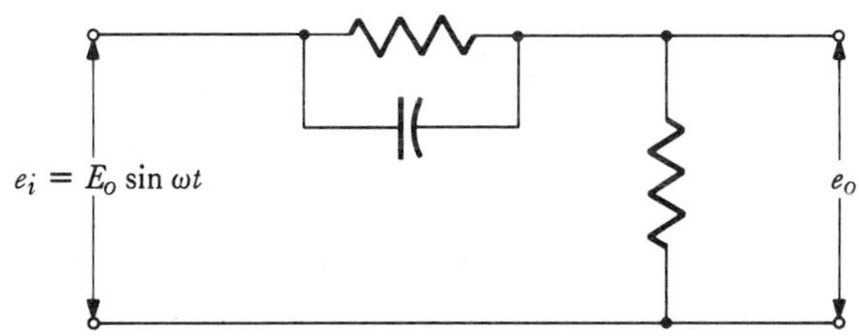

Fig. 6.5

which when supplied with a sinusoidal input of frequency 150 rad/sec will produce an output which leads the input by a phase lead of 30°. A circuit as shown in Fig. 6.5 has been proposed. Will it work, and, if so, what should be the values of the electrical components?

Definition of the Problem This is not a difficult problem to define in specific terms. It is again simply a problem of translation of a verbal statement into more exact symbolic notation, the resistors and capacitor being labeled as shown. Note that the input voltage as a sinusoidal function is also shown. The problem, therefore, is to determine e_0 in terms of E_0, ω, R_1, C, and R_2. With this accomplished, the next step will be to examine the result to see if there are values of R_1, C, and R_2 which will allow e_0 to lead e_i by the desired 30° phase angle.

Model Formation Here is a case where, in defining the problem, the model has essentially already been formed. Lumped parameters are used. The resistance of the connecting leads, junctions, etc., is neglected. Constant electrical properties are assumed.

Before going on now, the student should attempt to write the equations, based on the circuit principles involved, which govern this problem.

Application of Physical Principles The electric circuit principles needed are Kirchhoff's voltage and current laws. As a start it is necessary to define the currents in the various branches of the circuit. This is

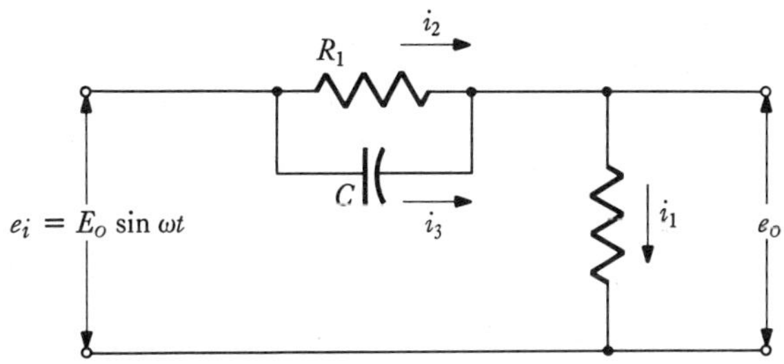

Fig. 6.6

shown in Fig. 6.6. Then, neglecting the current delivered to the output, the current law gives

$$i_1 = i_2 + i_3$$

The voltage laws provide three equations, because there are three loops:

$$e_i = i_2R_1 + i_1R_2$$

$$e_i = \frac{1}{C}\int i\,dt + i_1R_2$$

$$e_0 = i_1R_2$$

There is also $$e_i = E_i \sin \omega t$$

This set of equations describes the behavior of the model circuit. They must be solved for e_0 in terms of R_2, R_1, C, and time.

Before going on now, the student should attempt to set up the differential equation for i_1 from the above equations.

Computation Since $e_0 = i_1B_2$, if i_1 can be determined, e_0 is easily found and can be compared with e_i. Differentiating the second loop equation gives

$$\frac{de_i}{dt} = \frac{i_3}{C} + R_2\frac{di_1}{dt}$$

Now i_3 can be eliminated using $i_1 = i_2 + i_3$, and then the i_2 which appears is eliminated using $e_i = i_2R_1 + i_1R_2$. The result is

$$\frac{de_i}{dt} = \frac{i_1}{C} - \frac{1}{R_1C}(e_i - i_1R_2) + R_2\frac{di_1}{dt}$$

Note that $e_i = E_i \sin \omega t$ and hence

$$\frac{de_i}{dt} = E_i\omega \cos \omega t$$

This can be rewritten as

$$\frac{di_1}{dt} + \frac{R_1 + R_2}{R_1R_2C}i_1 = \frac{E_i\omega}{R_2}\cos \omega t - \frac{E_i}{R_1R_2C}\sin \omega t$$

Before going on now, students should attempt to solve this equation.

If one neglects the transient portion of the solution, the solution is

$$i_1 = A \cos \omega t + B \sin \omega t$$

where $A = \dfrac{E\omega}{R_1R_2C}\dfrac{(1 + R_1Ca)}{(\omega^2 + a^2)}$

$$B = \frac{ER_1C\omega^2 - aE}{R_1R_2C(\omega^2 + a^2)}$$

$$a = \frac{R_1 + R_2}{R_1R_2C}$$

Rewriting the current i_1 as

$$i_1 = \sin \phi \cos \omega t + \cos \phi \sin \omega t$$

it follows that $\qquad \sin \phi = \dfrac{E\omega(1 + R_1Ca)}{R_1R_2C(\omega^2 + a^2)}$

$$\cos \phi = \frac{ER_1C\omega^2 - aE}{R_1R_2C(\omega^2 + a^2)}$$

Hence $\qquad \tan \phi = \dfrac{E\omega(1 + R_1Ca)}{ER_1C\omega^2 - aE} = \dfrac{A}{B}$

From trigonometry, the above expression for i can be written alternatively as

$$i_1 = \cos(\omega t - \phi)$$

Since $e_0 = i_1R_2$, the phase lead angle ϕ should be 30° and hence R_1, R_2, and C must be such that

$$\tan 30° = \frac{A}{B} = \frac{E\omega(1 + R_1Ca)}{ER_1C\omega^2 - aE}$$

Some simplification gives

$$\tan 30° = \frac{\omega R_1C(2R_2 + R_1)}{\omega^2R_1^2R_2C + R_1 - R_2}$$

The conclusion is, therefore, that the proposed circuit will work and that the resistors and capacitors chosen must satisfy the above equation for e_0 to lead e_i by 30°. In this case, a communication on the above work is left for the student.

6.13 EXAMPLE: THE CONTRARY PUSHROD

Statement of the Problem In a certain fast-acting cycling machine, a pushrod is returned to its "home" position after an operation by a hammer. See Fig. 6.7*a*. The pushrod and hammer are essentially free

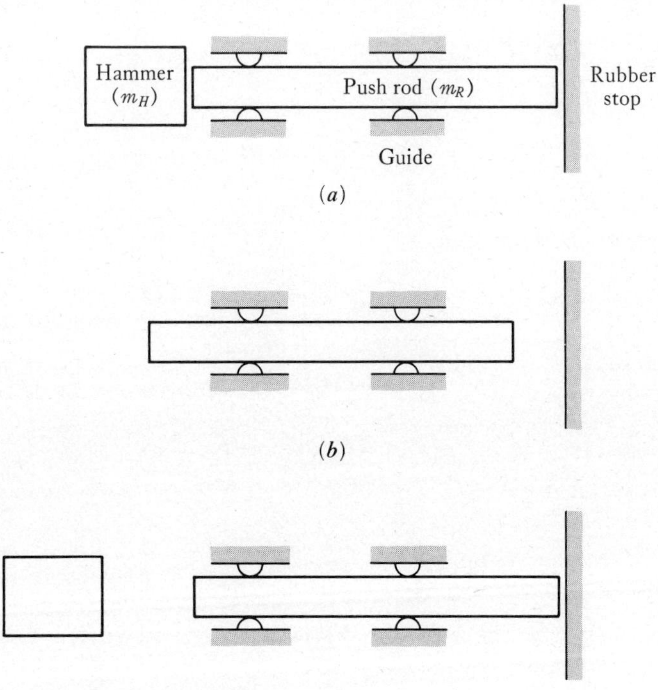

Fig. 6.7 (*a*) Pushrod and hammer in "home" position. (*b*) Motionless pushrod after its operation, awaiting hammer to drive it "home." (*c*) After new design, pushrod ends up at "home" but hammer does not.

floating, but are constrained to move along the line which coincides with the axis of the pushrod.

The whole process had been working well, but since a recent redesign of the hammer to make it lighter, it has been found that the hammer is not moving home with the pushrod as it should. See Figs. 6.7*b* and 6.7*c*.

What is wrong, and what do you recommend?

Before going on now, the student should define the problem.

Definition of the Problem To define the problem in this one is a big portion of the solution. Why doesn't the hammer return along with the pushrod? At first one might say that it does but for some reason it bounces back since it has been made lighter. This is a possibility, of course, but because the stop worked before and because the pushrod stays put, it doesn't seem highly probable. Before analyzing that unlikely possibility, one should look for some alternative explanations.

In addition to the contact of rod and hammer with the stop, the other interaction which takes place is between the rod and hammer when the hammer strikes. Maybe this is the trouble spot! Maybe the new lighter hammer doesn't "overpower" the rod and carry it along, but bounces back instead. Now this seems quite possible and is deserving of analysis.

Before proceeding, however, the engineer sees another possibility: perhaps the new design reduced the velocity as well as the mass of the hammer. He was assured, however, by the designer that the hammer speed was not altered.

His decision, therefore, is to analyze the hammer–pushrod collision to see what effect hammer mass could have on the resulting motions. Referring to Fig. 6.7, he defines his problem specifically as follows: Given M_H, M_R, V_{H1}, $V_{R1} = 0$, and l, determine V_{H2} and V_{R2}, magnitude and direction, in terms of the given parameters. Establish the effect of changes in M_H from the result.

Making a Guess The following is a true story. The engineer working on the hammer-pushrod problem had a pool table in his basement. He noticed one evening that if a direct collision is made between a moving ball (with no spin on it) and a stationary one, the stationary one moves but the moving one stops. Could the same effect be stopping the hammer? What is there about the two billiard balls that produces this effect?

Model Formation In the model formation, the engineer must assume something about the kind of collision which takes place between hammer and pushrod. Are elastic traveling waves within the hammer and/or pushrod important? Perhaps, but it seems more appropriate to try the simpler motion first, the one suggested by the billiard-ball phenomenon mentioned above. That is, consider that the hammer and pushrod act as rigid bodies and assume that the collision is a *conservative* one in which energy is not dissipated irreversibly. Such an assumption may fail to explain the observed phenomena, of course, but still it is a reasonable place to begin. Thus the model is that of a conservative collision between two rigid bodies, one of which is moving and one of which is initially at rest.

Before going on now, the student should consider which physical principles are significant in this problem.

Application of Physical Principles The problem is to find the final velocities of the hammer and pushrod following their collision in terms of their masses and initial velocities. Before the collision, there is

After the collision,

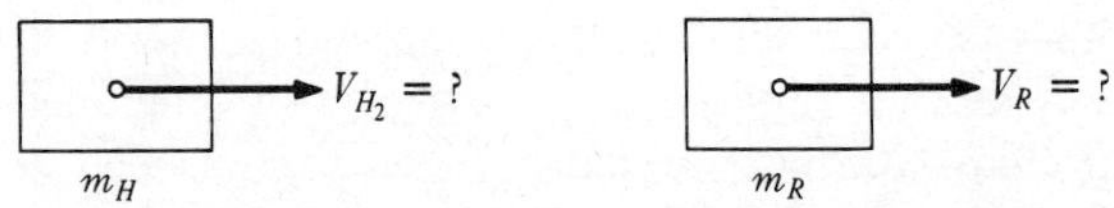

The principles which are significant in this problem are conservation of momentum and conservation of energy. (Did you get both?) Applying conservation of momentum, and noting that there are no external forces in the model and that all motion is in a single direction, there results

$$m_H V_{H1} = m_H V_{H2} + m_R V_{R2}$$

Conservation of energy gives

$$\tfrac{1}{2} m_H V_{H1}^2 = \tfrac{1}{2} m_H V_{H2}^2 + \tfrac{1}{2} m_R V_{R2}^2$$

Thus there are two independent equations and two unknowns V_{H2} and V_{R2}.

Computation Simultaneous solution of the above two equations for the velocities of the rod and hammer after impact gives

$$V_{R2} = \frac{2}{1 + m} V_{H1}$$

$$V_{H2} = \frac{1 - m}{1 + m} V_{H1}$$

where m equals the ratio of the masses:

$$m = \frac{m_R}{m_H}$$

During the solution, both m and V_{R2} were used as divisors, so that the results do not apply if either m or V_{R2} is zero.

Evaluation and Generalization The pushrod-hammer mechanism worked fine until the mass of the hammer was reduced. Examination of the above results shows that if $m_H \gg m_R$ (m small), then $V_{R2} \to 2V_{H1}$ and $V_{H2} \to V_{H1}$. In this case, the hammer will continue its motion after the collision. On the other hand, if $m_H \ll m_R$ (m large), $V_{R2} \to 0$ and $V_{H2} \to -V_{H1}$. In this limiting case, the hammer simply bounces back off the much heavier pushrod. Recalling the billiard-ball situation where $m = 1$, we get the results

$$V_{R2} = V_{H1}$$

and

$$V_{H2} = 0$$

This appears to be precisely what is happening in the machine. The engineer checks the masses and finds that as a result of the new design, they are nearly equal. Hence his report (not included here) recommends that the hammer be made heavier or the pushrod lighter in order to avoid the "billiard-ball effect."

6.14 EXAMPLE: ACCELERATING A WHEEL

Statement of the Problem For on-the-car dynamic balancing of automobile wheels, a small powered rotating cylinder is used in contact with the tire as shown in Fig. 6.8. In using this equipment, it is common

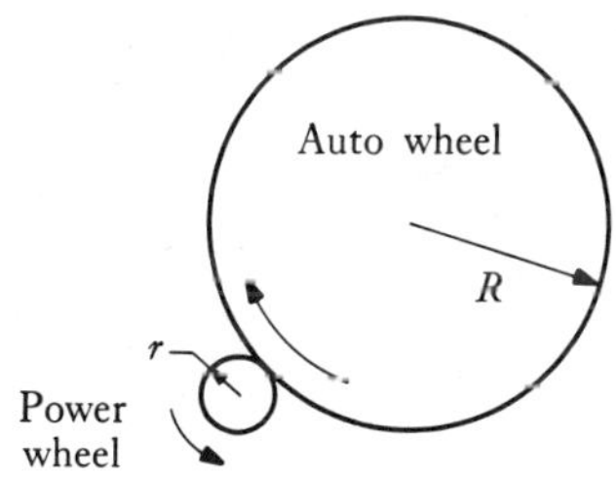

Fig. 6.8

practice in garages to bring the power wheel up to the desired speed before contacting the tire. Then the power wheel is jammed tight against the tire. For design purposes it is necessary to have an estimate of the energy dissipated as heat when this practice is followed.

Before proceeding, the student should make an attempt to define the problem.

Definition of the Problem This is certainly a vague problem statement as it stands. However, it is not difficult to define an engineering problem in this case, if only some symbols are defined to represent the important parameters in the problem situation (see Fig. 6.9). The problem, then, is to determine the energy dissipated as heat when the auto wheel is accelerated from rest to a speed ω_w by a power wheel running at w_p in terms of R, r, I_w, I_p, ω_w, and ω_p. Before going on, the student should now

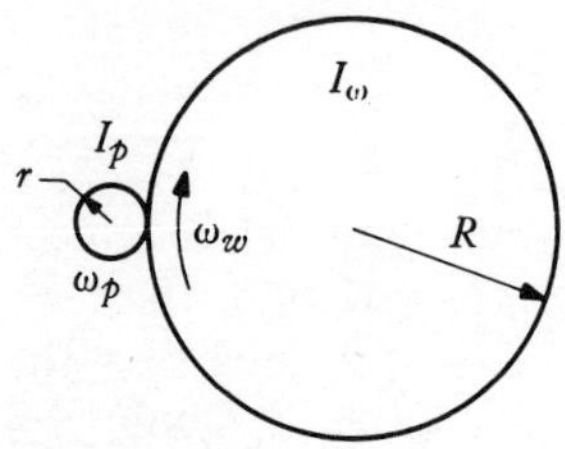

Fig. 6.9

consider what assumptions are appropriate in forming a model of this situation and what principles are likely to be useful.

Model Formation There is the possibility that the power wheel will slow down somewhat when it is required to accelerate the wheel, but if it is driven by an a-c induction motor (which is most likely), such speed reduction will be very small. In any event, the conclusions will not be affected much by assuming falsely that ω_p is a constant, so this will be done.

Where else could energy go besides dissipation as heat? Some energy, of course, is used to start the auto wheel and is stored as rotational kinetic energy of the wheel. Friction in the wheel bearings will absorb a little, but this should be negligibly small compared with the rotational energy. Also, elastic deformation of the tire will absorb some energy. This latter energy is recovered, however, when the tire returns to its normal shape, and any dissipation involved will certainly be negligible. The model then is this: A constant-speed drive wheel accelerates the nondeforming tire and wheel in such a way that energy is either absorbed as kinetic energy of the wheel or dissipated as heat because of the slippage between the two wheels.

In applying physical principles to this problem, it is necessary to derive a differential equation to find the *total* work done by the power wheel as it accelerates the auto wheel. The portion of this work dissipated as heat is then found by subtracting from the total the portion stored as kinetic energy in the auto wheel. Before proceeding the student should try to set up and solve the differential equation for the total work. (Note: The coefficient of friction between the tire and power wheel is *not* needed.)

Application of Physical Principles The work done by the power wheel goes to two places: to rotational kinetic energy of the wheel and to dissipation as heat. The amount stored as rotational energy is $\frac{1}{2}I_w\omega_w^2$. If the total work can be found, then the amount dissipated must be the difference. To find the total work done by the power wheel requires the setting up of a simple differential equation. Consider a moment during the process when the auto wheel has a speed ω (between 0 and its final speed ω_w), as shown in Fig. 6.10. The rate of power being consumed is

$$P = \text{force} \times \text{velocity} = Fr\omega_p$$

In time dt, the wheel is accelerated to $w + dw$ and the work done is

$$dW = P\,dt = Fr\omega_p\,dt$$

The force F here can be found if the normal force between the wheels and the coefficient of friction are known. But it can also be found in another way, by relating it to the acceleration of the auto wheel with Newton's

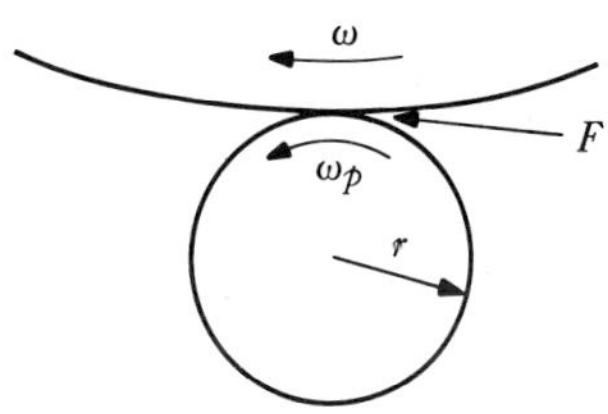

Fig. 6.10

second law, in this case as $\tau = Ix = I(d\omega/dt)$. The torque on the auto wheel is

$$\tau = FR = I_w \frac{d\omega}{dt}$$

Substituting this into the equation for work gives

$$dW = \frac{I_w}{R} r\omega_p \frac{d\omega}{dt}\,dt$$

Simplifying and integrating from $\omega = 0$ to $\omega = \omega_w$ gives

$$W = \frac{I_w}{R} r\omega_p = \int_0^{\omega_w} dW = \frac{I_w}{R} r\omega_p\omega_w$$

Now it is also true that

$$r\omega_p = R\omega_w$$

and so

$$W = I_w\omega_w^2$$

Since $\frac{1}{2}I_w\omega_w^2$ is absorbed as rotational energy of the wheel, the amount dissipated is

$$D = I_w\omega_w^2 - \tfrac{1}{2}I_w\omega_w^2 = \tfrac{1}{2}I_w\omega_w^2$$

With this information, designers can now estimate the heat-transfer problems associated with removal of this energy. A one-page report of the results is left as an exercise for students.

6.15 SUMMARY AND PREVIEW

The purpose of this chapter has been to provide students with a review of and a new perspective on physical principles. The idea that basic concepts should be used whenever possible in problem solving has been stressed. The examples have illustrated this point, as do all the examples in the book. Students who will develop understanding of the concepts themselves and who will also learn to set up and derive differential equations will be richly rewarded with new confidence in their ability to handle engineering problems.

The next chapter deals with the role of checking in engineering analysis. As the reader will see, checking is an important skill involving considerably more than just redoing the arithmetic.

SUGGESTED READING

Haberman, Charles M.: "Engineering Systems Analysis," Charles E. Merrill Books, Inc., Columbus, Ohio, 1965. A good place for study for those who have trouble setting up their differential equations.

Hughes, William, and Eber Gaylord: "Basic Equations of Engineering Science," Schaum Pub. Co., New York, 1964. A vast collection of formal equations. In order to use them, though, you will have to know what the symbols mean in physical reality and what all the unwritten conditions and assumptions are!

Shames, Irving: "Mechanics of Fluids," McGraw-Hill Book Company, New York, 1962. This book contains an excellent treatment of the thermal-vector approach to systems and control volumes using conservation of mass and Newton's second law.

Shapiro, Archer H.: "The Dynamics and Thermodynamics of Compressible Fluid Flow," The Ronald Press Company, New York, 1953, vol. 1. This book presents a clear and rigorous development of control-volume equations for mass, momentum, energy, and entropy from the basic system laws.

Tribus, Myron: "Thermostatics and Thermodynamics," D. Van Nostrand Company, Inc., Princeton, N.J., 1961. This book treats work in a general way and deals with many processes not often covered by engineers.

Ver Planck, D. W., and B. R. Teare: "Engineering Analysis," John Wiley & Sons, Inc., New York, 1959. This book was mentioned before but it is worth repeating. It has some good material on setting up differential equations.

PROBLEMS

6.1 The beam equation $EI(d^2y/dx^2 = M)$ is typically used in strength-of-material problems. Discuss the principles upon which it is based, the inherent assumptions, and the technique of its correct application.

6.2 What is surface tension?

6.3 Suppose you wanted to train a child to have an "intuitive" idea of the Coriolis effect. How would you go about it?

6.4 What is a viscoelastic fluid?

6.5 Suppose you measured in a laboratory experiment an emissivity of 1.2. What principle does this number violate? Would you believe the principle or your number? Why?

6.6 Springs are often assumed massless, though of course they are not. Consider the free vibration of a real spring by itself. Now consider the vibration of a massless spring with an attached mass. What should the attached mass be to give the same natural frequency of vibration as the real spring?

6.7 Why cannot Newton's second law be applied directly in a noninertial frame of reference?

6.8 What are the inherent assumptions in the Navier-Stokes equation? In Bernoulli's equation?

6.9 Many problems can be worked by assuming that the center of mass of a distributed or disjointed system does not move. What principle is used in such cases? Solutions of this type often seem "tricky," because students do not think to use the principle when two or more parts of a collection of things are involved in some action. What can be done to help people think of this possibility (without getting them to apply it indiscriminately)?

6.10 A cannon on a 1,000-lb railroad car fires a 50-lb cannon ball a half mile. How far will the railroad car move?

6.11 A helium-filled balloon is tied by a long string to the floor of a car. Which way, relative to the car, does the balloon move when the car accelerates?

6.12 A rocket burns fuel at a constant rate as it blasts off. Write the equation of motion.

6.13 What principle applies to the flow of fluids through porous media?

7
CHECKING

7.1 INTRODUCTION

Checking doesn't take place at any one point in a problem-solving situation. Checking permeates the whole procedure. However, in a textbook of this kind it must be considered in one location, though the reader will note that all the example solutions stress checking. If the chapter on checking is put first, the discussion would be somewhat sterile because the reader would not be familiar with the type of situations being referred to. If it is put later, there is no opportunity for the student to use the ideas on checking in any of the examples. As a compromise, then, it is being put near the middle.

Checking in engineering problem solving means a great deal more than common arithmetic and algebraic checking of addition, subtraction, etc. Such routine checking is important and it should not be belittled or ignored as students do who feel that accurate computation is beneath their dignity. The concept of checking as it will be discussed here, however, is a broader one in which arithmetic or algebraic checks are only one part. In particular, in Fig. 7.1 is shown a table which lists three types of checks:

1. Arithmetic or algebraic checks
2. Higher mathematical checks
3. Engineering-sense checks

As the figure indicates, each of these kinds of checks can be applied to either numerical or analytical steps or results. In the remainder of this chapter,

Types of checks	*Numerical results*	*Symbolic results (equations)*
Arithmetic or algebraic	Repeat check Reverse check Alternative-method check	Repeat checks Reverse checks Alternate-method check
Higher mathematical	Repeat check Reverse check Alternate-method check	Repeat check Reverse check; e.g., solution satisfies differential equation and boundary conditions. Alternate-methods check; e.g., formal and conceptual methods
Engineering sense	Consistent with experience? Consistent with practical? Order-of-magnitude checks	Dimensionally consistent? Limit checks Trend checks Sign checks All relevant and no irrelevant factors Order-of-magnitude checks

Fig. 7.1 Types of checks and their application to numerical or symbolic results.

the three types of checks are discussed in more detail, and then two problem examples are presented.

7.2 ARITHMETIC AND ALGEBRAIC CHECKS

Arithmetic and algebraic checks are the usual kinds of routine checks performed to ensure against inadvertent errors in arithmetic and algebraic operations. These are important checks, for a wrong answer is no less wrong in practice because it was due to, say, an addition error than if it had been due to a more basic error. Hence, this is a kind of check which should be made routinely at every step of an analysis involving either numbers or manipulation of equations.

There are at least three ways to make this check. One way is simply to repeat the original computations; hence it will be called a *repeat check* here. Repeat checking is the least reliable method of checking because of the human tendency to make the same error over and over again.

A second method of making an arithmetic or algebraic check is to reverse the order of the original computation. In the simple case of adding, for example, this means adding the column from top to bottom if the original computation was from bottom to top. Algebraic symbolic equations can also be manipulated in reverse order to give the original equations. This type of check is called a *reverse check*.

Finally, the third kind of check in this category is to find an *alternative method*. For example, if two simultaneous equations are solved for x and y by first eliminating x and then solving for y, the result can be checked by first eliminating y and then solving for x.

Nothing is more embarrassing to an engineer than a senseless error in a simple calculation. These arithmetic and algebraic checks should be made habitually and routinely.

7.3 HIGHER MATHEMATICAL CHECKS

Higher mathematical checks are still mathematical in nature (as distinguished from the more physical checks discussed in the next section) but they involve higher mathematical skills than do arithmetic or algebraic checks. Examples of higher-mathematical checks are (1) checking to ensure that a solution to a differential equation is correct, (2) checking the convergence of a numerical solution to a differential equation, (3) checking the correctness of a matrix inversion, (4) checking the derivation of a differential equation, etc.

Higher mathematical checks can also be subdivided into repeat, reverse, or alternate-method categories. In checking the solution of a differential equation, it is possible to repeat the original calculations, or to reverse the procedure by ensuring that the result satisfies the original equation and the boundary conditions, or to find an alternate method of determining the solution. The derivation of a differential equation can be checked, for example, by using both the formal and conceptual approaches. (This procedure has great educational value, too!)

7.4 ENGINEERING-SENSE CHECKS

The final kind of check on engineering analysis work is called *engineering sense* rather than *common sense* for good reason: common sense indicates that the world is flat! Not everything that is done or said in the name of common sense is sensible. Common sense burned "witches," persecuted Galileo, and laughed at the steam engine. (What is it doing today?) It's better to have ethical sense and engineering sense than common sense. (End of anti-common sense sermon.)

What, then, is meant by engineering sense? In the case of numerical results, it means knowing whether numbers are *consistent with experience* and are *practical*. A case of a young engineer who had designed an ordinary gas pipeline in such a way that the gas velocity was Mach 3 has recently been exposed. The calculations were all correct and the material costs had been cleverly minimized. However, he had failed to check the solution with his engineering sense.

In the case of symbolic or equation results, checking with engineering sense involves more than judgment or experience. There are five explicit aspects to the checks which can be made. *First and foremost, is the equation dimensionally consistent?* This simple check is so easy and yet so fantastically valuable that it should definitely be number one in any list of engineering analysis "ten commandments." This is a check which, like the arithmetic and algebraic checks, *must* be made a habit.

Second there are *limit checks*. Does the equation give sensible (i.e., consistent with experience) results when the various parameters are imagined to approach a limit, such as zero, infinity, or some boundary value? Limit checks are most important in determining the sensibleness of engineering results.

Third, there are *trend checks*. Parameters need not be imagined to go to some limit to obtain a sense check. They can be imagined either to increase or to decrease, and the equation can be examined to ensure that the remaining parameters respond in the expected way.

Fourth, there are *sign checks*. Do factors that should add complement one another, or influence the result in the same way, or have the same sign? That is, are the signs of the various terms consistent with what is expected?

Finally, a good check on problem-solving effort is to use engineering judgment to determine that *all the relevant factors appear in the equations*. A fluid-drag equation, for example, that did not involve the fluid viscosity would be worthy of some investigation before going on.

The remainder of the chapter contains two example problems in which the role of checking is emphasized. As with all the example problems, however, the ones here are complete examples of the engineering analysis process in action.

7.5 EXAMPLE: A CHEMICAL PROBLEM

Statement of the Problem Data on the performance of a large furnace burning a gaseous fuel called a producer gas have been leading to seemingly absurd results. The fuel is being burned with 25 percent excess air and contains

3.0 percent CO_2
26.0 percent CO
1.0 percent O_2
12.8 percent H_2
57.2 percent N_2

An Orsat analysis of the stock gas has been taken and it shows

11.25 percent CO_2
0.84 percent CO
2.46 percent O_2
85.45 percent N_2

Because of the awkward physical arrangement of the equipment, the Orsat sample was taken at a considerable distance from the furnace. It has been suggested (and note the inventiveness of the suggestion) that an air leak into the stack could be the cause of the strange data. Also, there has been difficulty with condensation in the stack. See what can be found out about this situation.

Definition of the Problem Assigned to this problem, an engineer has some problem definition to do for himself. He has been told only to "find out about this situation." He must redefine this vague assignment into meaningful and answerable (by engineering analysis) questions. What should these questions be? Form answers before going on to the next page.

Below are suggested problem definitions.

1. Determine the magnitude of the (presumed) air leak. (One possible magnitude is zero!)
2. Determine the dew point temperature of the stack gases before and after the (presumed) air leak.

It is absolutely necessary that the engineer choose a chemical *basis* on which to work a problem of this type. He could choose any of the following as a basis: 1 lb of fuel, 1 lb of dry stack gas, 1 lb-mole of fuel, 42.6 lb-moles

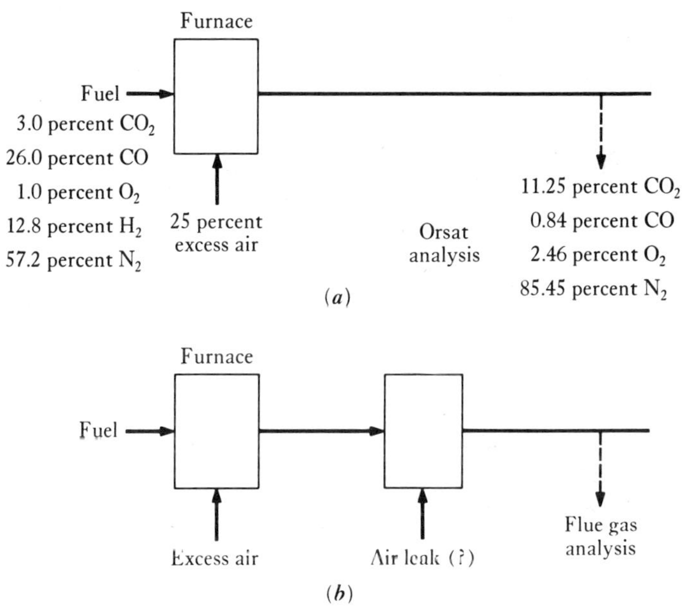

Fig. 7.2 (*a*) Given information. (*b*) Model.

of fuel, 0.293 lb-moles of dry stack gas, etc. On the other hand, some *available* choices lead to very difficult calculations. (Suppose he chooses 1 lb of the combustion air or 100 lb-moles of the leaking air, for example.) He plans, however, to take as a basis 1.0 lb-moles of fuel, and he will work everything on that basis. He further plans to make a material balance to see whether the input equals the output, or whether air from a leak is needed to make the balance. Then he will compute the percent of water vapor (on a mole basis) in the gases before and after the leak, if there is one. This will give him the partial pressure of water vapor from which he can determine the dew points.

Model Formation The problem as stated is to determine the magnitude of a postulated air leak in a furnace flue system. The engineer has sketched Fig. 7.2*a* to show the given information. What does he need to assume about the combustion? About the Orşat analysis? Others? Think it over before going on to the next page.

He can't very well make the common assumption of "complete" combustion because he sees there is carbon monoxide in the products. But he does need to say something about the Orsat analysis and what it measures. Note that there is no hydrogen listed among the products, but that the fuel is 12.8 percent H_2. What happened to it? One needs to remember here that the Orsat measures percentages in terms of the *dry* flue gas. The reason is that most water will be condensed out as the gas sample is cooled in the test apparatus. Hence, it is common to assume that the entering hydrogen leaves as H_2O, which does not show up in the Orsat results.

Finally, it is convenient to redraw the schematic sketch as shown in Fig. 7.2*b*. This will enable the engineer to hypothesize x units of air leaking in at the point shown. Then if his plans for a material balance are fruitful, he can compute x. Notice how far he has progressed from the initial assignment to "find out about this situation." Now he has a workable engineering problem.

Application of Physical Principles Given the model of Fig. 7.2*b*, the task is to determine the magnitude of the air leak. The first principle to be applied is conservation of mass to determine how much, if any, air must leak in to make things balance.

To make a material balance, the engineer must first select the basis for his computation. A good basis is one which makes computation easy, and it must not be a variable in the problem. For example, he would not want to choose, say, a mole of air leaking in as a basis in this problem. Nor would he want to choose a mole of products. However, a mole of *fuel* makes a convenient basis because everything in the problem can be easily referred to this basis and all the calculations will be understood to be *per mole of fuel entering*.

Once the basis is selected, the principle of conservation of mass is easily applied. Certain elements are easily followed through the process. Carbon is one because it enters only with fuel—none comes in with the air. But finally, of course, *all* the elements must balance.

First of all compute the amount of combustion air and excess air entering. If there were just the right amount of air for complete combustion, there would be

0.130 moles of O_2 for the 0.26 moles of CO to CO_2
0.064 moles of O_2 for the 12.8 moles of H_2 to H_2O
0.194 moles of O_2 for complete combustion

Since there is 25 percent excess air, $1.25 \times 0.194 = 0.243$ moles of O_2 (per mole of fuel understood) must be added. The numbers are easily checked by the "reverse" method.

At this point the engineer needs an assumption not made earlier. This will frequently happen. He has to specify what he means by *air*. The usual assumption, and the one used here, is that air is a dry mixture consisting by moles of 21 percent oxygen and 79 percent nitrogen. This is equivalent to saying that for each mole of O_2 in air, there are 3.77 moles of N_2. (How could you check this?) He thus neglects water vapor and other elements in the air. In this problem there are 0.243 moles of O_2, so there are also $0.243 \times 3.77 = 0.916$ moles of N_2 introduced as part of the combustion and excess air. Thus,

Furnace

0.03 moles CO_2
0.26 moles CO
0.01 moles O_2
0.128 moles H_2
0.572 moles N_2 → [Furnace] → 1.488 moles N_2
0.29 moles C

↑
0.243 moles O_2
0.916 moles N_2

Now since no carbon enters except with the fuel, there must be 0.29 moles C in the final mixture. Thus $11.25 + 0.84 = 12.09$ percent of the final mixture is equivalent to 0.29 moles. The moles of final mixture are

$$\frac{100}{12.09} \times 0.29 = 2.40 \text{ moles}$$

It must not be forgotten, however, that the Orsat analysis is a dry analysis. Therefore, it must be assumed that all the hydrogen is burned to form H_2O. Since there are 0.128 moles of hydrogen in the fuel, there will also be 0.128 moles of H_2O in the gases. The *total* moles of final mixture is thus $2.40 + 0.128 = 2.528$. Check this figure.

If it is now noted that 85.45 percent of the dry Orsat mixture of 2.40 moles was nitrogen, the moles of nitrogen leaving can be found:

$$\text{Moles nitrogen leaving} = 0.8545 \times 2.40 = 2.05$$

But only 1.488 moles of nitrogen leave the furnace! There seems to be an air leak of $2.05 - 1.488 = 0.562$ moles of nitrogen! That would mean an additional $0.562/3.77 = 0.149$ moles of oxygen as well. So, progress has been made, to this point:

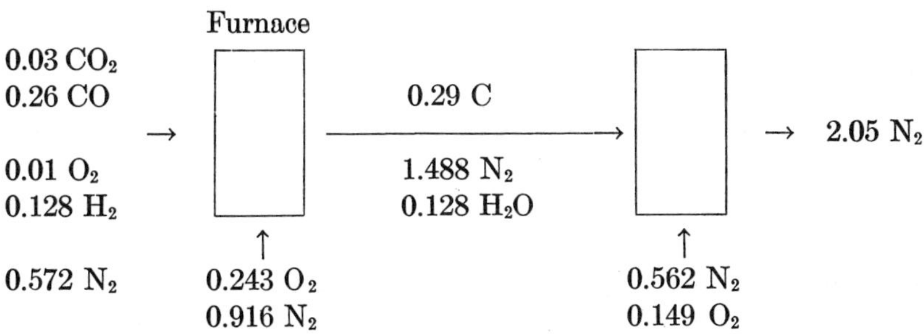

That is half the job. This work requires complete checking before the engineer goes on. To save space here, it will be assumed that appropriate checks (which ones?) have been made and that all is well. The other half of the job is to find the dew point of the mixture before and after the air leak.

The dew point is the temperature at which condensation occurs in a mixture if it is cooled at constant total pressure. It is, in thermodynamic jargon, the saturation temperature corresponding to the partial pressure of water vapor. Thus the mole fraction of water vapor is $0.128/2.528 = 0.0506$, and the partial pressure of water vapor is

$$0.0506 \times 14.7 = 0.795 \text{ psia}$$

From the Steam Tables one finds the corresponding saturation temperature to be a little over 94°F. Since the flue wall temperature could easily be below 94°F, the condensation reported seems very reasonable.

After the leak, moles of the various gases are as follows:

CO_2:	0.1125×2.40	$= 0.270$
CO:	0.0084×2.40	$= 0.020$
O_2:	0.0246×2.40	$= 0.059$
N_2:	0.8545×2.40	$= 2.050$
H_2O:		$= 0.128$
		2.527

At the leak, there are 0.149 moles of O_2 and 0.562 moles of N_2. There seems to be less free O_2 *after* the leak than before, indicating that additional combustion is taking place in the stack! The moles of nitrogen prior to the leak are 2.050 − 0.562 = 1.488. The total moles cannot be computed at this point because the chemical form of the constituents is not known. There is no law of conservation of *moles*—only of elements. As an estimate the engineer might note that the nitrogen undoubtedly makes up about 80 to 85 percent of the mixture. This works out as follows:

Assumed percent N_2	*Total moles*	*Mole fraction* H_2O	*Partial pressure* H_2O, *psia*	*Dew point*, °*F*
80	1.488/0.80 = 1.86	0.128/1.86 = 0.0689	0.0689 × 14.7 = 1.01	102
85	1.488/0.85 = 1.75	0.128/1.86 = 0.0733	1.08	104

Thus it is clear that regardless of the exact number of moles, the leak reduces the dew point by about 10°F. This may well contribute to the condensation problem. It looks as if the company had better find and repair that leak!

7.6 EXAMPLE: A DISPLACEMENT-GAGE PROBLEM

Statement of the Problem For use in measuring the thickness of small flat parts, a gage using a liquid gage as shown in Fig. 7.3 has been proposed. Investigate the feasibility of this device.

Definition of the Problem Before any conclusions about the usefulness of the proposed device can be made, something about its performance must be determined. At this point, however, the engineer does not

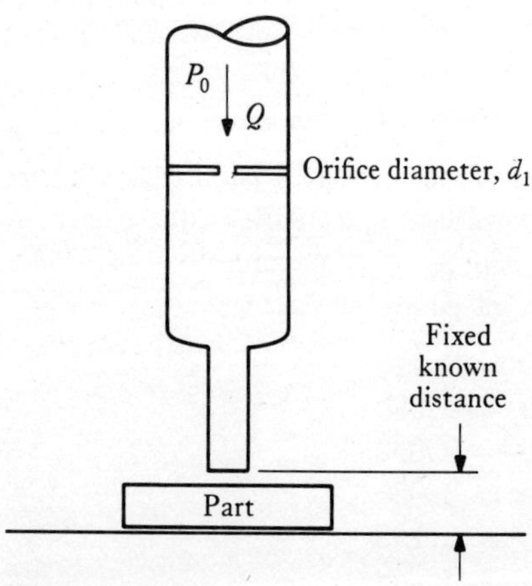

Fig. 7.3 Schematic of the air-thickness gage.

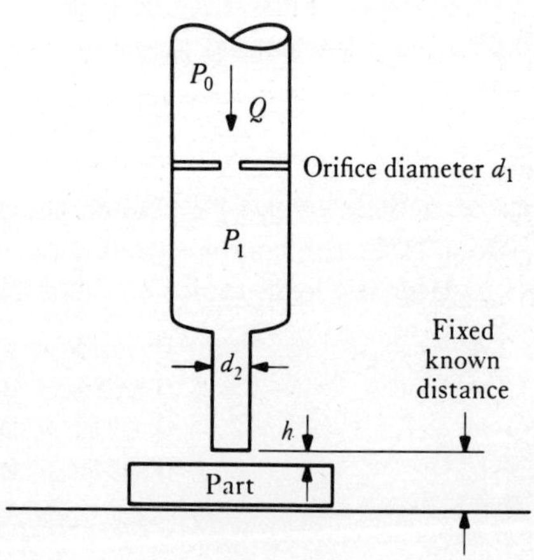

Fig. 7.4

even know how the thing works. Before going on, the student should study the device and decide how it could be used to determine the thickness of the part as shown.

Here is how it works (refer to Fig. 7.4): P_0 is a constant supply pressure, but P_1 will vary as a function of the spacing h. As h approaches zero, P_1 will approach P_0. As h becomes quite large, P_1 will approach atmospheric pressure (zero gage pressure). In between, P_1 will be a function of h, and measuring P_1 will give an indication of h.

As a first step in the determination of feasibility, then, it is desirable to analyze the device to find a relationship among the parameters in the problem. In particular, the analysis should relate P_1 to h in terms of P_0, d_1, d_2, and h. To find this relationship is the problem.

The student should now consider the assumptions to be made.

Model Formation Since a liquid has been specified as the fluid to be used, it is reasonable to consider the flow to be incompressible and isothermal. Also, except for pressure drops through the orifice at d_1 and at the exit, frictional effects will be negligibly small. P_0 will be maintained constant, and steady conditions are assumed. With this model, it is now possible to begin the application of physical principles.

Application of Physical Principles Since the flow is steady and incompressible, conservation of mass indicates that the volume flow through d_1 equals the volume flow through the exit. The momentum principle applied to either the orifice or the exit yields only information about the forces in the orifice plate. In the case of the steady flow of a frictionless, incompressible fluid, the energy equation reduces to the familiar Bernoulli equation. (Newton's second law in a lagrangian system also gives the same result.) The general steady-flow energy equation is

$$\dot{Q} - \dot{W} = \left(h_2 + \frac{V_2^2}{2} + gz_2\right)\dot{m} - \left(h_0 + \frac{V_0^2}{2} + gz_0\right)\dot{m}$$

where the subscript 2 refers to conditions at the throat of the orifice. Gravity terms are negligible, and if one follows a particular system of matter through the device, the first law gives

$$Q - W = u_2 - u_0$$

Combining these gives

$$0 = \frac{P_2}{\rho} + \frac{V_2^2}{2} - \frac{P_0}{\rho} - \frac{V_0^2}{2}$$

Now since $V_2^2 \gg V_1^2$, the final result is

$$V_2^2 = \frac{2(P_0 - P_2)}{\rho} \qquad V_2 = \sqrt{\frac{2(P_0 - P_2)}{\rho}}$$

The flow rate is $Q = AV$ and so

$$Q = \frac{\pi d_1^2}{4}\sqrt{\frac{2(P_0 - P_2)}{\rho}}$$

To account for the various factors which are neglected in the analysis, it is common practice to introduce an orifice coefficient C_1 into this equation and to use the downstream pressure P_1 instead of the unmeasurable orifice pressure P_2. Thus,

$$\dot{Q} = C_1\frac{\pi d_1^2}{4}\sqrt{\frac{2(P_0 - P_1)}{\rho}}$$

Orifice coefficients are well known for certain standard orifices and vary from about 0.60 to about 0.80. This equation should be checked. First the units:

$$\dot{Q} \rightarrow \text{ft}^3/\text{sec}$$

$$\frac{\pi\, d_1^2}{4}\sqrt{\frac{2(P_0 - P_2)}{\rho}} \rightarrow \text{ft}^2\,\frac{\text{lb}_f^{1/2}\ \text{ft}^{3/2}}{ft\ \ \text{lb}_m^{1/2}}$$

They do not check! What is needed is the conversion factor $g_0 = 32.2(\text{ft})(\text{lb}_m)/\text{lb}_f)(\text{sec}^2)$. This gives

$$\frac{\pi\, d_1^2}{4}\sqrt{\frac{2(P_0 - P_2)g_0}{\rho}} \rightarrow \text{ft}^2\,\frac{\text{lb}_f^{1/2}\ \text{ft}^{3/2}}{\text{ft}\ \ \text{lb}_m^{1/2}} \rightarrow \frac{\text{ft}^{1/2} - \text{lb}_m^{1/2}}{\text{sec} - \text{lb}_f^{1/2}} \rightarrow \frac{\text{ft}^3}{\text{sec}}$$

In other respects, the equations seem satisfactory.

Now if the exit is also treated as an orifice, the result is

$$\dot{Q} = C_2\pi\, d_2 h\sqrt{\frac{2(P_1)g_0}{\rho}}$$

where the orifice area is taken as $\pi\, d_2 h$, the exit orifice coefficient is C_2, and P_1 is taken as gage pressure.

Using conservation of mass the $\dot{Q}$'s must be equal:

$$C_2\pi\, d_2 h\sqrt{\frac{2(P_1)g_0}{\rho}} = C_1\,\frac{\pi\, d_1^2}{4}\sqrt{\frac{2(P_0 - P_1)g_0}{\rho}}$$

Simplifying gives

$$P_1 = P_0\,\frac{1}{1 + 16(C_2^2\, d_2^2h^2/C_1^2\, d_1^4)}$$

The units of this equation can be checked by inspection. Also note that as $h \rightarrow 0$, $P_1 \rightarrow P_0$, which is as expected. (Limit check.) Also, as $h \rightarrow$ large, $P_1 \rightarrow 0$, as expected. (Limit check.) The equation, therefore, seems quite reasonable.

The orifice coefficients are a function of the orifice to pipe-area ratio, but as the ratio becomes quite small, the orifice coefficient approaches a constant value near .60. If both coefficients are equal, then, the equation becomes, in dimensionless form,

$$\frac{P_1}{P_0} = \frac{1}{1 + 16(d_2^2h^2/d_1^4)}$$

A plot of this result is shown in Fig. 7.5. The equation can be used to design d_2, d_1, and P_0 for desired ranges of h, and this completes the assigned task.

As a final check, an observant engineer might question the fact that fluid properties (especially density) do not enter the equation or affect the result. Does it seem reasonable that this result should be independent of the fluid used in the gage?

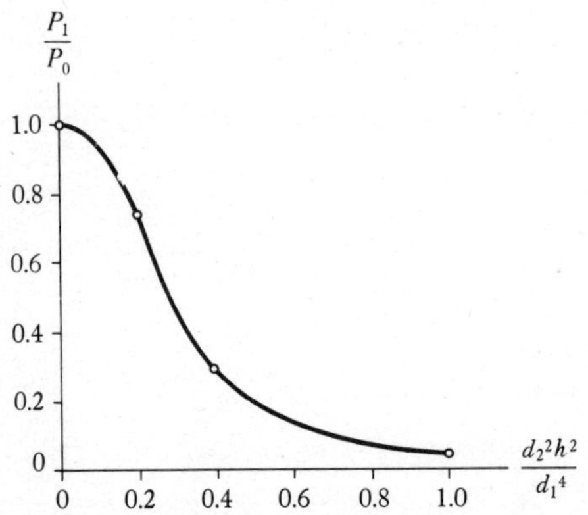

Fig. 7.5 Results of the analysis of the air-thickness gage.

7.7 SUMMARY AND PREVIEW

Checking, whether arithmetic, mathematical, or by engineering sense, should permeate the engineering analysis process. In this chapter, the various kinds of checks have been cataloged and discussed briefly. However, a student engineer can be told about checks but he cannot be made to use them—remember the horse and water story. Careful and skillful checking, however, is evidence of maturity in an engineer. *Check your work!*

In the next chapter, attention is directed to computation in engineering analysis. Routine arithmetic and algebraic computations are not discussed, nor is the solution of differential equations treated. These are matters amply dealt with in regular courses. Instead, the situation is discussed in which the engineer must have a numerical answer but where conventional analytical methods fail. In such cases, either numerical analysis (perhaps using an analog or digital computer) or graphical analysis is needed. Hence, numerical and graphical solution techniques are discussed.

SUGGESTED READING

Ver Planck, D. W., and B. R. Teare: "Engineering Analysis," John Wiley & Sons, Inc., New York, 1954. This book also has a chapter on checking.

PROBLEMS

7.1 If x is feet and t is time, what are the dimensions of the following differential expressions (*a*) dx/dt, (*b*) d^2x/dt^2, (*c*) d^5x/dt^5, (*d*) $d/dt\, x^2t^3$?

7.2 Comment on the possible connection between improved perceptiveness, discussed in Chap. 2 in connection with inventiveness, and the ability to make good engineering analysis sense checks.

7.3 Discuss the dimensions of $F = ma$ and $\oint d'Q = \oint d'W$.

7.4 Go to the library and get several technical journals in your field of interest (e.g., *Journal of Fluid Mechanics*, *Journal of Applied Mechanics*, etc.) and check the dimensions of equations you find in the papers there. Also check the equations using trend and limit checks.

7.5 Discuss ways and means of making yourself perform dimensional, limit, and trend checks on the equations you derive.

8
COMPUTATION

8.1 INTRODUCTION

The engineering analysis process was being considered in step-by-step fashion—problem definition, model formation, application of principles—before we went out of the way a bit in Chap. 7 in order to discuss checking. Checking, as was said, occurs within and between every step. Now it is time to return to the stepwise procedure. The next step is computation. Once a model has been formed and physical principles applied, there are generally equations to be solved for desired unknowns. The equations may be differential equations, integral equations, or a variety of types of algebraic equations. The calculations may require simple arithmetic, quite difficult algebra, or differential or integral calculus. In many cases exact solutions are not possible and various approximate techniques, such as graphical or numerical methods, are necessary. In this chapter the analytical solution of differential and integral equations will be discussed only briefly because this is a topic covered in mathematics courses. Graphical methods and numerical methods will be discussed in more detail, however. The extensive and important use now being made of computers, both digital and analog, is making numerical methods increasingly important. Up-to-date engineers simply must be able to make good use of this exciting new computational tool.

8.2 ANALYTICAL CLOSED-FORM SOLUTIONS

Systems of algebraic equations and differential or integral equations often have exact analytical closed-form solutions. By this is meant that it is possible to write down exact and explicit equations for the unknowns. For example, $\int_a^b \sin x\, dx = \cos a - \cos b$ and $\int_a^b dx/x = \ln (b/a)$. Or given $2x + 2y = a$ and $x + y = b$, one can solve and find $y = 2b - a$ and $x - a - b$. Also, given the differential equation $d^2y/dx^2 + by = 0$, one can find the solution of the form $y = Ae^x$. All of these, and many others like them or more complex, are solved in courses on differential equations, calculus, and algebra. Therefore time and space will not be spent on them in this book. Instead, ways of getting numerical answers to problems when exact analytical methods fail to provide a closed-form solution will be considered.

8.3 ALGEBRAIC EQUATIONS: GRAPHICAL AND TRIAL-AND-ERROR SOLUTIONS

A system of n independent equations which contains n unknowns has a unique solution. However, it is not always possible to solve the system explicitly for the unknowns. For example, consider this simple equation in one unknown: $\sin x = x^2$. Here is one equation in one unknown, but the solution cannot be written in a form which permits solution for x directly. Similar situations can occur with two or more unknowns. For example, the following simultaneous equations have no direct solutions:

$$x^4y + xy^2 = 1$$

$$x^3y^2 + x^2y^3 = 2$$

The heating-duct problem in Chap. 5 gave a similar result, too. Equations or systems of equations with this contrary characteristic are called *tran-*

scendental equations. Their solution requires the use of graphical techniques, or the use of numerical trial-and-error methods, or both. The appearance of transcendental equations in real engineering work is too common to be considered an exception, and hence it is important for an engineer to be able to solve them with facility.

Sometimes situations arise where direct and explicit solutions are *possible* but not practical. That is, there are circumstances where a graphical or numerical trial-and-error solution is quicker and easier than the exact analytical solution. This occurs most often when the equations involved are quite long or complex. It should be pointed out that if more than two unknowns are involved, graphical methods become generally impractical because of the unavailability of three- (or more) dimensional paper. Trial-and-error solutions are also inappropriate for cases of many unknowns, unless a calculating machine or computer is used and a scheme is devised for iterating quickly to a solution. Generally speaking, the work required in trial-and-error solutions grows in proportion to 2^n where n is the number of unknowns.

Graphical solutions to transcendental equations are very common. The reasons for student reluctance to undertake graphical solutions are mysterious. It is probably a combination of (1) lack of graph paper, (2) "set" on analytical methods, (3) belief that graphical solutions take a long time and are a lot of "dog-work," and (4) belief that graphical solutions are not accurate. There are answers to these objections. First, lack of graph paper limits accuracy but usually does not prevent an approximate solution. Besides, a simple remedy is the purchase of a package of graph paper! Second, if the situation is recognized as transcendental, then this should break the "set." Third, it simply isn't true that graphical solutions are time-consuming. If they are approached directly and sensibly, it takes only a few minutes to plot the solution. Fourth, slide rule accuracy is readily obtained using ordinary $8\frac{1}{2}$- by 11-in. graph paper in many problems. Larger paper will permit greater accuracy.

As an example, consider again the equation $\sin x = x^2$. A graphical solution requires that one plot $\sin x$ vs. x and x^2 vs. x on a single graph. The intersection of the curves will be the solution. Figure 8.1 shows a freehand sketch made for the solution. It gives a solution of $(1.13)(\pi/4) = 0.886$ on regular $8\frac{1}{2}$ by 11 graph paper. Using $x = 0.886$ as a solution, we get $\sin x = 0.774$ and $x^2 = 0.786$, which is not bad agreement for a freehand graph of this size. The time consumed in doing the original graph was less than 10 min.

The same problem is also easily solved by numerical trial and error. It is always best in these cases to begin by setting up a table, as shown in Fig. 8.2. Generally it is necessary to have *some* idea of the order of magnitude of the solution. Often a quick, qualitative, freehand graph will give this information. In other cases engineering sense will indicate a reasonable first guess. Once a starting guess is made, the next objective should be to get, as quickly as possible, another trial value which is on the opposite side of the initial guess. That is, if the initial guess is too small, then one wants to find *quickly* another guess that is too large, and vice versa. It is desirable to get the answer *bracketed* in this manner as early in the process as possible, because much time can be wasted in creeping up to a solution from one side. (It may be a *very* long creep in some cases!) Once a value is obtained which is too big and another that is too small, centering down on the answer is quickly done. So a rule to remember in numerical trial and error solutions is this: Bracket the answer quickly!

A neat way to bracket answers quickly, especially when very little is known about the magnitude of the answer, is to select a starting value

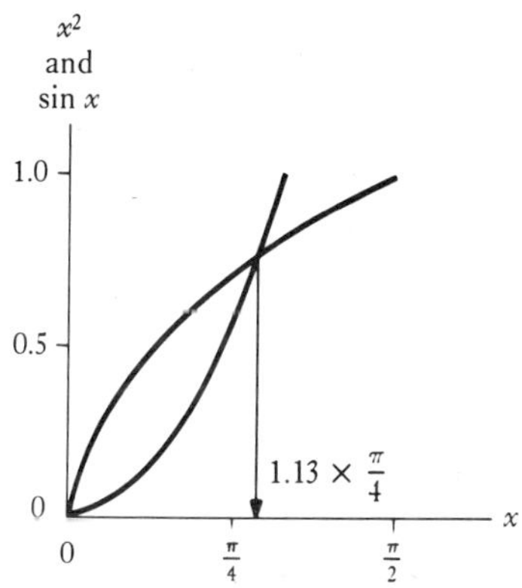

Fig. 8.1 Graphical solution of $x^2 = \sin x$.

x	x^2	$\sin x$	$x^2 - \sin x$
0.1	0.0100	0.0996	−0.0896
0.2	0.0400	0.1982	−0.1582
0.4	0.1600	0.3883	−0.2283
0.8	0.6400	0.7157	−0.0757
1.8	3.2400	0.9996	+2.2404
1.2	1.4400	0.9311	+0.5089
1.0	1.0000	0.8410	+0.1590
0.9	0.8100	0.7815	+0.0225
0.85	0.7225	0.7501	−0.0276
0.875	0.7660	0.7672	−0.0012

Fig. 8.2 Tabular solution of $x^2 = \sin x$.

which is felt to be definitely on the low side. Then for each new guess, double the preceding value. In the above example, for instance, one might start with $x = 0.1$, as shown in Fig. 8.2. The second guess would then be $x = 0.2$, the next $x = 0.4$, and so on. (This sort of thing is easily done on the computer, too.) Note that in column 4 in the figure, we have listed the difference $(x^2 - \sin x)$, which should be made to go to zero. When this difference changes sign, the answer is bracketed. Thus in the example it is clear that $x = 0.8$ is too small and $x = 1.6$ is too large. At this point convergence on the answer is done very quickly by selecting each new guess halfway between the known brackets. The distance between the bracket values will thus decrease by one-half at each step. In our example, the sequence of guesses is thus 0.8, 1.6, 1.2. The results of the $x = 1.2$ calculation determine whether the next guess should be 1.0 or 1.4. This repeated halving of the unknown range can go on as long as desired, presumably until the desired accuracy is obtained. Using only a slide rule for the given example we got $x = 0.875$, compared with $x = 0.886$ by the graphical method.

The example worked here is a very simple one, but the method used and the techniques discussed will serve well in other, more complex situations. The doubling or halving techniques are quite useful. The use of a table similar to Fig. 8.2 is essential if one is not going to waste a lot of time, effort, and paper in writing and rewriting equations. Of course, special time-saving methods can often be developed for any given problem. Most complaints by students about the "dog-work" of graphical and numerical trial-and-error solutions are the result of the use of very inefficient techniques. Most of these inefficiencies could be corrected by an open mind and a little clever forethought. For instance, some unnecessary work has been done in the $x^2 = \sin x$ example. Since the difference between x^2 and $\sin x$ is

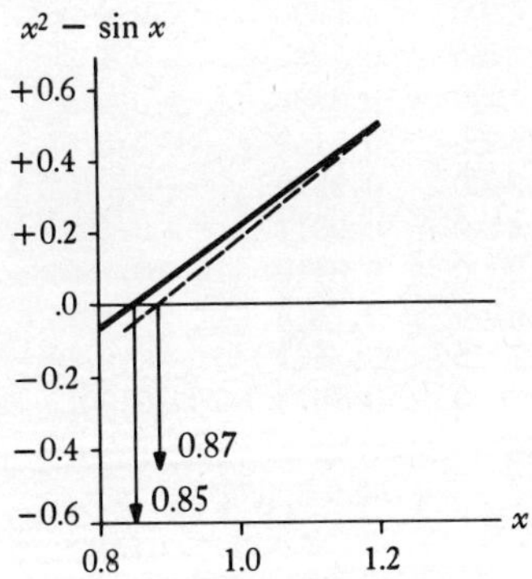

Fig. 8.3 Graphical solution of $x^2 = \sin x$.

-0.0757 at $x = 0.8$ and $+2.2404$ at $x = 1.6$, it seems reasonable to assume that the answer is closer to $x = 0.8$ than to $x = 1.6$. A better guess than $x = 1.2$ might therefore have been made. Alternatively, the numerical work could have been supplemented with a graph of $x^2 - \sin x$ vs. x, as shown in Fig. 8.3. Suppose this graph is started at the time following computation of the $x = 0.8$ and $x = 1.2$ cases. Assuming a linear relationship between $x^2 - \sin x$ and x, the solution would now be $x = 0.85$, as shown by the solid line. This is now used as the next guess. Doing so gives $x^2 - \sin x = -0.0246$. Again a linear relationship is assumed, a new intercept obtained (see the dashed line in Fig. 8.3), and a new value computed. This time the result is $x = 0.87$, a value which is recognized as quite accurate. To get this answer there was the added work of the graph, but there were only two computations for $x^2 - \sin x$. In this example, the total work is not much affected, but in cases where lengthy computation is involved, this simple linear interpolation scheme can save a lot of time. It is easily handled on the computer also.

Obviously, there are a number of time- and energy-saving schemes for solving transcendental equations. Most are simply the clever inventions of men who need to get a good numerical answer quickly. Some simple schemes have been shown here. Others are discussed in the literature or in courses on numerical analysis.

8.4 INTEGRAL EQUATIONS: GRAPHICAL SOLUTIONS

Frequently engineers are confronted with the necessity for obtaining a numerical answer for the value of an integral I

$$I = \int_{x=a}^{x=b} f(x)\, dx$$

where $f(x)$ is such that closed-form analytical solution is not possible. Evaluation of such integrals can be done either numerically or graphically.

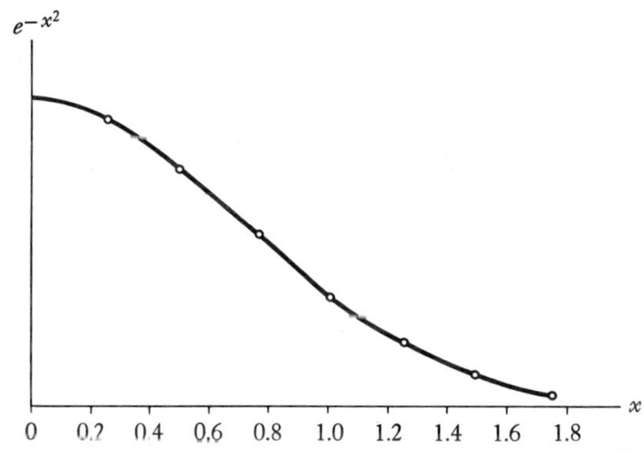

Fig. 8.4

In this section graphical methods are discussed, and in the next section numerical methods are considered.

Graphical evaluation of an integral $\int_a^b f(x)\, dx$ requires that $f(x)$ be plotted as a function of x. The area under the curve between a and b is then obviously the desired value. As an example, consider the integral

$$I = \int_0^1 e^{-x^2}\, dx^*$$

A plot of e^{-x^2} is shown in Fig. 8.4. The area under the curve can be ob-

* The integral $I = (2/\sqrt{\pi})\int_0^x e^{-x^2}dx$ is called the "error function" and is denoted by erf 1 = 0.8427. Hence the answer is $(\sqrt{\pi}/2)$ 0.8427 = 0.747. This information can be used to check up on the accuracy of the graphical and numerical solutions found here.

tained by counting squares. On regular $8\frac{1}{2}$ by 11 graph paper, using the small squares, the grand total is 7,460. Each 100 squares represents an area of $0.1 \times 0.1 = 0.01$, so the integral is equal to $7{,}460/100 \times 0.01 = 0.746$.

To three decimal places, the exact analytical answer is 0.747.* The time consumed to get the graphical answer was about 10 min. There is, therefore, absolutely no reason for anyone to be reluctant to use graphical integration on the basis of either time consumed or accuracy.

8.5 INTEGRAL EQUATIONS: NUMERICAL SOLUTIONS

Solutions for integrals of the form $I = \int_a^b f(x)\,dx$ can also be obtained by a variety of numerical techniques. The most common technique consists simply of approximating the curve by a series of steps. The interval $a \rightarrow b$

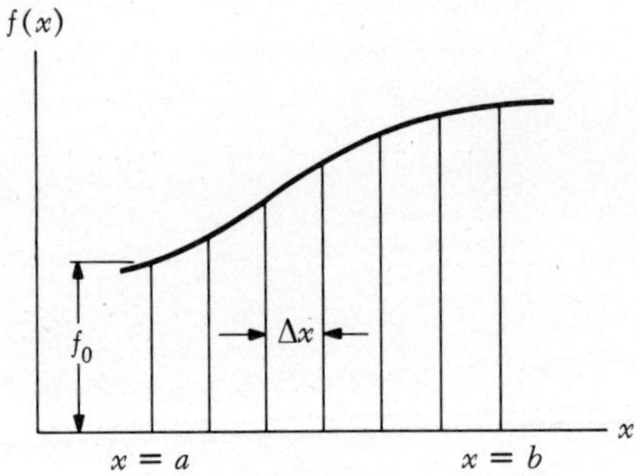

Fig. 8.5

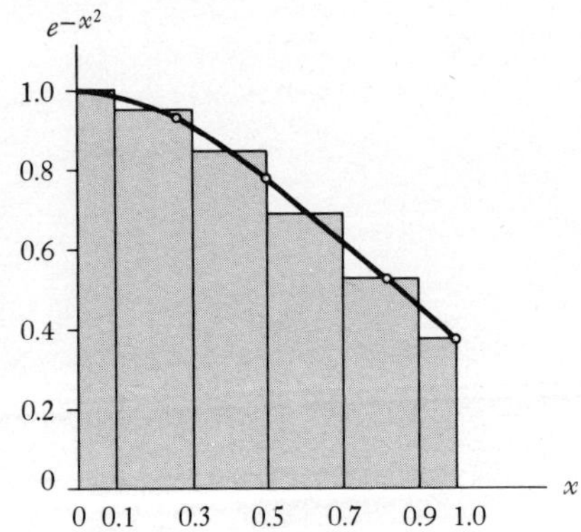

Fig. 8.6

is first divided into a convenient number of small intervals each Δx wide (see Fig. 8.5). The function $f(x)$ is then evaluated at each of the intervals. Use the notation f_0 to mean the value of $f(x)$ at $x = a$, f_1 to mean the value of $f(x)$ at $x = a + \Delta x$, f_2 to mean the value of $f(x)$ at $x = a + 2\,\Delta x$, etc. The area under the curve between a and b is then given by

$$I = \int_a^b f(x)\,dx = f_0\frac{\Delta x}{2} + f_1\,\Delta x + f_2\,\Delta x + \cdots + f_k\frac{\Delta x}{2}$$

where k is the number of intervals used. Figure 8.6 shows the area computed by this method for the integral $\int_0^1 e^{-x^2}\,dx$. Using only five intervals ($\Delta x = 0.2$)

* Believe it or not, this very good answer was actually obtained with a freehand sketch and by "counting" squares mostly by eye. To be honest, the accuracy is not usually quite this good!

the numerical result is

$$I = 1\frac{0.2}{2} + 0.961(0.2) + 0.852(0.2) + 0.698(0.2) + 0.523(0.2) + (0.368)\frac{0.2}{2}$$

$$= 0.744$$

The accuracy is excellent. The time involved is only a few minutes, and one does not really need the graph once the technique is understood. This method will give *exact* results when the "curve" being integrated is a straight line. The good results just obtained are due in part to the fact that the function e^{-x^2} does not depart drastically from a straight line in the interval $0 \to 1$.

When more rapidly changing functions are being integrated, better numerical approximations are available. The most common is called Simpson's rule and it will give exact results when the curve is a parabola. Consider the parabolic curve shown:

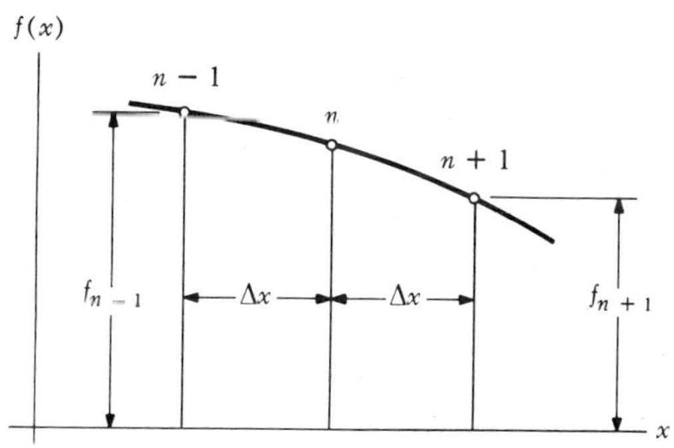

The shaded area under the curve will be exactly given by

$$\left(\frac{f_{n-1} + 4f_n + f_{n+1}}{6}\right) 2\,\Delta x = \frac{\Delta x}{3}(f_{n-1} + 4f_n + f_{n+1})$$

Now given a curve as shown in Fig. 8.7, one can use the above equation to estimate the area if it is first divided into k intervals, where k must now be an even number. Then the total area under the curve is given by

$$I = \int_a^b f(x)\,dx = \frac{\Delta x}{3}(f_0 + 4f_1 + f_2) + \frac{\Delta x}{3}(f_2 + 4f_3 + f_4) + \frac{\Delta x}{3}(f_4 + 4f_5 + f_6) + \cdots + \frac{\Delta x}{3}(f_{k-2} + 4f_{k-1} + f_k)$$

$$= \frac{\Delta x}{3}(f_0 + 4f_1 + 2f_2 + 4f_3 + 2f_4 + \cdots + 4f_{k-1} + f_k)$$

Returning to the previous example, $I = \int_0^1 e^{-x^2}\,dx$; if $k = 4$, then $\Delta x = 0.25$ and

$$I = \frac{0.25}{3}[1.0 + 4(0.9395) + 2(0.7795) + 4(0.5705) + 0.368]$$

$$= 0.748$$

Note that the accuracy is now a little better, even though fewer points were used. The parabolic approximation requires more work per interval but will give the same accuracy with fewer intervals. Hence it is generally preferred. The number of intervals selected is arbitrary and depends on the accuracy desired and the shape of the curve. Intervals should be small

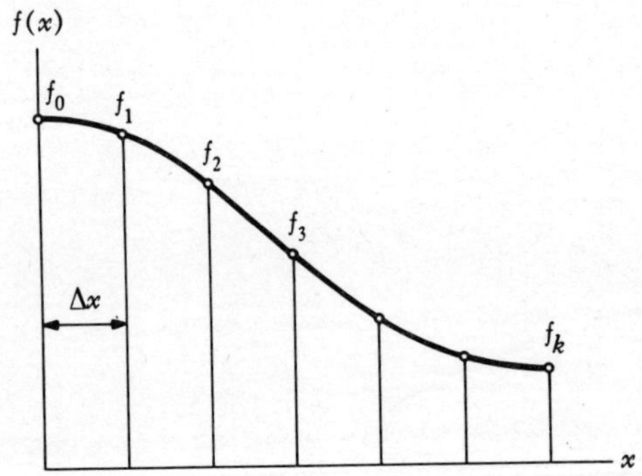

Fig. 8.7

enough so that each interval can be reasonably approximated by a parabola. Otherwise, the fewer the better, so as to save computation time.

8.6 DIFFERENTIAL EQUATIONS: NUMERICAL SOLUTIONS

Differential equations can be classified in two groups: those which are initial-value problems, and those which are boundary-value problems. An initial-value problem requires a solution which "marches" forward from a known region into an unknown region. A boundary-value problem requires a solution within a region which is bounded by known conditions. Figure 8.8 shows a one-dimensional schematic view of these concepts. Numerical procedures for solving the two types of problems will now be discussed.

Initial-value Problems The Taylor-series expansion is the basis for many useful numerical methods of solving initial-value problems. Suppose it is required to find y, given that $dy/dx = f(x,y)$ and given an initial condition $y = y_0$ at $x = x_0$. Taking a small interval Δx, the Taylor-series expansion is

$$y(x + \Delta x) = y(x) + \frac{dy}{dx}\frac{\Delta x}{1!} + \frac{d^2y}{dx^2}\frac{\Delta x^2}{2!} + \cdots$$

If it is noted that $x_n = x_0 + n\,\Delta x$ and $y_n = y(x_0 + n\,\Delta x)$, the above equation becomes

$$y_{n+1} = y_n + \left(\frac{dy}{dx}\right)_n \frac{\Delta x}{1!} + \left(\frac{d^2y}{dx^2}\right)_n \frac{\Delta x^2}{2!} + \cdots$$

Thus if y and its derivatives at the nth point are known, the above equation can be used to find y and its derivatives at the $(n + 1)$th point. Thus the

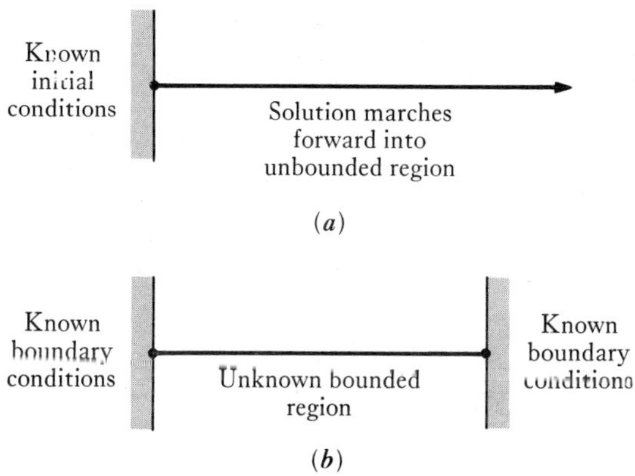

Fig. 8.8 **(*a*) Schematic view of one-dimensional initial-value problem. (*b*) Schematic view of one-dimensional boundary-value problem.**

solution can march forward. In practice, when making numerical approximations one does not retain the whole series but drops many of the higher-order terms, since Δx is usually taken as a small number.

As an example, suppose it is desired to solve for $y(x)$ from the differential equation

$$\frac{dy}{dx} - 2y = -x^2$$

with $y = 0$ at $x = 0$ as the initial condition. Δx is selected as, say, 0.1.

For use in the Taylor series the derivatives are

$$\frac{d^2y}{dx^2} = 2\frac{dy}{dx} - 2x = 4y - 2x^2 - 2x$$

$$\frac{d^3y}{dx^3} = 2\frac{d^2y}{dx^2} - 2 = 8y - 4x^2 - 4x - 2$$

$$\frac{d^4y}{dx^4} = 2\frac{d^3y}{dx^3} = 16y - 8x^2 - 8x - 4$$

$$\frac{d^5y}{dx^5} = 2\frac{d^4y}{dx^4} \qquad \frac{d^6y}{dx^6} = 2\frac{d^5y}{dx^5} \qquad \text{etc.}$$

The Taylor series gives us

$$y_{n+1} = y_n + \left(\frac{dy}{dx}\right)_n \frac{\Delta x}{1!} + \frac{d^2y}{dx^2}\frac{\Delta x^2}{2!} + \cdots$$

and so for y_1 there is

$$y_1 = y_0 + \left(\frac{dy}{dx}\right)_0 \Delta x + \left(\frac{d^2y}{dx^2}\right)_0 \frac{\Delta x^2}{2} + \cdots$$

At the starting point $n = 0$, $x = 0$, $y_0 = 0$

$$y_0 = 0 \qquad \left(\frac{dy}{dx}\right)_0 = 0$$

$$\left(\frac{d^2y}{dx^2}\right)_0 = 0 \qquad \left(\frac{d^3y}{dx^3}\right)_0 = -2$$

$$\left(\frac{d^4y}{dx^4}\right)_0 = -4 \qquad \left(\frac{d^5y}{dx^5}\right)_0 = -8 \qquad \text{etc.}$$

Thus for y_1 the result is

$$y_1 = 0 + 0(0.1) + 0\frac{0.01}{2} - 2\frac{0.001}{6} - 4\frac{0.0001}{24} - 8\frac{0.00001}{120}$$

$$y_1 = 0.000333 - 0.000167 - 0.0000067 - \cdots$$

$$y_1 = -0.0005067$$

Now with this value of y_1 and with $x_1 = \Delta x$ all the derivatives at point

$n = 1$ can be computed, and y_2 computed using the Taylor series once again, this time as

$$y_2 = y_1 + \left(\frac{dy}{dx}\right)_1 \frac{\Delta x}{1!} + \left(\frac{d^2y}{dx^2}\right)_1 \frac{\Delta x^2}{2!} + \cdots$$

This procedure will normally produce quite accurate results on a desk calculator, but it can be quite time-consuming if a large range of x is to be covered. The method is easily extended to differential equations of higher order. Though the above method is sometimes practical for hand or desk calculator computations, the digital computer is now the primary tool for numerical solution of differential equations.

Boundary-value Problems Taylor-series expansions are a standard method of marching out into an unknown region from given initial conditions. When the problem is of the boundary-value type, one of the most common methods used to obtain an approximate numerical solution is called *collocation*. Collocation involves assuming a solution in terms of a number of undetermined parameters. The parameters are then determined by requiring that the assumed solution exactly satisfies the differential equation at a number of specific points equal to the number of parameters. As an example, consider the differential equation

$$\frac{d^2y}{dx^2} = \sin x$$

with the boundary conditions

$$y = 0 \qquad \text{at } x = 0$$

$$y = 0 \qquad \text{at } x = \frac{\pi}{2}$$

The exact analytical solution to this system is

$$y = -\sin x + \frac{2x}{\pi}$$

If it is assumed that an exact analytical solution is not possible, the collocation method can be used and the answer compared with the exact solution. Any form of solution with undetermined parameters may be assumed. A polynomial is convenient in this case so assume

$$y = ax^3 + bx^2 + cx + d$$

From the boundary conditions, $d = 0$ and $c = -\pi/2(a\pi/2 + b)$. There are, therefore, two undetermined parameters (a,b) remaining, and these are found by requiring that the assumed equation exactly match the correct equation at two arbitrary points. The hope is, then, that the assumed solution will not stray far from the correct solution at other points. It is reasonable to select the points for exact matching at equally spaced intermediate points, in this case at $x = \pi/6$ and $x = \pi/3$. The assumed solution gives

$$\frac{d^2y}{dx^2} = 6ax + 2b$$

When this is set equal to sin x, as required by the given equation $d^2y/dx^2 =$

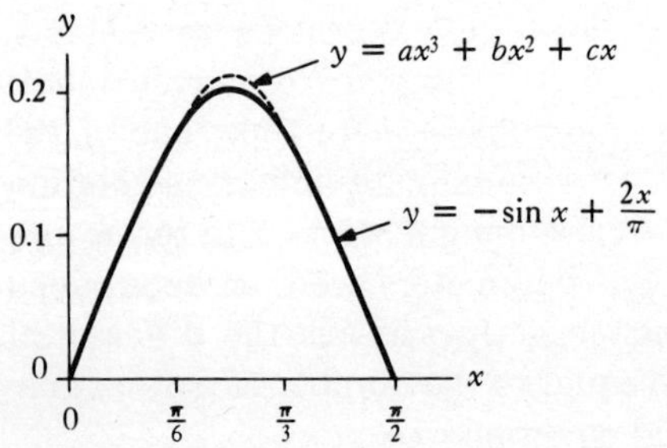

Fig. 8.9

sin x, then at $x = \pi/6$ and $x = \pi/3$ the two equations in a and b are as follows:

$$6a\frac{\pi}{6} + 2b = \sin\frac{\pi}{6}$$

$$6a\frac{\pi}{3} + 2b = \sin\frac{\pi}{3}$$

Solution for a and b, and hence for c, also gives

$$a = 0.1165 \qquad b = 0.067 \qquad c = -0.393$$

so that
$$y = 0.1165x^3 + 0.067x^2 - 0.393x$$

This is the approximate solution obtained by collocation. The exact solution is $y = \sin x + 2x/\pi$. These two are compared in Fig. 8.9. As can be seen, the agreement is excellent.

The Taylor series and collocation are but two of the numerical methods available for solving differential methods. Others are covered in the literature or in courses on numerical analysis.

8.7 INTRODUCTION TO FINITE-DIFFERENCE CALCULUS

The advent of the digital computer has greatly increased the importance of a branch of mathematics known as *finite-difference calculus.* A differential equation which cannot be solved by traditional analytical methods can often be handled by using the computer and a finite-difference approximation to the equation. Basically, the method is this: Instead of taking the limit, as is done in the regular calculus procedure, the equation is left or rewritten in terms of the original finite-difference form. A derivative, dy/dx for example, is the limit of the ratio of two *finite* differences, $\Delta y/\Delta x$, as Δx goes to zero. Left in finite-difference form, the computation can be done on the computer.

The major question that arises is how small must Δx be in order to give results which are close to the true value of dy/dx? The answer would seem to be as small as possible, thereby getting Δx as close to zero (where it belongs) as possible. However, other considerations arise. The smaller the Δx chosen, the larger the number of calculations necessary to cover a given range of x. This involves, then, the question of the cost of computer time and also the question of accuracy as a result of round-off errors. Obviously, even a large computer can keep track of only a certain number of significant figures. Also, the difference equations themselves involve some error due to omitted terms, as with a Taylor series.

On the other hand, of course, if Δx is too large, a poor approximation to dy/dx will be obtained. This is called the problem of *convergence.* Also, if Δx is too large, problems of *stability* may occur; that is, situations can arise where the results will be completely meaningless, oscillating from very large to very small, somewhat like a shaft which has exceeded its critical speed. The size of the finite difference Δx must be chosen small enough for convergence and stability but large enough to avoid too much round-off error.

Mathematical procedures are available for studying the stability of finite-difference equations but are beyond the scope of this book. Usually, an unstable equation is perfectly obvious from a glance at the results. Convergence is usually ascertained by reducing Δx to a smaller value (say, half) and showing that the same results are obtained. If the results are independent of the size of Δx, then Δx is small enough, and the result is said to have converged. Detailed studies of round-off error are available in the literature. If reasonable care is taken, round-off is not usually a serious problem in most routine problems. The major danger comes when subtracting two numbers that are very nearly equal. In this process, it is easy to lose many significant figures. If the answer from such a subtraction is then used (multiplied or divided by other numbers) in further calculations, the final result may well become quite inaccurate. Since, in working with complicated expressions, points where subtraction of nearly equal numbers take place may be hard to detect, every subtraction sign in a computer program should be examined carefully.

Let us return now to finite differences. The simplest finite-difference approximation for the first derivative is obtained by eliminating all but the first two terms of the Taylor-series expansion. Rearranging gives

$$\left(\frac{dy}{dx}\right)_n = \frac{y_{n+1} - y_n}{\Delta x}$$

As written, this is called a *forward-difference approximation*, as distinguished from the equation

$$\left(\frac{dy}{dx}\right)_n = \frac{y_{n+1} - y_{n-1}}{2\ \Delta x}$$

which is called a *central-difference approximation*. As an example of the use of this equation, reconsider the differential equation

$$\frac{dy}{dx} = 2y - x^2$$

with initial condition $y = 0$ at $x = 0$. Using the forward-difference approximation for the derivative gives

$$\frac{y_{n+1} - y_n}{\Delta x} = 2y_n - x_n^2$$

Rearranging and including $x_n = n\ \Delta x$ gives

$$y_{n+1} = y_n + (2y_n - n^2\ \Delta x^2)\ \Delta x$$

If $x = 0.1$ as before, at the starting point $n = 0$, $x_0 = 0$, and $y_0 = 0$, and so

$$y_1 = 0 + (0 - 0)0.1 = 0$$

The equation is now repeatedly applied, an easy task for the computer, but not impossible by hand in this simple case:

$$\begin{aligned}
y_2 &= 0 + (0 - 0.01)0.1 = -0.001 \\
y_3 &= -0.001 + (-0.002 - 0.04)0.1 = -0.0052 \\
y_4 &= -0.0052 + (-0.0104 - 0.09)0.1 = -0.01524 \\
y_5 &= -0.01524 + (-0.03048 - 0.16)0.1 = -0.034288
\end{aligned}$$

Typically for these solutions, the accuracy is not good at the beginning. At $x = 0.5$ the exact analytical answer is $y_5 = -0.055$. Accuracy will improve, however, at larger values of x, where the starting error becomes less significant. The central-difference approximation is generally more accurate; however, since it cannot be used to obtain y_1, the forward-difference approximation can be used to start, and the central difference equation used thereafter. Using this technique and a digital computer, the following results were obtained for $x = 0.50$:

Δx	y
0.10	−0.0906
0.02	−0.0619
0.01	−0.0582

The correct value for y at $x = 0.5$ is -0.055. For $x = 2.0$, the following results were obtained:

Δx	y
0.10	−12.7
0.01	−10.6

The correct value here is $y = -10.4$; thus, the value obtained with $\Delta x = 0.01$ is correct to within 2 percent. Note that the answer obtained with $\Delta x = 0.10$ has not yet converged to the correct solution. Stability problems are seldom encountered in simple problems of this kind. Round-off error due to neglected terms in the Taylor series is a small contributor to the lack of convergence at the larger values of Δx. The neglected terms are of the order of $(\Delta x)^2$. If $\Delta x = 0.10$, to find y at $x = 2.0$ requires 20 calculations, each with a potential error of $(\Delta x)^2 = 0.01$, for a total possible cumulative error of 0.20. The total error experienced is about ten times this, indicating that lack of convergence is responsible for most of the error. More accurate, but also more complex, approximations for the first derivative have been developed, and these are discussed in detail in most texts on numerical analysis.

Finite-difference approximations also can be developed by various schemes for the higher derivatives. For the second derivative, there is

$$\frac{d}{dx}\left(\frac{dy}{dx}\right)_n = \frac{d}{dx}\,\frac{y_{n+1/2} - y_{n-1/2}}{\Delta x}$$

$$\left(\frac{d^2y}{dx^2}\right)_n = \left(\frac{y_{n+1} - y_n}{\Delta x} - \frac{y_n - y_{n-1}}{\Delta x}\right)\frac{1}{\Delta x}$$

$$= \frac{y_{n+1} - 2y_n + y_{n-1}}{(\Delta x)^2}$$

If y_n and y_{n-1} are known, the above equation gives y_{n+1}. Repeated application will thus give all values of y. For example, consider the equation

$$\frac{d^2y}{dx^2} = x + y$$

with the conditions that $y = 0$ at $x = 0$ and $dy/dx = 0$ at $x = 0$. Using the above approximation for the second derivatives, this equation becomes

$$\frac{y_{n+1} - 2y_n + y_{n-1}}{\Delta x^2} = x_n + y_n$$

Solving for y_{n+1}:

$$y_{n+1} = 2y_n - y_{n-1} + (n\ \Delta x + y_n)\ \Delta x^2$$

To get the solution started, note that $x_0 = y_0 = 0$, and since $(dy/dx)_0 = 0$, the first derivative forward difference equation gives $y_1 = y_0 = 0$. Thus

$$y_2 = 2y_1 - y_0 + (\Delta x + y_1)\ \Delta x^2$$

$$y_3 = 2y_2 - y_1 + (2\ \Delta x + y_2)\ \Delta x^2$$

and so on. Other more accurate, and more complex, approximations also have been developed for the second-order derivative. Approximations for the higher derivatives are also available in books on numerical methods. The finite-difference equations above serve most engineering purposes, and their use is illustrated in one of the example problems later in this chapter.

Numerical solution of a differential equation always involves, as a first step, making the appropriate finite-difference approximation for the derivatives. As a general example, consider the equation:

$$\frac{d^2y}{dx^2} + a\frac{dy}{dx} + by = f(x)$$

For the derivatives,

$$\frac{d^2y}{dx^2} = \frac{y_{n+1} - 2y_n + y_{n-1}}{\Delta x^2} \quad \text{and} \quad \frac{dy}{dx} = \frac{y_{n+1} - y_{n-1}}{2\ \Delta x}$$

Thus

$$\frac{y_{n+1} - 2y_n + y_{n-1}}{\Delta x^2} + a\frac{y_{n+1} - y_{n-1}}{2\ \Delta x} + by_n = f(x_n)$$

Solving this for y_{n+1} gives

$$y_{n+1} = \frac{2y_n - y_{n-1} + (a\ \Delta x/2)y_{n-1} + f(x_n) - by_n\ \Delta x^2}{1 - (a\ \Delta x/2)}$$

Thus if one knows x and y at position n, the above equation can be used to get x and y at position $n + 1$. Repeated application of the equation gives the complete x and y relationship. Equations such as the above which are used repeatedly in this fashion are called *recursion equations.*

Note that two points (n and $n - 1$) must be known in order to compute a new point ($n + 1$). This raises the question of how to get the process started. In initial-value problems, initial conditions prescribed at $x = 0$ will provide the necessary information, but there are too many possible situations to discuss them all. As one example, however, assume in the general equation given above that it is given that at $x = 0$,

$$y = d \qquad \frac{dy}{dx} = e$$

Now from the differential equation, there results

$$\left(\frac{d^2y}{dx^2}\right)_0 = f(0) - bd - ae$$

Now if it is assumed that

$$\left(\frac{dy}{dx}\right)_0 = \frac{y_1 - y_0}{\Delta x}$$

then it can be shown that

$$y_1 = y_0 + \frac{\Delta x}{a}\left[f(0) - by_0 - \left(\frac{d^2y}{dx^2}\right)_0\right]$$

This enables calculation of y_1, and now the general recursion formula given above can be used for future points. Computer solution of this type of problem is, of course, essential.

When the problem is a *boundary*-value problem instead of an *initial*-value problem, it is not possible to get the finite-difference recursion equation started as outlined above. One method of solving such problems using a finite-difference technique is by trial and error. With one boundary condition given at each end of the region of interest, the method involves *guessing* at a second boundary condition (say, the first derivative) at one side, using this trial value to get the recursion equation started, and then determining whether the result satisfies the other boundary condition. All the remarks made earlier about trial-and-error solutions of transcendental equations apply to this situation also.

8.8 REMARKS ON ANALOG COMPUTATION

The foregoing discussion has been devoted largely to *numerical methods* as distinguished from *analog methods*. Numerical methods are best suited for use with the digital computer. Analog methods are also very powerful, however, when numerical answers are required in otherwise intractable situations. The basis for analog computation is the fact that the governing differential equations for many kinds of problems are analogous to the equations governing d-c electric circuits. Problems can thus be formulated as electrical problems, run on the analog computer, and the results obtained then interpreted in terms of the analogous original situation. Because of space limitations, this oversimplified view of analog computation cannot be expanded upon here. Interested readers are referred to the books in the Suggested Reading at the end of the chapter.

8.9 EXAMPLE: THE FLOATING TOMATO

Statement of the Problem In Chap. 3, the problem of separating green from ripe tomatoes was discussed from an inventiveness point of view. One of the schemes suggested there depended upon the different densities of ripe and green tomatoes. A little experimental trial will show

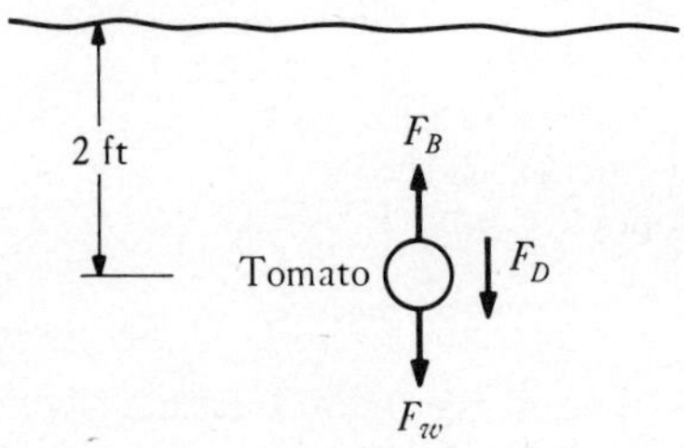

Fig. 8.10 Forces on the submerged tomato.

that green tomatoes will float whereas ripe ones sink in water. In the design of a device to utilize this principle, it is necessary to compute the time required for a green tomato to rise to the surface of a tank of water when it is released from rest 2 ft below the surface. In particular, a sample tomato is to be used as a test of the procedure developed. The tomato obtained was green, weighed 4.8 oz, and had a volume of 0.00500 ft^3. How long will it take for this tomato to rise the 2 ft, as described above?

Definition of the Problem Given a tomato which weighs 4.8 oz, has a volume of 0.00500 ft^3, and is released from rest 2 ft below the surface in a water tank, how much time is required for the tomato to reach the surface?

Model Formation Most of the questions relating to the model in this problem are trivial, except for the shape of the tomato. Tomatoes are not round. The assumption that the tomato is a sphere will surely not be correct, but it seems a reasonable place to begin. One would not want to assume anything else certainly, in the absence of specific or statistical information about the actual shapes of tomatoes. The model, then, will be that of a spherical object in a quiet, incompressible fluid.

Application of Physical Principles Newton's second law must be applied to the tomato. Refer to Fig. 8.10. Initially, the external forces on the tomato are its weight (downward) and the buoyant force of the water (upward). The buoyant force F_B is given by

$$F_B = \rho_w g \mathcal{V}$$

where ρ_w = density of the water, slugs/ft^3
g = acceleration of gravity, ft/sec^2
$\mathcal{V}$ = volume of the tomato

After the tomato begins to move, another external force, due to fluid drag, enters the problem. In terms of the fluid-drag coefficient C_D the drag force is

$$F_D = C_D \frac{\rho_w V^2}{2} A$$

where V is the tomato velocity and A is the cross-sectional area of the tomato. Applying Newton's second law gives:

$$F_B - F_w - F_D = ma$$

$$\rho_w g - \rho_T g - C_D \frac{\rho_w V^2}{2} A = \rho_T \mathcal{V} a$$

where ρ_T is the density of the tomato and a is its acceleration. Checking units:

$$\underbrace{\frac{\text{slugs}}{\text{ft}^3}\frac{\text{ft}}{\text{sec}^2}\text{ft}^3}_{\text{lb}_f} - \underbrace{\frac{\text{slugs}}{\text{ft}^3}\frac{\text{ft}}{\text{sec}^2}\text{ft}^3}_{\text{lb}_f} - \underbrace{\frac{\text{slugs}}{\text{ft}^3}\frac{\text{ft}^2}{\text{sec}^2}\text{ft}^2}_{\text{lb}_f} = \underbrace{\frac{\text{slugs}}{\text{ft}^3}\text{ft}^3\frac{\text{ft}}{\text{sec}^2}}_{\text{lb}_f}$$

A little algebraic manipulation gives

$$a = \left(\frac{\rho_w}{\rho_T} - 1\right) g - \tfrac{3}{4} C_D \frac{\rho_w}{\rho_T} \frac{1}{D} V^2$$

Since $a = dV/dt = d^2x/dt^2$, solution of this equation will give the desired position-time relation for the tomato.

Model Formation Before we go on to computation, some additional assumption making is suggested. Often particles released for free fall or rise in a fluid reach a terminal velocity very quickly. Thus most of their journey takes place at this constant terminal speed ($a = 0$). It is not illogical to assume that this circumstance is possible in this case. Alternatively, it might be assumed that drag is a small and negligible factor. If either of these assumptions is made, it would have to be justified, of course. As a first trial, let it be assumed that drag is negligible. The following computations are then based on this assumption.

Computation If drag is negligible, the above equation becomes

$$\left(\frac{\rho_w}{\rho_T} - 1\right) g = a = \left(\frac{62.4}{60.0} - 1\right) 32.2 = 1.25 \text{ ft/sec}^2$$

For constant acceleration and $x = 2$ feet,

$$t = \sqrt{2x/a} = \sqrt{4/1.25} = 1.79 \text{ sec}$$

Now to show that the drag is truly negligible, its order of magnitude can be estimated. The average velocity for $x = 2$ ft and $t = 1.79$ sec is 1.12 fps.

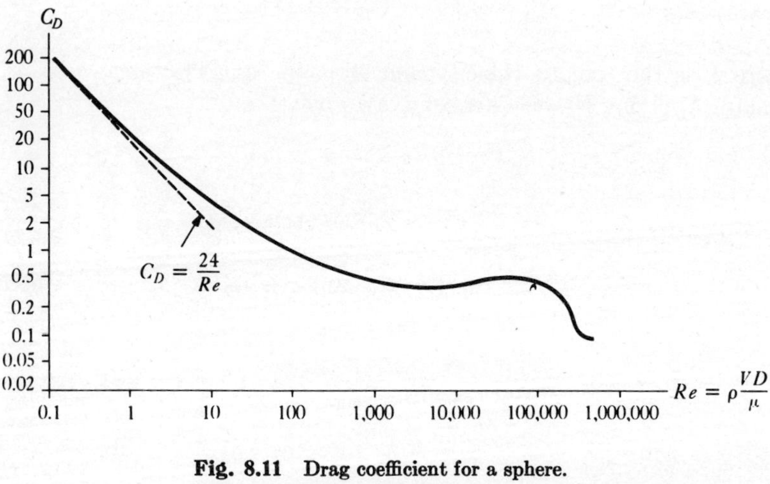

Fig. 8.11 Drag coefficient for a sphere.

To get the drag coefficient C_D the Reynolds number is needed (see Fig. 8.11). The Reynolds number $\rho VD/\mu$ is to be based on the diameter of the (assumed) spherical tomato. A sphere with a volume of 0.0050 ft³ has a diameter of 0.212 ft. Hence the Reynolds number is

$$Re = \frac{\rho VD}{\mu} = \frac{(1.94)(1.13)(0.212)}{(1.3 \times 10^5)} = 0.358 \times 10^5$$

$$\frac{\text{slugs}}{\text{ft}^3}\,\frac{\text{ft}}{\text{sec}}\,\frac{(\text{ft})(\text{ft}^2)}{\text{lb}_f\text{-sec}} = \frac{(\text{slug})(\text{ft})}{\text{lb}_f\text{-sec}^2} = 1$$

At this value of the Reynolds number, $C_D = 0.50$. The *average* drag force would thus be

$$F_D = C_D \frac{\rho V^2}{2} A = 0.50 \frac{1.94(1.12)^2}{2} \frac{(.212)^2}{4}$$

$$F_D = 0.0216 \text{ lb}_f$$

The net force due to buoyant and weight terms is

$$g(\rho_w - \rho_T) = \frac{62.4 - 60.0}{32.2} (32.2)0.005 = 0.012 \text{ lb}_f$$

Obviously, then, the drag forces are *not* negligible, and a new and more meaningful model must be formed.

Model Formation A second simplifying assumption that might be made is that the tomato reaches a terminal speed very quickly. In such a case the time and distance traveled during the initial period of acceleration from rest to a terminal speed would be neglected. The following computation is based on this assumption.

Computation When steady state is reached, there is a balance of forces and $a = 0$. The equation of motion is thus

$$\left(\frac{\rho_w}{\rho_T} - 1\right) g = \tfrac{3}{4} C_D \frac{\rho_w}{\rho_T} \frac{1}{D} V^2$$

Apparently C_D cannot be computed until V is known, and vice versa. However, note that C_D is nearly constant in the range of the Reynolds number expected in this problem. Hence a C_D can be assumed, V computed, and then C_D checked against the graph. If necessary, then, a second guess can be made. In this case, a guess of $C_D = 0.45$ gives $V = 0.880$ fps. The Reynolds number with $V = 0.880$ fps is 2.8×10^4. The corresponding C_D from the graph is .45. Hence, this is the solution.

A constant terminal velocity of 0.88 fps gives $t = 2.27$ sec for the time required to rise 2 ft. The assumption that the acceleration time is negligible must now be checked. The initial force on the tomato is merely the difference between the buoyant force and the weight and equals 0.012 lb_f. As the tomato begins to move, the net upward force on the tomato decreases because of the drag. However, for estimation purposes, suppose the maximum force (0.012 lb_f) acts fully during the accelerating period. Then the time required to accelerate the tomato from rest to the terminal velocity of 0.88 fps would be given by

$$a = \frac{F}{m} = \frac{0.012}{0.300} \times 32.2 = 1.29 = \frac{\Delta V}{\Delta t} = \frac{0.88}{t}$$

$$t = 0.68 \text{ sec}$$

This is a *minimum* and is hardly negligible compared with the 2.27-sec answer. Thus the accelerating time cannot be neglected unless an answer accurate to about 25 percent is satisfactory.

Model Formation Apparently none of the possible simplifying assumptions are accurate in this situation (see also Prob. 8.3). It is necessary to return to the complete equation

$$a = \left(\frac{\rho_w}{\rho_T} - 1\right) g - \tfrac{3}{4} C_D \frac{\rho_w}{\rho_T} \frac{1}{D} V^2$$

Putting in the available numerical values, this is

$$a = 1.29 - 3.68\, C_D V^2$$

Because V appears on the right, it is convenient to write a as dV/dt. The result is then a differential equation for the velocity-time relation:

$$\frac{dV}{dt} = 1.29 - 3.68\, C_D V^2$$

If the velocity is known as a function of time, then since $x = \int V\,dt$, the distance-time relationship can also be found, and that will give the desired answer.

To get velocity vs. time from the above equation, a graphical solution will be necessary. For this purpose note that

$$\int \frac{dt}{dV}\, dV = t$$

and so the area under the curve on a plot of $dt/dV = 1/(dV/dt)$ vs. V will be time. The procedure is this: (1) assume V in equal intervals starting with 0 up to the terminal velocity $V = 0.88$ fps; (2) for each V, compute Re and determine C_D from the graph; (3) use the equation above to compute dV/dt; (4) plot dt/dV vs. V; (5) count squares to obtain the time as a function of V for the various velocities (the time that the tomato has a velocity V_1 is the area under the curve between $V = 0$ and $V = V_1$); (6) with V vs. t now known, plot V vs. t and find $x = \int V\,dt$.

V	Re	C_D	dV/dt	dt/dV
0	0	0	1.29	0.776
0.1	0.32×10^4	0.40	1.28	0.782
0.2	0.64×10^4	0.37	1.24	0.806
0.3	0.96×10^4	0.40	1.16	0.863
0.4	1.28×10^4	0.40	1.06	0.944
0.5	1.60×10^4	0.42	0.90	1.11
0.6	1.92×10^4	0.45	0.69	1.45
0.7	2.24×10^4	0.45	0.48	2.08
0.8	2.56×10^4	0.45	0.23	4.35
0.88	2.80×10^4	0.45	0	—

These figures are shown plotted in Fig. 8.12. Also shown there are the

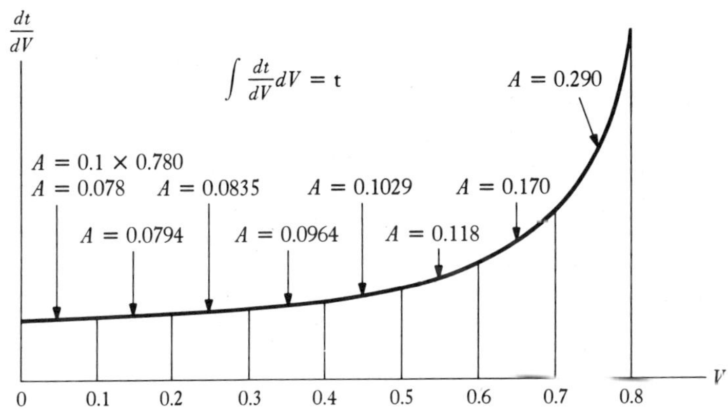

Fig. 8.12

areas corresponding to time. The results are:

V, fps	*t, sec*
0	0
0.1	0.078
0.2	0.157
0.3	0.241
0.4	0.331
0.5	0.434
0.6	0.552
0.7	0.722
0.8	1.012
0.886	

Plotting this data is shown in Fig. 8.13. Counting squares to get the area,

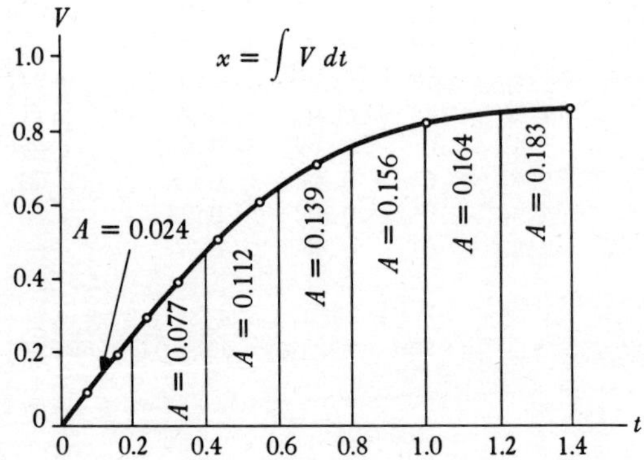

Fig. 8.13

and hence the distance, gives the following results:

t, sec	*x, ft*
0	0
0.2	0.024
0.4	0.096
0.6	0.208
0.8	0.347
1.0	0.403
1.2	0.567
1.4	0.735

After 1.4 sec the velocity is 0.84 fps and is approaching 0.88 fps asymptotically. Little error will be introduced, therefore, if the velocity is assumed constant at, say, 0.87 fps beyond this point. The remaining distance is $2.0 - 0.735 = 1.265$ ft, which at $V = 0.87$ fps requires 1.46 sec. The total time for $x = 2$ ft is therefore $1.46 + 1.40 = 2.86$ sec.

Evaluation and Generalization The above results are about as accurate as can be obtained using the drag coefficient for a sphere. Obviously, a tomato is a poor sphere, and one would expect the actual drag coefficient to be a little larger. The computed time of 2.86 sec would seem, therefore, to be a minimum.

Communication of the results of this analysis is left as an optional exercise for students. Students are also referred to Probs. 8.2, 8.3, and 8.6.

8.10 EXAMPLE: A BEAM PROBLEM FOR OPTIMIZATION

Statement of the Problem A cantilever beam is to be designed to carry a static-distributed, linearly decreasing load, as shown in Fig. 8.14. Manufacturing limitations restrict the design to solid cross sections in every plane. The depth of the load into the paper is L and is large compared with the length l, so that the problem is two-dimensional. The beam is to be designed so that the product of the beam weight times the maximum beam deflection is a minimum.

Definition of the Problem The problem is fairly well defined by the problem statement, a situation which is not normally true. The assignment is to determine the shape of a cantilever beam to carry the load shown in the figure such that $\delta_m W$ is a minimum, where δ_m is the maximum deflection and W is the total weight of the beam. (This is an unusual criterion, but we engineers must not be thrown by the unusual.)

Fig. 8.14

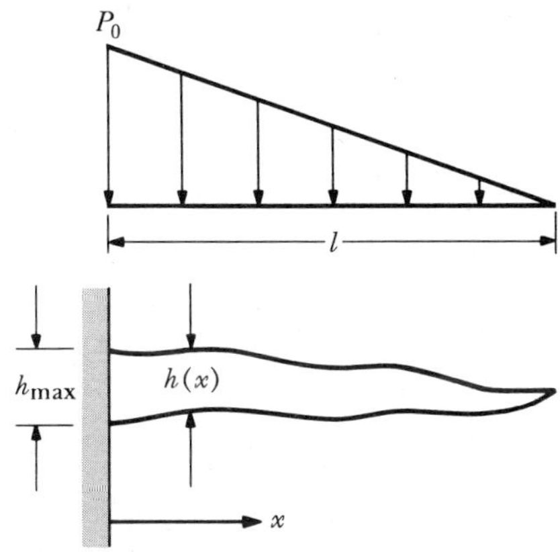

Model Formation Since h_0 is limited to a small percentage of l, the assumptions of ordinary elastic beam theory are appropriate. No other assumptions appear necessary at this time.

Application of Physical Principles Refer to Fig. 8.15. The basic equation for beam deflection is

$$EI\frac{d^2y}{dx^2} = M$$

The bending moment M for the triangular load at a location x is

$$M = \frac{P_0L(l-x)}{l}\frac{(l-x)}{2}\frac{(l-x)}{3} = \frac{P_0L(l-x)^3}{6l}$$

The moment of inertial I for a rectangular cross section of height h is

$$I = \tfrac{1}{12} Lh^3$$

Thus the equation for δ_m will be

$$\delta_m = \int_0^l \int_0^x \frac{d^2y}{dx^2}\,dx\,dx = \frac{1}{E}\frac{P_0L}{6l}\frac{12}{L}\int_0^l\int_0^x \frac{(l-x)^3}{h^3}\,dx\,dx$$

$$= \frac{2P_0}{E1}\int_0^l\int_0^x \frac{(l-x)^3}{h^3}\,dx\,dx$$

The boundary conditions are $x = 0$, $y = dy/dx = 0$. The weight of the

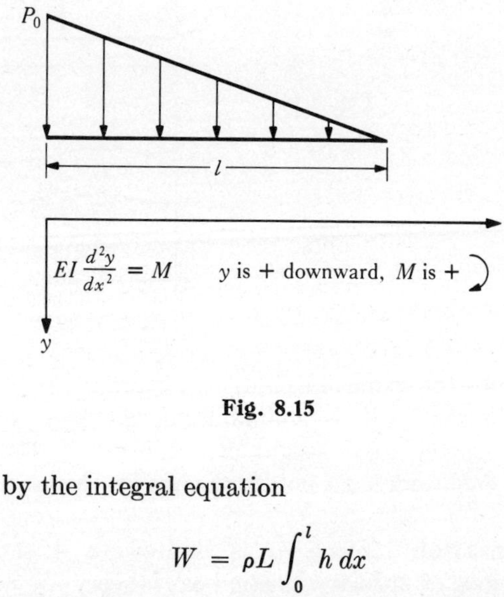

Fig. 8.15

beam is given by the integral equation

$$W = \rho L \int_0^l h\,dx$$

The problem is to find h such that $\delta_m W$ will be a minimum.

Computation Often in complex problems, especially in areas not fresh in the mind of the problem solver, it is helpful to work out a solution for a simple case before tackling the complete problem. This technique enables the problem solver to gain insight into the problem, its solution, and the method of solution, without having to consider all the complicating factors at once. In this present case, for example, it might be useful to consider a beam with a constant cross section first, just to get a better feel for the situation. In this case $h = h_0 =$ constant, and the equations for δ_m and W are easily integrated. The results are

$$\delta_m = \frac{2}{5}\frac{P_0 l^4}{L h_0^3 E} \qquad W = \rho h_0 l L$$

The product $\delta_m W$ is

$$\delta_m W = \frac{2}{5}\frac{P_0 l^5}{E h_0^2}$$

Checking units,

$$\text{ft-lb}_f = \frac{(\text{lb}_f)(\text{lb}_f)(\text{ft}^5)(\text{ft}^2)}{(\text{ft}^3)(\text{lb}_f)(\text{ft}^2)(\text{ft})} \rightarrow \text{ft-lb}_f$$

A slightly more complex case can be solved if a triangular beam is assumed, that is, if

$$h = h_0 \frac{l - x}{l}$$

In this case, also, the integrations called for to get δ_m and W are easily done. The results are

$$\delta_m = \frac{P_0 l^4}{E L h_0^3} \qquad W = \tfrac{1}{2}\rho h_0 l L$$

$$\delta_m W = \frac{1}{2}\frac{\rho P_0 l^5}{h_0^2 E}$$

At this point an alert problem solver may note that a somewhat general solution is possible if the beam height is assumed to be a power function

$$h = h_0 \frac{(l - x)^n}{l}$$

where n is an unknown factor to be found by optimizing $\delta_m W$. Using this form of solution does not give an absolute optimum, but it gives the optimum power function, which is real progress. It is often not possible to find *the* optimum in engineering problems, and engineers have to settle for *an* optimum which is the best they can do at the time. Using the power-function equation for h gives

$$\delta_m = \frac{2P_0 l^4}{E L h_0^3}\frac{1}{5 - 3n}$$

where $n \neq \frac{4}{3}$ and $n \neq \frac{5}{3}$. (Why?) For the weight, the result is

$$W = \frac{\rho h_0 l}{n+1} \qquad n \neq -1$$

The product to be minimized is

$$\delta_m W = \frac{2P_0 l^5}{Eh_0^2} \frac{1}{(5-3n)(n+1)} \qquad n \neq -1, \tfrac{4}{3}, \tfrac{5}{3}$$

$\delta_m W$ will be a minimum when $(5-3n)(n+1)$ is a maximum. Setting

$$\frac{d}{dn}(5-3n)(n+1) = 0$$

gives $n = \frac{1}{3}$. Thus the optimum power-function shape is

$$h = h_0 \left(\frac{l-x}{l}\right)^{1/3}$$

The resulting product $\delta_m W$ is

$$\delta_m W = \frac{3}{8} \frac{P_0 l^2 \rho}{Eh_0^2}$$

Evaluation and Generalization An optimum has been found, and though perhaps not *the* optimum, it certainly gives a good idea of the best shape to satisfy the constraints. The result obtained is the best power function, and that covers a lot of ground. The shape of the solution is shown in Fig. 8.16.

Examination of the result yields some additional information. For $\delta_m W$ to be a minimum, the quantity $P_0 l^2 \rho / Eh_0^2$ should also be as small as possible. P_0 and l are fixed. ρ and E are functions of the material chosen, and it can be concluded that the best material will have a minimum value

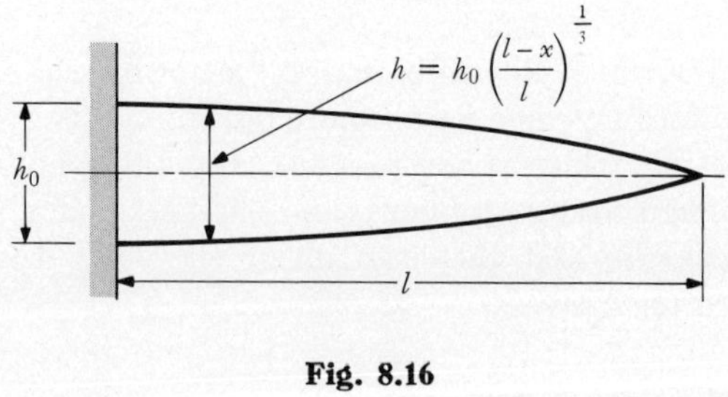

Fig. 8.16

Material	E/ρ(in. × 10^6)
Beryllium	670
Al_2O_3	390
TiC	260
Aluminum	110
Steel	110
Stainless Steel 301	100
Copper	60
Lead	15

Fig. 8.17 Values of E/ρ for some materials.

of ρ/E or a maximum value of E/ρ. Figure 8.17 shows approximate E/ρ values for a few materials.

The beam thickness h_0 should be as *large* as possible in order to minimize $P_0l^2\rho/Eh_0^2$. Thus h_0 will be made as large as the h_{max} constraint will

Problem: Find the shape of the solid cross-sectional cantilever beam to support the load shown such that $\delta_m W$ is a minimum.

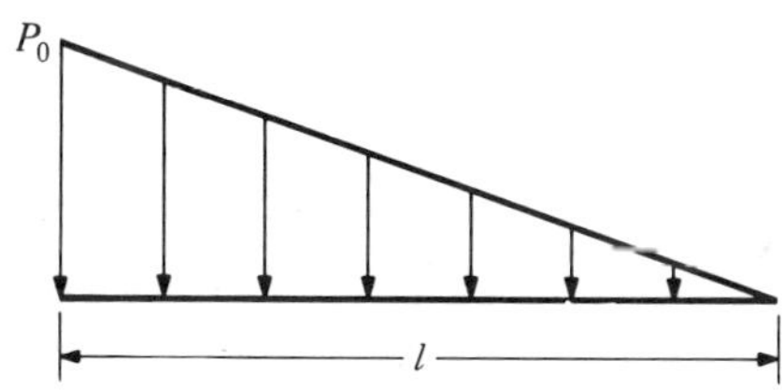

Results: The product to be minimized is

$$\delta_m W = \left[\frac{2P_0\rho L}{El}\int_0^l\int_0^x \frac{(l-x)^3}{h^3}\,dx\,dx\right]\left[\int_0^l h\,dx\right]$$

Assuming a power function shape $h = h_0[(l-x)/l]^n$, the optimum result is $n = \frac{1}{3}$ and

$$\delta_m W = \frac{3}{8}\frac{P_0l^2\rho}{Eh_0^2}$$

Material which satisfies the relation $S_{max} > P_0l^2/h_0^2L$ and has a minimum value of ρ/E should be chosen.

Fig. 8.18 Communication of results for the beam problem.

permit. This defines a limitation on the yield stress of the material. The maximum bending moment is at $x = 0$, where $M_0 = P_0l^2/6$. Thus the maximum stress will be $S_{\max} = M_0h_0/2I = P_0l^2/Lh_0^2$. The material chosen must at least be large enough to satisfy this condition.

A one-page communication of the results of this analysis is shown in Fig. 8.18.

8.11 SUMMARY AND PREVIEW

This chapter has been an introduction to numerical analysis—that is, to the techniques of getting numerical answers when they are needed and when the usual analytical techniques will not work. Graphical, numerical, and computer methods should be looked upon as being as prominent and useful as any other methods. In fact, real-life engineering situations are more likely to result in equations which require numerical solution than in equations which are amenable to closed-form analytical solution. Students interested in being at the forefront of their profession should seriously consider choosing an elective course or should do some self-study in numerical analysis. Some books which will help are listed in the Suggested Reading.

After computation in the engineering analysis process, the results must be evaluated for their significance to the original question. Also, the results must be studied for whatever new information they may imply. They should be generalized and learned from. Finally, of course, to be of any value at all, they must be effectively communicated. In the next chapter, then, attention is directed to the topics of evaluation, generalization, and communication.

SUGGESTED READING

Crandall, Stephen H.: "Engineering Analysis," McGraw-Hill Book Company, New York, 1956. An excellent advanced treatment of the science of numerical analysis.

Dusinberri, G. M.: "Heat Transfer Calculations by Finite Differences," International Textbook Company, Scranton, Pa., 1961. A good example of numerical analysis as applied to the field of heat transfer.

Hildebrand, Francis B.: "Introduction to Numerical Analysis," McGraw-Hill Book Company, New York, 1956. An excellent book on numerical analysis.

James, M. L., G. N. Smith, and J. C. Wolford: "Analog and Digital Methods of Engineering Analysis," International Textbook Company, Scranton, Pa., 1965. A good introductory book on both analog and digital computation.

Pennington, R. H.: "Introductory Computer Methods and Numerical Analysis," The Macmillan Company, New York, 1965. An excellent introduction to numerical methods of computation with the computer.

Truitt, Thomas D., and A. E. Rogers: "Basics of Analog Computers," John F. Rider, Publisher, Inc., New York, 1960. An elementary introduction to analog computation.

PROBLEMS

8.1 Use the digital computer to find the solution of $\sin x = x^2$.

8.2 Use the digital computer to solve the equation of motion of the tomato in the example in this chapter.

8.3 Stokes' law for the drag of a sphere applies at Reynolds numbers less than about 1. The drag is then given by $24/Re$. Assume the tomato in the example in this chapter satisfies the Stokes' law condition and solve the equation of motion. (It now has an analytical solution.) Show that the assumption of Stokes' law is invalid.

8.4 Use the digital computer to find the optimum value of n in the beam example problem in the chapter.

8.5 Use the digital computer to find the maximum beam deflection for a triangular beam, $h = h_0 [(l - x)/l]$. For this phase use the finite-difference approximation on the second derivative that is given in the chapter.

8.6 A tomato of the size and weight used in the example in the chapter was used in an actual experiment. It required 4.3 sec to rise to the surface from rest 2 ft below the surface. Calculate the drag coefficient for the tomato.

8.7 Find the value of the integral $\int_0^2 (x^3 - 2x^2)\,dx$ by (*a*) analysis, (*b*) graphical methods, (*c*) numerical analysis using the computer.

8.8 Find the solution to the equation $\ln x = x$ using (*a*) analysis, (*b*) graphical method, (*c*) digital computer.

8.9 Solve the following differential equations using numerical methods of your choice:

a $dy/dt = 2y^2 + y; y = 1$ at $t = 0$; find $y(t)$ for $0 < t < 2$

b $d^2y/dt^2 = y^2 + 2y; y = 1$ at $t = 0$; $dy/dt = 0$ at $t = 0$; find $y(t)$ for $0 < t < 1$

c $2\,d^2y/dt^2 + dy/dt = 4ty; d^2y/dt^2 = dy/dt = 0$ at $t = 0$; find $y(t)$ for $0 < t < 4$

9 EVALUATION, GENERALIZATION, AND COMMUNICATION OF RESULTS

9.1 EVALUATION

At the conclusion of the computation stage, the desired numerical or symbolic (equation) answer has been obtained. The answer applies strictly to the model. It is the answer *for the model* to the operationally defined problem which was formed near the beginning of the engineering analysis process. The question at this stage is: What does the answer mean with respect to reality?

There are two aspects to this question. First, it is necessary to evaluate the answer in terms of how well the answer for the model matches the answer for reality. Short of an actual experiment, of course, there is no way to determine the true correspondence. However, often the numerical answer does shed light on the appropriateness of some of the assumptions made, and once a numerical answer is obtained, it is wise to review the various assumptions to ensure that the answer is at least consistent with them. Even though the assumptions are poor, of course, the answer may still be consistent with them. But there are also many cases where answers and assumptions turn out to be inconsistent. This kind of error should be detected in the evaluation stage.

The second part of the evaluation question has to do with the meaning of the answer in terms of the original vague, nonoperational question (see

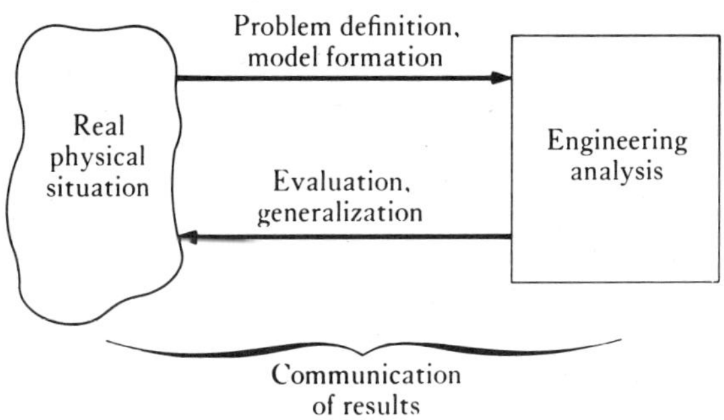

Fig. 9.1 Schematic of the analysis process.

Fig. 9.1). Earlier, questions such as "Will it work?" and "Will it be too hot?" had to be translated into answerable questions such as "How long will it be?" and "Given —, at what temperature will it operate?" By now, presumably, the engineer has found the length, the temperature, etc. He must now interpret his answer to determine what it means in terms of "Will it work?" or "Will it be too hot?" That is, if the length is 12 in., does that mean it will or will not work? If the temperature is 500°F, is that too hot? Etc.

The engineer at this stage is a decision maker. Part III of this book deals specifically with decision making, but it is appropriate to point out here that not all factors in decision making are objective. Many decisions are based at least partly on subjective factors. (*Objective* here means completely impersonal and external to the decision maker. *Subjective* means that personal factors enter.) There can be style and character in subjective decisions. One man or company may prefer to take more chances than another, for example. Sometimes the best that can be done in answering the vague but all-important questions like "Will it work?" is to answer in terms of probabilities, and hence probability and statistics are important decision-making tools.

9.2 GENERALIZATION

Generalization is learning. At this stage in the engineering analysis process, an answer to a specific problem has been obtained. To make the most of what has been done, it is useful to consider the general implications of the specific solution obtained. The idea is to see what can be learned that will apply to other similar problems and to new and different situations. Always remember to take a few minutes to *learn from your work!*

9.3 COMMUNICATION AND SEMANTICS

The remaining and major portion of this chapter is to be devoted to the subject of communication. Communication is almost always symbolic. Words, either written or spoken, and pictures (graphs, drawings, sketches, etc.) are the common symbols. But in some circumstances, bodily motions, facial expressions, tone of voice, manner of wording, and so on, also are used for communication, either consciously or unconsciously.

There is a tendency in discussing and studying communication to concentrate on the properties of the message. This is especially true of the field of study and research called information theory. In this book, however, communication will be viewed as an *activity* involving *people*. The important elements of communication will be considered to be a *sender*, a *message*, and a *receiver*. In general the sender and the receiver can each be either a man or a machine. In this book, the concern is with communication when both sender and receiver are human and when the message is either verbal or graphic. It is of tremendous importance in this kind of communication that both the sender and receiver understand that such communication includes two human beings—with all their associated human frailties—as well as the message itself. Undue concentration on the message itself without appropriate consideration of the effects which caused the message to be sent and of the effects the message will have on the receiver can lead to good messages but poor communication.

An understanding of the basic notions of the field of study called general semantics is essential to effective communication. Do not think that semantics is just "a matter of defining terms." Semantics is the study of the interaction among words, thought, and behavior. That is, it is the study of the way in which words affect how people think and behave, and vice versa. It is the science of human communication in the most general sense. This is far from the simple matter of definition, though as might be expected, semantics has some things to say about definitions too. In the remainder of this section, some of the elementary ideas of semantics will be outlined. Students who wish to pursue the subject further (a *very* good idea!) should consult the Suggested Reading.

A first principle of semantics is that words are symbols; the word is *not* the thing. Semanticists like to use the analogy of a map and a territory, pointing out that the map is not the territory. In the same way, the *word* "chair" is *not the physical object* (used by people to sit on) referred to. The thing referred to is called the *referent* in the jargon of semanticists. Thus the map is not the territory and the word is not the referent. It is perfectly obvious that maps can be drawn which are poor descriptions of the territory they are supposed to represent. In fact, maps can be drawn for which there is *no territory at all!* What about words?

Words may or may not stand for something. And they mean different things to different people. The symbol is *not* the thing symbolized. It is just a symbol.

Another basic principle of semantics has to do with the matter of definitions. The classical concept of the way to define the meaning of a word is to (1) place it in a general class of similar things and (2) distinguish it from the other members of its class. Thus a dog is (1) an animal (2) with fur, four legs, etc. This sort of definition is all right as far as it goes. The trouble is that it is possible to define something in this way which has absolutely no referent! Thus a dolevert is an animal with seven legs, red-, white- , and blue-striped skin, etc. The fact that doleverts do not exist is immaterial to such a definition.

Semanticists have a somewhat different concept of definition. They recognize that a definition (one kind of message) serves the purpose of telling some human being (a receiver) the meaning of the symbol involved. Thus semanticists say that a definition is a kind of communication and that the message or definition used must take into account the ability of the receiver to understand the words used in the definition. This leads directly to the conclusion that definitions must be made in terms of, or with reference to, things in the direct experience of the receiver. In fact, this kind of statement can be made about all messages and communication situations: The words used, references made, and implications introduced must be within the experience—or explained in terms of the experience—of the receiver.

It is to be noted, therefore, that people who have not shared similar experiences cannot communicate at all! The sounds made by one will have no meaning to the sounds made by the other—even though they may speak the same language. Keep this in mind the next time you hear an argument between, say, a labor leader and a company executive, a religious believer and an atheist, or a scientist and a humanist.

A good technique for illustrating the point of view of semanticists is to use the so-called abstraction ladder. As an example, consider the abstraction ladder of thermodynamic terms shown in Fig. 9.2. At the lowest level there is experience and observation of the physical world. At this level verbal definitions are not possible except in terms of activities. Consider,

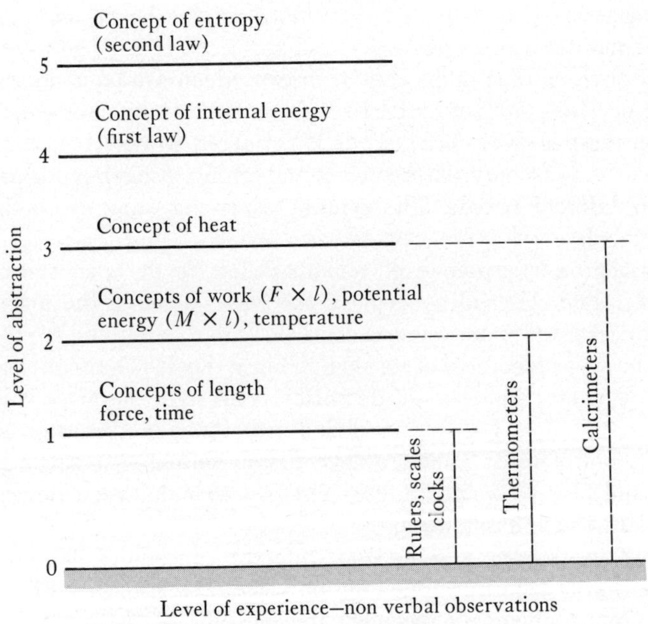

Fig. 9.2 Thermodynamic abstraction ladder.

for example, the definition of length. No quantity or clever selection of words will ever describe length as a property. To define it one can only describe the procedure for measuring it. The procedure for measuring length is an activity or, as it is usually called, an *operation*. Definitions of this kind—that is, in terms of activities—are called operational definitions. Operational definitions are the connection between the verbal-symbolic world and the physical world.

People develop a concept of length as they measure it, observe its effects, etc., and the word (symbol) stands for this concept. Backing up the concept and its symbol is the operational definition, without which the whole idea is meaningless. The concepts of mass, force, and time have similar definitions on the operational, or nonverbal, level.

This picture is complicated, however, by the fact that man's great intelligence has enabled him to abstract and generalize to a higher level. Thus there are words *about words*, and concepts which are at least once removed from any operational definition involving experience. Potential energy (see thermodynamic abstraction ladder) is a good example. Potential energy has an operational definition: it is the paper-and-pencil operation of multiplying the weight by its height (length) above a reference plane. The definition of potential energy is operational and, furthermore, it is connected with experience through the operational definitions of length and weight. Suppose, however, that I define an alavert as the product of the cervalosis and the boktock. Alavert is operationally defined (a paper-and-pencil operation), but since cervalosis and boktock have no meaningful definitions operationally and the whole chain has no ultimate referent in experience, it is all meaningless.

Notice how the thermodynamic abstraction ladder has a large number of steps, and how many steps separate entropy from experience. (This is the reason entropy seems like such a mysterious concept when first encountered.) On any level of abstraction, if the terms are to be meaningful, they must be operationally defined with words below, *not* above, them on the abstraction ladder (that is, less abstract words or words closer to experience), *and* the ladder must ultimately connect with experience where there are real referents.

Western thought has generally been founded on the basis of aristotelian logic. It has been the lot of the semanticist to point out that all the effects of aristotelian thinking are not good. It is ingrained in Western thought processes that no statement can be both true and false at the same time. That is, a thing cannot have property "A" and property "not A" at the same time. What is not nearly so well ingrained, however, is that this type of logic applies *only to certain kinds of statements or things*. In particular, it applies only when there are agreed-upon operational definitions of "true" and "false" or "A" and "not A." Misuse of the aristotelian law of the excluded middle (a thing cannot be both A and not A) leads to great difficulties, such as the tendency to think in either-or terms when it is not appropriate to do so. Consider, for example, the statement: "You are either for me or against me." The semanticist knows that the person being addressed may have a third view, or may be sometimes for and sometimes against, or for on some points and against on others, or unconcerned, or undecided, or at some point halfway or partway between "for" and "against."

As a final point in this brief introduction to semantics, the tendency that people have of making things anthropomorphic (anthropomorphic means "in the image of man") should be pointed out. The "economy," for example, is spoken of as if it were a living, breathing person about to be sick or well or to stumble over a stick. The "engineering profession" is also given anthropomorphic qualities by some writers and speakers. The anthropomorphic analogy has its limitations, however, and even with such a cursory examination of semantics as has been given here, students should find referents and operational definitions for words used in the anthropomorphic sense before spending a great deal of time in meaningless discussion.

9.4 FACTORS IN EFFECTIVE COMMUNICATION

With the general introduction to semantics in the preceding section serving as an introduction, it is now possible to set down some more specific rules or factors for effective communication. These are not profound and are fairly obvious to anyone who considers the problem. However, they are often neglected.

1. Specify the purpose of the communication. Is it to convey specific information, to persuade, to entertain, to stimulate thought, etc.? Is it even necessary?

2. Construct the message so that it is *appropriate for the audience (receivers)*. Is the audience interested, and willing and able to listen? Is the length appropriate? Are the words and other symbols within the experience of the audience? Does it proceed from what the audience knows to what it (he) does not know? Is it clear what the audience is supposed to do as a result of the message?

3. Make the mechanics of the message as efficient as practical. Are sentences short, concise, and to the point? Are graphs and pictures clear and easily read or interpreted? Are words and other symbols meaningful in the sense described in Sec. 9.3? Has the best method—written, oral, graphic, etc.—been selected? Are facts, opinions, recommendations, etc., clearly distinguished? Is an introductory overview provided which covers the high point or points, so that the busy audience (reader) can get the essence without going into full detail? A thousand questions could be asked in the same vein, but the general idea should be obvious by now. All that remains is practice—which suggests another "rule":

4. Evaluate your communications with hindsight to see where and why they succeeded or failed. This sort of self-evaluation will provide a continual check on the effectiveness of communications made and will, with time, result in highly effective communication.

One last rule:

5. Learn to be a good listener. Since not all messages received will be from people who are careful and competent communicators, it is often necessary to use ingenuity to discover the purposes of the message, the high points, the operational meanings, the expected actions, etc. Good listening involves (1) listening to only one message at a time, and (2) listening for the purpose of understanding the message before thinking about a reply.

Engineering students do a great deal of communicating with their teachers via tests, homework, and reports. A great deal can be gained from thinking of such assignments and tests as messages in a communication system. Consider, for example, the purpose of the message. (Students will have to decide this for themselves. It will vary from instructor to instructor!) If students will examine these messages *as pieces of communication* both before and after they are graded, much of benefit can be learned that will improve understanding—and grades, too!

9.5 COMMENTS ON WRITTEN COMMUNICATION

The belief that college students generally, and engineering students particularly, are poor writers is a widespread one. There is some substance to and evidence for such an opinion. "Why can't Johnny write?", or something similar, is the title of articles which appear in educationally oriented journals in many fields. The *Saturday Review* has over the years carried a number of articles on the subject, and some of the ideas that follow in this section appeared in that journal on February 20, 1965, in an article by E. R. Clapp entitled "Why the Devil Don't You Teach Freshmen to Write?"

In considering a piece of written communication—an informative rather than artistic communication—there are two distinguishable aspects of concern. One is called *literacy* and the other *competence*. Literacy is the correct use of grammar and spelling. It is also having a reasonable vocabulary. Errors in literacy are easily detected and are a common source of complaints about poor writing. Spelling and grammar are not difficult skills, so that poor performance in them is often taken as a sign of poor education.

Competence is a higher-level skill than literacy and hence is harder to judge. By competence is meant the ability to express thoughts, trains of thoughts, logic, ideas, argments, etc., in a clear, concise, understandable, and readable fashion. Competence has to do with the subject matter being presented, whereas literacy is independent of the subject matter. Competence is the ability to use language to express thoughts in a way that others can understand. Note that a competent writer may be a terrible speller (he probably can't be *too* bad on grammar, however) and that a completely literate person who knows all the mechanical rules of language may be quite incompetent. Competence in no way excuses poor literacy, of course, but it is important to see that literacy does not *necessarily* indicate competence, and vice-versa.

How can engineering students ensure their literacy and competence in engineering communications? An awareness of the principles of semantics is essential to communication competence. Also, some of the basic rules outlined in the preceding section will help in specific situations. Having each piece of writing checked for both literacy and competence by a friend or colleague is always a safe procedure. A good secretary can often check for literacy but cannot be expected to help much on competence.

In the long run, engineers and engineering students will do well to realize the relationship among reading, thinking, and writing. The failure of engineering students to read no doubt contributes a great deal in the long run to a low level of writing competence. Reading novels as well as good expository articles is a good therapy for both poor literacy and writing incompetence. The connection between writing and thinking is obvious. Some go so far as to say that they are one. Certainly a poorly written (incompetent) communication is commonly considered to be poor thinking, but it seems more plausible to assume that there are some who understand their subject, but because of writing incompetence cannot articulate their thoughts.

As a final thought, consider this: You cannot write writing! Before you can write, you must have someting to say and someone to say it to.

9.6 COMMENTS ON ORAL COMMUNICATION

Most of the principles of semantics, rules of good communication, and remarks on written communications apply as well to oral communications. There are some obviously different features of oral reports, however. For example, the audience may be able to talk back, or ask questions. The audience may be captive, and so the length of the message may be critical. [Note that the receiver of a written communication can (and will!) put it aside if he gets bored or restless. An audience can hiss, boo, or walk out, but, out of courtesy, seldom does. An employer may well become exasperated and say, "Get to the point," however!] Finally, an oral communication has the added dimension of person-to-person contact. Mannerisms, facial expressions, body motions, and the like are thus part of the message. The writer can use language (flowery, sharp, curt, kind, etc.) to do some of the same things, but the written communication is inherently less personal than the oral.

Most schools offer courses in speaking and these are popular with students, for good reasons. As with writing, literacy and competence can be distinguished in an oral report. The added dimension of the manner of presentation (analogous to handwriting in written communications) is also very important in oral reporting, however. The practice gained in a good speaking course that stresses all three—literacy, competence, and presentation—should be considered invaluable.

The principles of semantics, the general rules of good communication, and the rules of grammar apply to oral as well as to written reports. Nevertheless, written and oral reports are usually quite different in style. The oral report has the added dimension of vocal and visual expression. Hence reading aloud a report intended for readers is almost always very poor communication. *Never read aloud to an audience a report that was prepared for readers.* Good communication demands that the oral and written messages be prepared and transmitted differently, each in a manner appropriate to its special characteristics. Repeat: *Never* read a written report aloud to an audience! If you are inexperienced at delivering oral reports, then prepare and read, if necessary, an *oral* report. But do not simply read a written version aloud.

9.7 COMMENTS ON GRAPHICAL COMMUNICATION

Illiteracy in graphical communication is a very important sin—more so than in written work, where incompetence is most significant. The rules of graphical communication—the conventions—are the very reason that it works. This is the reason the rules have been taught so thoroughly in the past. In recent times, however, as engineering design activity has moved further from the drafting boards, there is less and less need that these rules be taught in the context of actual mechanical drawing. The engineer must be graphically literate, but he need not be a draftsman any more than he need be a typist. Knowledge of the rules of orthographic projection, of dimensioning, and the like is necessary for graphical literacy but they can be learned quickly through sketching.

Sketching is coming to be appreciated as a valuable method of communication for engineers. It seems that, just as some good thinkers look poor because of writing incompetence, so others look poor because of graphical incompetence. A good clear sketch is analogous to a good, clearly written message. The ability to sketch can be learned—or at least improved—just as competence in writing can be learned or improved: by reading and by practice. Several of the books in the Suggested Reading have good sections on sketching, for those students who are interested in improving their graphical skills of communication.

Figure 9.3 shows in table form the relationship among some of the concepts being discussed here in terms of graphical and verbal communication. There are several levels shown: technical skill, literacy, and compe-

Type of communication	Level of communication skill			
	Technical	*Nontechnical*	*Literacy*	*Competence*
Graphical	Drafting	Sketching	Rules of Graphics	Obtaining ideas from sketches; expressing ideas as sketches
Verbal (written)	Typing	Handwriting	Grammar Spelling Sentence structure	Principles of semantics Knowledge of subject Organization of material Clarity of expression
Verbal (oral)		Speaking (use of voice, expression)	Grammar Sentence structure	

Fig. 9.3

tence. Notice that knowledge of the rules of graphics (e.g., orthographic projection) is analogous to knowledge of the rules of grammar, spelling, sentence structure, etc., in verbal communication. An engineer must be able to read drawings, just as he must be able to read verbal messages. Note too that there are two levels of skills, called here *message skills* and *technical skills*. Engineers need facility with message skills—sketching, speaking, writing. These are the basic staff of everyday communication. In the case of speaking, it is also important on special occasions such as giving a paper to a technical audience or discussing plans in an engineering or management group. There seems to be little need for engineers to be skilled draftsmen any longer. This is not universally true, but it is clearly the lowest in importance in the communication system. The most important communication skills are on the level of those shown under *competence* in the table.

9.8 SUMMARY AND PREVIEW

This essentially concludes the separate discussions of the steps in engineering analysis. Design was separated into three parts at the outset: inventiveness, engineering analysis, and decision making. The engineering analysis process was then further broken down into problem definition, model formation, application of physical principles, computation, checking, evaluation, generalization, and communication of results. It is nearly time, therefore, to move on to decision making. But first, and as a kind of bridge to that topic, the next chapter will present another case study. This one has to do with the design—inventiveness, analysis, and decision making—of an air-water toy rocket. It was written with the cooperation of the manufacturer.

SUGGESTED READING

Hayakawa, S. I.: "Language in Thought and Action," 2d ed., Harcourt, Brace & World, Inc., New York, 1964. This is a good introduction to the principles of semantics.

Johnson, Wendell: "People in Quandaries," Harper & Row, Publishers, Incorporated, New York, 1946. A very good book on the use of semantics in everyday life.

Rosenstein, A. B., Rathbone, R. R., and Schneerer, W. F.: "Engineering Communications," Prentice-Hall, Inc., Englewood Cliffs, N.J., 1964. This contains an introduction to information theory in communication. The authors do not stress semantic problems in communication. The section of the book on graphics and sketching is excellent. There are also good sections on written, oral, and graphic reports. Available in paperback.

PROBLEMS

9.1 The example problems worked in Chap. 7 do not have written reports presented. Prepare one-page reports for the results.

9.2 If I say, "I have a pain in my arm," how do you know what I mean or how I feel?

9.3 Write an essay on Newton's second law (or some other technical topic) for (*a*) a coed in history, (*b*) your mechanics instructor, and (*c*) your friend who missed two weeks of class when this was discussed. Analyze how and why these are different.

9.4 For your last summer (or some other project of interest) prepare two reports for your class: (*a*) written, and (*b*) oral. Analyze how and why these are different.

9.5 Analyze and criticize yourself as a skilled communicator in terms of the items and levels shown in Fig. 9.3. Make specific plans for strengthening your weaknesses. There is "nobody else but you" to do it!

10
A CASE STUDY

10.1 INTRODUCTION

This chapter discusses certain aspects of the design of an air-water two-stage toy rocket (Park Plastics Company's Product No. 507).[1] A single-stage forerunner of this product was first produced and marketed in the United States in 1954 by the Park Plastics Company under a German patent. Excerpts and pictures from the United States patent are shown on pages 220–221. To the German inventor must go the credit for the creativity of the basic idea. Also to him went a considerable sum of money, for he was paid royalties by the United States company for a number of years. Finally, the patent was purchased outright, so that it is now solely owned by Park Plastics Company.

[1] The author wishes to acknowledge the cooperation and assistance of Mr. Arthur O. Lange, president of Park Plastics Company, Linden, New Jersey, in the preparation of this chapter. Mr. Lange gave freely of his time and knowledge in order that this case study could be brought to engineering students in a realistic manner.

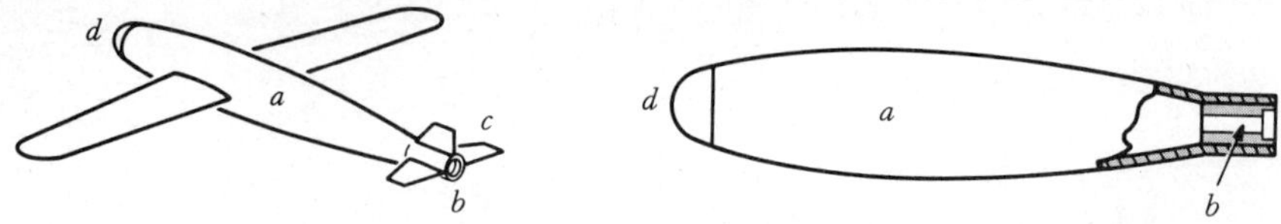

Fig. 10.1 Toy rocket.

Jet Driven Aerial Toy

2,732,657
Adam Krautkrämer, Budenheim (Rhine), Germany
Application November 22, 1952, Serial No. 322,077
5 Claims. (Cl. 46–76)

This invention relates to a driving mechanism for toys.

The development of modern industry has been reflected in many ways in the toy industry. In particular, model airplanes in the most varied embodiments have awakened the keen interest of children.

It is an object of the present invention to provide a new toy of this kind. This, and still further objects will become apparent from the following description, read in conjunction with the drawing, in which:

Figure 1 shows a perspective view of a toy airplane in accordance with the invention.

Figure 2 is a side elevation of the fuselage of the plane shown in Fig. 1, partially in section.

In operation, the hollow fuselage member *a* is partly filled, for example, with water, with a pipette, syringe, funnel or the like, or through an inlet *i–g*. Thereupon, compressed air is introduced into the remaining part, as by means of an ordinary bicycle pump, so that the air and thus also the water introduced are under a pressure of a few atmospheres. When introducing the compressed air, the fuselage body is placed on an air inlet *i–g* in such a manner that the end of the fuselage provided with an opening points downward, and a dependable seal between the inlet and the outlet opening is obtained. In this way the water collects in the lower part of the pressure container, the water being under the pressure of the air which has been blown in. As soon as a restraining device *k* which presses the hollow body in a pressure-tight manner against the air inlet and holds it fast, is loosened, the water discharges under the air pressure from the opening, and the reaction force produced thereby acts in an accelerating manner on the airplane. Depending on the energy content of the compressed air in conjunction with the substances exerting the reaction, the toy will be propelled upward to a greater or lesser height. After the cessation of the driving action, the airplane, for example, when developed as a glider, will gradually glide to the earth.

The invention relates to an airplane which is not shot off as heretofore customary by means of some type of catapult or thrown out or up in some other known manner, but rather compressed air or other gas is utilized as the driving force. The air acts on solid substances or liquids in such a manner that they are expelled from the pressure container which is provided with an outlet opening. The jet reaction produced thereby is utilized to drive the airplane.

The special advantage of this embodiment lies in the elimination of any danger of fire, inasmuch as there does not enter into question in connection with the drive any lighting of any powder train or the like, but merely the inflation with a gas and preferably air, as, for example, by means of a bicycle pump.

I claim:

1. In combination, a toy projectile comprising a hollow, pressure-tight body, a jet outlet opening defined at one end of said body in communication with the hollow interior, and a launcher therefor carrying a fluid inlet check valve dimensioned to freely detachably fit said jet outlet opening for the pressure closure thereof, and manually operable releasable holding means carried by said launcher for releasably engaging the rear end of said projectile for holding said fluid inlet check valve in closure contact with said jet opening.

2. Device according to claim 1, in which said hollow body is formed as at least a portion of a toy flying device.

3. Device according to claim 2, in which said hollow body has a resilient impact cap at the forward tip thereof.

4. Device according to claim 1, in which said fluid inlet check valve has an adapter extension dimensioned to fit within said jet outlet opening.

5. Device according to claim 1, in which said hollow, pressure tight body is a tear-shaped body.

The concept of the *two-stage* air-water toy rocket originated in 1958. It was conceived, developed, and patented by Park Plastics Company. Its design will be discussed in detail later in this chapter. It is interesting to note that the marketing of the single-stage toy rocket preceded the launching of the Russian Sputnik and that the two-stage toy was already under development at the time Sputnik was announced. The public interest in rocketry and space that Sputnik precipitated created a huge boost in sales for these toy rockets. Sales went from roughly 60,000 in 1957 to nearly 700,000 in 1958, better than a tenfold increase! Of this 700,000, about 500,000 were of the two-stage variety. In 1964 over a million air-water toy rockets were sold in the United States.

The above data are presented here to illustrate the success of this product, a success which is based on a timely idea, competent engineering design, and economical production.

10.2 INVENTIVENESS AND PLASTIC TOYS

There are two levels of inventiveness to be discussed in connection with this case study. One is general and has to do with the creation of new ideas in plastic toys. This is important to the company that has the facilities and skills to mold and fabricate plastics. The other is specific and has to do solely with ideas for toy rockets.

Consider the general problem first. The chief engineer for one plastics company has said: "What am I supposed to do, just sit here and invent something?" The answer, though he might not agree, is "yes"! But this is no easy assignment, because the major dimension for evaluating an idea is "Will it sell?" Who knows whether a new product idea will sell or not? The toy rocket has done very well, but a "moon vehicle" flopped. One of the major costs to a company dealing in products for the public is the development of products which do not sell well once they are designed, produced, and marketed.

Despite the difficulties, most of the good ideas for new products come from a company's own engineering department. But not by just "sitting there." The most important source of ideas is by analogy to other products. Perhaps a metal or wooden toy can be made in plastic. Or, from a one-stage rocket, the idea for a two-stage rocket is conceived as an extension or modification. Analogies to real-life products are obviously important, and thus rockets, helicopters, earthmovers, and the like are popular. (See Probs. 10.1 and 10.2.)

A steady stream of private inventors can be counted upon to contact manufacturers with their ideas, but only a very small percentage of these turn out to have real commercial merit. A company has to rely primarily on its resources for new ideas. (See Prob. 10.3.)

The second aspect of inventiveness related to this case study has to do with toy rockets themselves. Suppose it is known that a toy rocket is wanted. What kind of rockets can be developed? It should be evident by considering the situation *before* the air-water concept (the pressurized air forces the water rapidly out the back, thus providing a thrust for accelerating the rocket) was developed, that the idea is a highly creative one. It is simple, it uses the most common of materials (air and water), and it works beautifully. Another good scheme used for propelling toy rockets uses a powder which produces gas when mixed with water, and the resulting pressure pushes out the water. What other schemes can be developed? (See Probs. 10.5 and 10.6.)

10.3 TECHNICAL FEASIBILITY OF THE AIR-WATER ROCKET

Once the air-water concept of propulsion for a plastic toy rocket is conceived, it is necessary to determine its technical feasibility. That is, will the concept work? This is, of course, essentially an engineering analysis question. As such, it must be translated into specific and objective questions which have quantitative answers. What are they?

In this case, the primary criterion of performance will be the height to which the rocket will go. At this stage an order-of-magnitude analysis will do. In fact, an order-of-magnitude analysis is definitely the first step, because should it produce an answer in, say, inches, the idea should be abandoned.

Before any direct answer, even an order-of-magnitude answer, can be obtained, more information about the rocket must be specified. Note how design and analysis interact here. The rocket concept cannot be analyzed, even in an order-of-magnitude way, until it is designed in a little more detail. (Design almost always keeps a step ahead of analysis!) For purposes

Fig. 10.2 Sketch of possible rocket design.

of obtaining an order-of-magnitude estimate of the height, a rough design can quickly be made and is shown in the freehand sketch of Fig. 10.2.

The sketch raises many questions, however. If a rocket of about the size shown is built, how thick shall the walls be made? What area discharge nozzle shall be used? To what pressure can it be pumped? What material shall be used? How much water shall be used? One wonders in fact whether a paper-and-pencil feasibility study should be continued, or abandoned in favor of an experimental program.

In the real-life development of the product, the feasibility study was entirely experimental. A model was built (at some expense!) and tried.

More can be determined for use in a paper-and-pencil order-of-magnitude analysis, however. It is assumed here that this development is taking place within a plastic-toy-manufacturing organization. Thus there is experience available that can be used to make estimates. It will be known, for example, that thicknesses less than about 0.045 in. are extremely difficult to mold. Hence the rocket thickness can be taken, as a first guess, to be 0.045 in. From this, and a knowledge of the approximate density of available plastic materials, the weight of the rocket itself can be estimated (about 0.060 lb). It should be emphasized here that only actual experience in molding plastics or a survey of other existing plastic products could possibly provide the number 0.045 in. It could not be calculated by any known means.

How much pressure can be used? It seems obvious that the higher the pressure, the higher the rocket will fly. Of course, if the pressure is to be generated by a pump used by a child (or his father!), human strength will limit the possible pressure. Also, safety is to be considered, but this is partly a materials-selection problem.

A company at this stage is most likely to select a material (and a manufacturing process) with which it has experience. Polystyrene plastic is inexpensive and strong, but it tends to shatter. A rocket of it would probably not withstand falls onto hard pavement. Butyrate, though considerably more expensive than polystyrene, does not shatter and is strong enough to withstand the necessary shocks. Butyrate also will deform (bulge) rather than shatter if the pressure inside the rocket becomes too high. Thus butyrate is the initial choice of materials.

Now the question has been reduced to: How much pressure can a rocket 0.045 in. thick made of butyrate withstand? An answer can be found in five ways: (1) through the experience of company personnel, (2) from suppliers of butyrate, (3) through the experience of professional consultants (who are paid well for their services), (4) by a study of the literature plus analysis, and (5) by an experiment with a model. In this case a phone call to the engineering department of the plastic suppliers establishes the fact that, though it depends upon the shape of the container, for small ones with no serious stress-concentration areas, a pressure of 40 to 50 psig is certainly feasible. Pumping practicalities dictate that pressures in excess of this are not possible anyway.

Finally, how much water should be put into the rocket? If no water is used, very little height can be expected, because there would be very little mass (hence little momentum) out of the nozzle. If a great deal of water is used, too much energy would be consumed in lifting the water in the rocket. Also, in this case, the air would expand to atmospheric pressure before all the water was ejected. There will, then, be an *optimum* amount of water which will produce a maximum height of the rocket. Finding the optimum is too long a task at this stage; the problem now is simply to determine whether or not the basic concept will enable the rocket to fly to interesting heights. For *this* purpose, the optimum is not necessary but only a reasonable value in the good operation range.

Assume the air pressure is 40 psig or 55 psia, and atmospheric pressure is 15 psia. Since it is the air-pressure difference which ejects the water, it seems that there should be a substantial Δp throughout the water-ejection period. In order to get an estimate, suppose the air pressure when the water is just gone is assumed to be 25 psia. The air thus is assumed to expand from 40 psia to 25 psia and from some unknown initial volume to a final volume equal to the total volume of the rocket. Now if the air expansion is of such speed that it may be considered reversible and adiabatic, the initial air volume (and hence initial water volume) can be estimated. For the initial air volume,

$$P_1\mathcal{V}_1^k = P_2\mathcal{V}_2^k$$

$$\mathcal{V}_1 = \mathcal{V}_2\left(\frac{P_2}{P_1}\right)^{1/k} = \mathcal{V}_2\left(\frac{25}{40}\right)^{1/1.4} = \mathcal{V}_2(.714)$$

It seems quite reasonable, therefore, to assume that the water initially occupies about one-fourth of the total rocket volume, and that the air expands to four-thirds of its initial volume.

These preliminary estimates are quite crude, but they are nevertheless very valuable. They are a kind of experience with the problem, and they help lead one to a better qualitative understanding of the problem. In this case, what others can be made?

The jet velocity can be related to the air pressure difference if Bernoulli's equation is used as an approximation. (A further discussion of the use of this equation in this problem is given later.) Δp varies from 40 to 10 psia. An average is $\Delta p = 25$ psia. For this value, the water-jet velocity estimate is

$$V_j = \sqrt{\frac{2\,\Delta p}{\rho}} = \frac{2 \times 25 \times 144}{1.94} = 60 \text{ fps}$$

Checking units:

$$\frac{\text{ft}}{\text{sec}} \rightarrow \left[\frac{\text{lb}_f}{\text{ft}^2}\,\frac{\text{in.}^2}{\text{ft}^2}\,\frac{\text{ft}^4}{(\text{lb}_f)(\text{sec}^2)}\right]^{1/2} \rightarrow \frac{\text{ft}}{\text{sec}}$$

The approximate volume of a circular rocket, as shown in Fig. 10.5, is 0.0080 ft³. If one-fourth of this is water (0.002 ft³), then the time required to eject the water can be estimated as a function of the nozzle area.

$$t = \frac{0.0020}{A_j(60)}$$

For example, a nozzle diameter of $\frac{1}{8}$ in. gives about 0.4 sec. A $\frac{1}{16}$-in.-diameter nozzle gives 1.6 sec. In any case, water ejection will be *very* rapid.

This information suggests two possible means for estimating the height a rocket might be expected to travel. One method is based on conservation of momentum. Suppose it is assumed that the water goes out very rapidly, so that the momentum of the water out the back of the rocket can be taken to be equal to the forward momentum of the rocket with the water gone. This neglects dissipative losses and, probably more important, neglects the fact that the momentum of the first water out is used partly to lift and accelerate the water remaining in the rocket. To account for this to some extent, a modified mass of the rocket can be used equal to the actual mass plus, say, one-half the mass of the initial water.

The total momentum of the water out the nozzle is

$$m_w V_j = (1.94)(0.002)(60) = 0.240 \text{ (slug)(ft)/sec}$$

If this is equal to the initial rocket momentum, the initial rocket velocity would be

$$V_{R0} = \frac{0.240}{m_R} = \frac{0.240}{(0.060/32.2) + 1.94(0.002/2)} = 60 \text{ fps}$$

Neglecting drag, a rocket with a vertical velocity at ground level of 60 fps will travel to a height of 56 ft. This is a promising height indeed!

Using conservation of energy, another rough estimate can be made. The only energy available is that of the expanding air in the rocket. If it is assumed that the expansion is reversible and adiabatic, and with volumes known, the mass and temperature change (hence energy change) of the air can be computed. For the numbers above, if the initial air temperature is 80°F, then the final temperature is found from

$$T_1\mathcal{V}_1^k = T_2\mathcal{V}_2^{k-1}$$

$$T_2 = T_1\left(\frac{\mathcal{V}_1}{\mathcal{V}_2}\right)^{k-1} = 540\ (\tfrac{3}{4})^{0.4} = 482°\text{R} = 22°\text{F}$$

The mass of air is

$$m_a = \frac{P_1\mathcal{V}_1}{RT_1} = \frac{(55)(144)(0.006)}{(53.35)(540)} = 0.0016\ \text{lb}_m$$

$$\text{lb}_m \rightarrow \frac{\text{lb}_f}{\text{in.}^2}\ \frac{\text{in.}^2}{\text{ft}^2}\ \frac{(\text{ft}^3)(\text{lb}_m)(°\text{R})}{(\text{ft})(\text{lb}_f)(°\text{R})}$$

The internal energy change of the air is

$$\Delta U_a = m_a C_v T = (0.0016)(0.17)(58) = 0.0158\ \text{Btu}$$

$$\text{Btu} \rightarrow \text{lb}_m\ \text{Btu}/(\text{lb}_m)(°\text{R})\ °\text{R}$$

Equating this to the potential energy of the rocket at its maximum height (this neglects any lifting of water before it is all ejected), the height is

$$E_{\text{rocket}} = E_{\text{air}}$$

$$m_R g h = 0.0158\ \text{Btu}$$

$$h = \frac{(0.0158)(778)}{(0.060)} = 205\ \text{ft}$$

This answer is a factor of 4 different from the estimate made above using momentum. However, it is the *same order of magnitude* and serves the purpose of this part of the analysis quite well. The conclusion must be that this air-water concept is at the least worthy of additional study and development.

10.4 ECONOMIC FEASIBILITY

Even though it appears that the proposed rocket will fly satisfactorily, further development is not justified unless the project appears to be economically feasible. Specifically, can such a product be manufactured and marketed at a price which will yield good sales and an adequate profit? This is a question in which engineers must participate, but in which others (market analysts, company accountants and economists, sales representatives, etc.) have a great deal to contribute. Often research or experimentation is necessary to determine costs, acceptable selling price, and expected sales. In the present case, the "go ahead" at this stage actually was made with very little specific or detailed information. Experience with producing other plastic products indicated that the rocket could be built inexpensively enough so that a modest market price could be established. Everyone concerned felt that the product would be popular.

The importance of economic feasibility should not be underemphasized. Manufacturing-company profits amount typically from about 3 to 10 percent, depending largely on how well the choices for new products have been made. That is, losses on products which do not sell well are a major factor reducing profits. Choosing the right products for development—based on technical and economic feasibility studies—is a vital activity.

At this point it might be well to note how retail toy prices reflect manufacturing costs. Typically, a toy selling for $1 has been sold by the manufacturer for 40 cents. Of this 40 cents, about 25 cents is actual cost of production—materials, labor, equipment, etc. The remaining 15 cents is general overhead, which includes such items as executive salaries, engineering, office expenses, and profit. The narrow margin of profit should make it clear why costs are a critical part of every design decision as well as of every production decision.

10.5 ENGINEERING ANALYSIS OF THE SINGLE-STAGE AIR-WATER TOY ROCKET

Assume now that the design decisions leading to the specifications shown in Fig. 10.3 have been made. (Also see Probs. 10.7 and 10.8.)

Overall length = 5.5 in.
Maximum diameter = 1.5 in.
Wall thickness = 0.045 in.
Total volume = 75 ml
Nozzle diameter = 16 in.
Weight = 16.7 g

Fig. 10.3 Specifications for small single-stage air-water toy rocket.

Maximum air pressure is limited to 50 psig. The following questions for engineering analysis are now presented:

1. For a given initial pressure and water volume, how high will the rocket fly?
2. What is the optimum amount of water for maximum height?

3. What is the optimum area for the nozzle?
4. How sensitive is the height of the flight to the air drag coefficient?
5. How sensitive is the height of the flight to nozzle design?

Students who wish a challenging engineering analysis problem should stop reading at this point and answer the above questions. Actual data for checking answers are provided later in the chapter. Those who read on will be presented with one possible engineering analysis for the above questions.

The momentum equation is basic to rocket-acceleration problems. If a

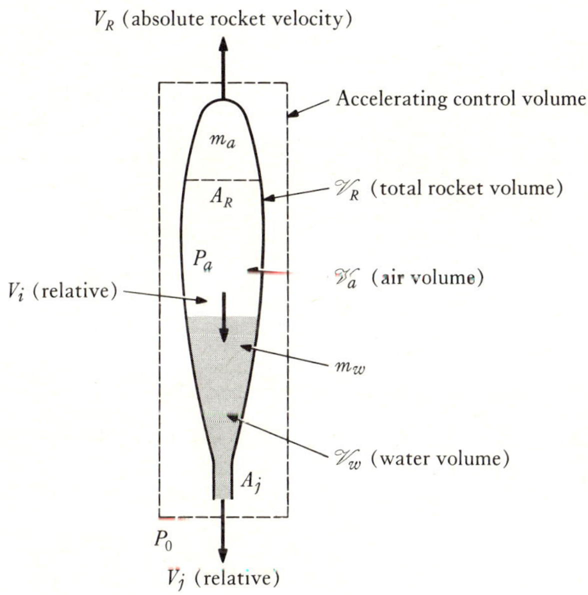

Fig. 10.4

control volume is attached to the accelerating rocket and symbols are defined as shown in Fig. 10.4, then the basic equation can be written as

$$\sum \mathbf{F} = \frac{d}{dt}\int_{\mathcal{V}} \rho V_a \, d\mathcal{V} + \oint_A V_a(\rho V_r n \, dA)$$

It is important to recall that V_a represents "absolute" velocity (velocity relative to an inertial or Newtonian frame, in this case the earth) and V_r represents velocity relative to the moving frame. The external forces on the rocket are its weight and the air drag:

$$F = -mg - C_D \frac{\rho_a V_R^2}{2} A_R$$

The coordinate direction is taken as positive upwards. Pressure forces have been neglected. The term $\int_{\mathcal{V}} \rho V_a \, d\mathcal{V}$ is the momentum within the control volume, so in this case

$$\frac{d}{dt} \int_{\mathcal{V}} \rho V_a \, d\mathcal{V} \cong \frac{d}{dt} (m V_R) = m \frac{dV_R}{dt} + V_R \frac{dm}{dt}$$

where m is the total mass:

$$m = m_R + m_w + m_a \cong m_R + m_w$$

Writing the momentum within the control volume as mV_R is a considerable oversimplification. It is correct only if all the material inside the control volume is moving with absolute velocity V_R, but the water in this case has a different velocity. The assumption will be made, however, that if this effect is neglected, sufficient accuracy will be attained for the purposes of this calculation.

The term $\oint V_a(\rho V_r n \, dA)$ represents the flux of momentum across the boundary of the control surface. $\oint_A \rho V_r n \, dA$ is the mass flow, which is also equal to $-dm_w/dt$, which is equal to $-dm/dt$. In fact, this is the conservation-of-mass concept for this problem:

$$\oint \rho V_r n \, dA = -\frac{dm_w}{dt} = -\frac{dm}{dt} = \rho_w A_j V_j$$

V_a is the absolute velocity of the water leaving the control volume. In this case it is constant and equal to $V_R - V_j$ when proper account of sign is made. Hence, the entire equation combined with continuity can be written as

$$-mg - C_D \frac{\rho_w V_R^2}{2} A_R = m \frac{dV_R}{dt} + V_R \frac{dm}{dt} - V_R \frac{dm}{dt} - V_j \frac{dm_w}{dt}$$

$$= m \frac{dV_R}{dt} - V_j \frac{dm_w}{dt}$$

This is the familiar equation for rockets, where the last term is called the *thrust*. Using the fact that $dm_w/dt = -\rho_w A_j V_j$ by continuity

$$-mg - C_D \frac{\rho_a V_R^2}{2} A_R + \rho_w A_j V_j^2 = m \frac{dV_R}{dt}$$

This gives a differential equation for the rocket velocity V_R as a function of time, but note that m is also a function of time and that the nozzle jet velocity V_j is still unknown. The mass flow and V_j are connected by

$$-\frac{dm}{dt} = \rho_w A_j V_j$$

V_j can be estimated by assuming that Bernoulli's equation can be used and by neglecting V_i^2 compared with V_j^2. The result is

$$V_j^2 = \frac{2(P_a - P_0)}{\rho_w}$$

The use of Bernoulli's equation here is wrong on several counts. First of all, it is a steady-state equation, and this is clearly an unsteady situation. Second, and perhaps most important, this is an accelerating system, but Bernoulli's equation applies strictly to an inertial system. However, in a first effort, these effects may be neglected (somewhat nervously!) and the above approximation used. If a nozzle coefficient is introduced, there is

$$V_j^2 = \frac{2C_N(P_a - P_0)}{\rho_w}$$

Now, however, P_a is a variable, since as the water is ejected, the air expands and its pressure falls. It can be related to the amount of water, however. The air expansion will be of such speed that a reversible adiabatic expansion can be assumed. Then for the air,

$$P_a \mathcal{V}_a^k = \text{const} = P_{a0} \mathcal{V}_{a0}^k$$

$$\mathcal{V}_a - \mathcal{V}_R - \mathcal{V}_w - \mathcal{V}_R - \frac{m_w}{\rho_w}$$

Theoretically, there are now enough independent equations so that one may solve for the variables V_R, m, P_a, and V_j as functions of time. Obviously, however, no closed-form analytical solution exists for this set of equations. Numerical solution will be necessary, and a digital computer will be required to perform the calculations.

Here is how it can be done. Writing the momentum equation in finite-difference form gives

$$(V_R)_{k+1} = (V_R)_k + A_j(V_j)_k^2 \left(\frac{1}{m_k}\right) \Delta t - g\,\Delta t - \frac{C_D \rho_a A_R}{2m_k} (V_R)_k^2$$

where the subscripts k and $k + 1$ refer to values at times k and $k + 1$ respectively. Given values of V_R, m, and V_j at time t (corresponding to k), the above equation makes it possible to compute V_R at $t + \Delta t$ (corresponding to $k+1$). $(V_j)_k$ can be found from

$$(V_j)_k^2 = \frac{2C_N(P_a)_k - P_0}{\rho_w} \tag{1}$$

if $(P_a)_k$ is known. $(P_a)_k$ can be found from

$$(P_a)_k\left[\mathcal{V}_R - \frac{(m_w)_k}{\rho_w}\right]^k = \text{const} \tag{2}$$

if $(m_w)_k$ is known. m_w can be found from

$$\frac{dm_w}{dt} = -\rho_w A_j V_j \tag{3}$$

by making it a finite difference equation too:

$$(m_w)_{k+1} = -(m_w)_k \rho_w A_j (V_j)_k \tag{4}$$

To get the iteration process starting, m_w, P_a, and V_R are initially known or specified. The constant in (2) is now computed and $(V_j)_k$ is found from (1). $(V_R)_{k+1}$ can be found from the momentum equation and a new value of m_w from (4). From (2), a new P_a is found and used in (1) to find the new V_j, which is used in momentum to find a new V_R and in (4) to find a new m_w. Etc. The iteration is continued until all the water is ejected. See Prob. 10.9.

After all the water is ejected, there is no more thrust (neglecting the little air which still remains under pressure). The equation of motion is then simply

$$-m_R g - C_D \frac{\rho_a V_R^2}{2} A_R = m_R \frac{dV_R}{dt} \tag{5}$$

This equation is readily solved by either numerical or analytical techniques, using the final velocity from the preceding numerical solution as the initial velocity for this part of the flight. See Prob. 10.10.

The following data are taken from an actual toy rocket:

Weight of rocket:	$m_R = 0.036$ lb
Volume of rocket:	$\mathcal{V}_R = 0.0026$ ft^3
Area of jet:	$A_j = 0.000184$ ft^2
Area of rocket:	$A_R = 0.0123$ ft^2

Other data needed to solve the above equations are the atmospheric pressure P_0, the nozzle coefficient C_N, and the drag coefficient C_D. As a

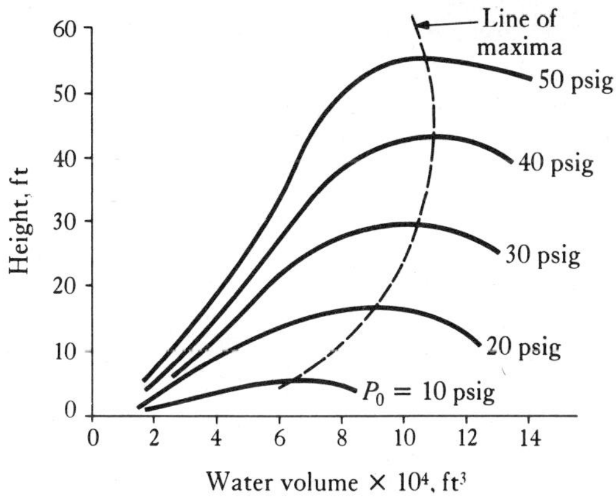

Fig. 10.5 Results of rocket analysis.

start, these may be taken as

Atmospheric pressure:	$P_0 = 15$ psia
Nozzle coefficient:	$C_N = 1.0$
Drag coefficient:	$C_D = 0.1$

The effect of variations in the latter two will be investigated.

Solution of the equations now makes it possible to answer the questions posed earlier in this section.

How high will the rocket fly for a given initial pressure and water volume? See Fig. 10.5. What is the optimum amount of water? Also see Fig. 10.5. What is the optimum area for the nozzle? See Fig. 10.6. How

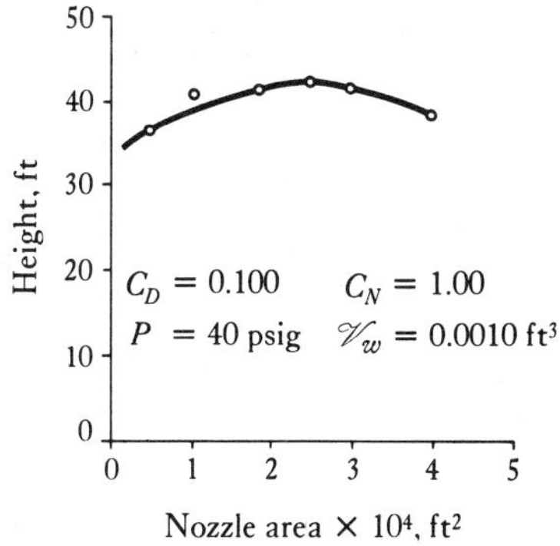

Fig. 10.6 Effect of nozzle area on rocket performance.

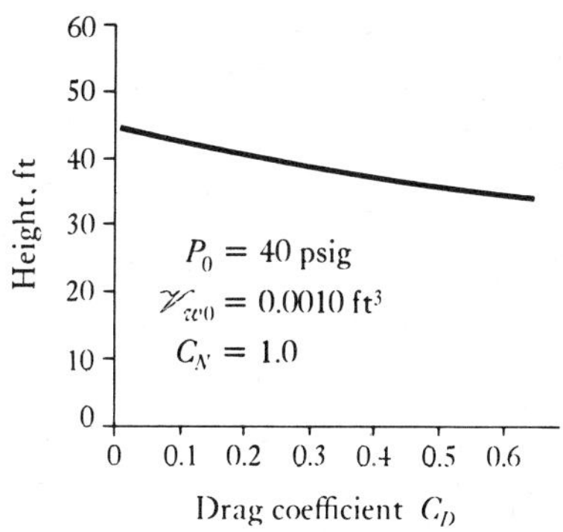

Fig. 10.7 Effect of drag coefficient on rocket performance.

sensitive is the height of the flight to drag coefficient? See Fig. 10.7. How sensitive is the height of the flight to nozzle coefficient? See Fig. 10.8.

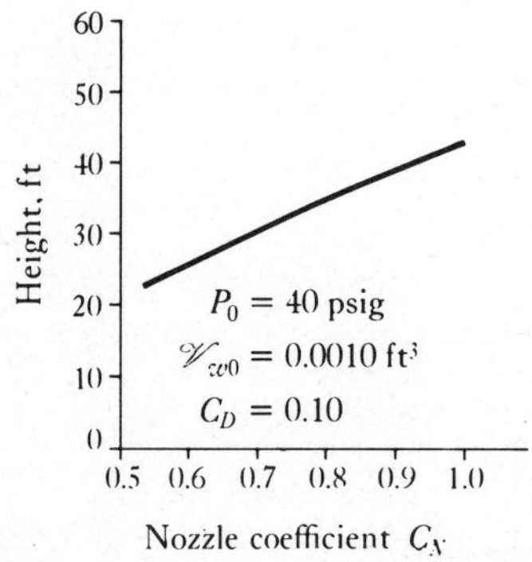

Fig. 10.8 Effect of nozzle coefficient on rocket performance.

Figure 10.9 gives a comparison of the actual performance of the rocket with the above analysis. Note that the agreement is quite good except in the 30-psig case. Whether this is theoretical or experimental error is not known. What do you think?

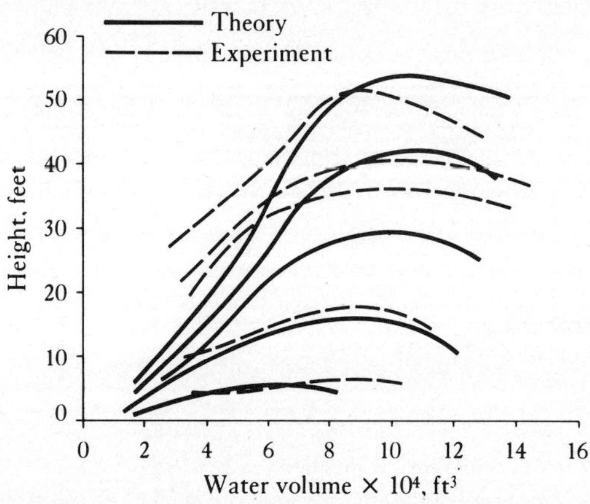

Fig. 10.9 Comparison of theoretical results with rocket performance.

10.6 INVENTING A TWO-STAGE TOY ROCKET

It is highly recommended that readers do Prob. 10.12, either alone or in a group, before proceeding.

There are again two levels of inventiveness with respect to the two-stage rocket. How is the idea for a two-stage rocket arrived at in the first place? And how can a two-stage toy rocket, once the idea is conceived, actually be built? Most likely the answer to the first question is: "By analogy with real rockets." The second question is a truly fine problem in engineering creativity (and is the subject of Prob. 10.12). There are many possible ways to do it, and all the procedures discussed in Chap. 2 are useful in getting ideas. The following list was compiled by one group of students in 20 min:

Exploding bolts.
A string connected to ground pulls lever that separates stages at specified height.
A timer, perhaps a spring, releases the stages.
Inertia; when the thrust of the first rocket falls off, a lever held by inertia moves to allow separation.
A timer based on small water flow from the second stage.
Radio control.
A pressure gage measuring first-stage pressure.
A velocity gage measuring velocity—when it slows down, second stage is fired.
A water-soluble diaphragm as a timer.
A diaphragm which ruptures when Δp between the two rockets is large.
An altimeter.

Readers should add their ideas to the above list and then evaluate the various ideas. See Prob. 10.13.

The concept which was chosen for actual development by the company was the timer based on a small amount of water flow out of the second stage. When asked why this particular concept was chosen, the company president said: "We knew we wanted to use the pressure difference between the stages and the water in the stage *because they were already there*." Note that this is a direct appeal for *simplicity*. If existing things could be used directly, very little extra would have to be added. See Prob. 10.14.

Figure 10.10 shows a schematic diagram of the mechanism that was developed. The mechanism works as follows: Water is charged into both rockets and held by the ball valves. The piston-plunger unit is initially in the up position, thus holding the split-end section of the second stage in place. The entire system is then pressurized with the hand pump. When the system is released, the pressure in the first stage falls as the rocket climbs. There is now a pressure difference between the rockets, and the tiny metering slot allows water to bypass the lower ball valve. As water escapes via the bypass, the piston plunger is pushed down by the high pressure in the second stage. When the piston plunger has moved far enough, the plunger portion no longer holds the split ends apart, and the second stage is free to blast off itself. The key to the timing of the separation is the size of the tiny bypass orifice around the lower ball valve. It is designed to be 0.008 in.[2] Flow rate through this orifice will be proportional to the pressure difference between the rockets (the first-stage pressure will be essentially atmospheric)

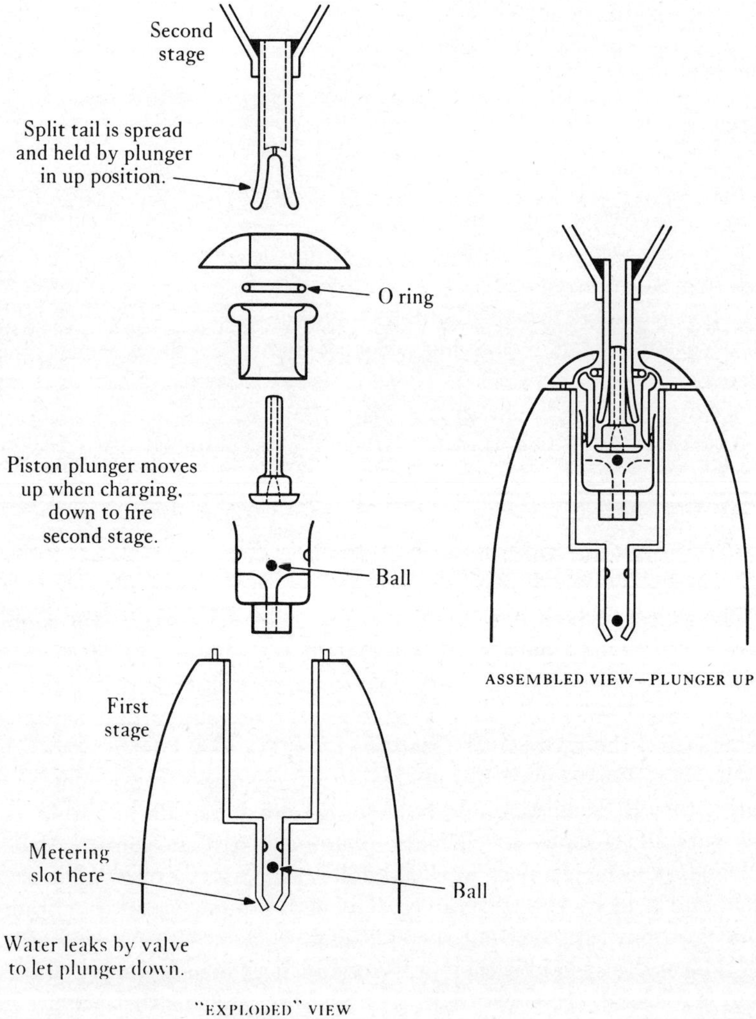

Fig. 10.10 Two-stage-rocket design.

and the orifice area. Note that the higher the initial pressure, the shorter will be the time to separation. Is this as it should be? Interested readers are referred to Prob. 10.16 for a challenging problem in engineering analysis.

10.7 DECISION MAKING IN THE DESIGN OF THE TWO-STAGE ROCKET

As was pointed out earlier in this book, decision making permeates the entire design process. Both economic and technical questions appear at every step, usually in some related fashion. There are simply too many of these decisions to discuss here. (See Prob. 10.17.) One technical-economic question is especially interesting, however: How big should each of the rocket stages be?

There is no technical way to establish the general range of the dimensions. That is, the approximate size or some starting dimension must be fixed by factors other than technical ones. In this case, of course, the fact that the product is a toy for children fixes the general size. It must be big enough to handle easily and to be seen in flight, but no bigger, so that material costs can be kept down. The fact that two single-stage rockets (one about 5 in. long and one about 7 in. long) are already in production turns out to be a major and specific determining factor in the decision on size. The reason is that the dies are already made and paid for for molding these rockets. Die costs range from \$20,000 to \$40,000 for a rocket. Thus if the present dies can be used, a considerable saving can be realized.

Clearly, then, one stage of the two-stage rocket will be the same as one of the existing single-stage rockets. Now the size of the second stage must still be treated as a decision-optimization problem. If the criterion selected for optimization is height, then the optimization will be based on technical factors. An analysis, or experimental program, can be made to determine the optimum. However, there are also other important criteria that must be considered. The most important is *cost*. By far the cheapest procedure to follow here is to use *both* the dies of the existing single-stage rockets, one for each stage of the new two-stage combination. This minimizes (optimizes) tooling cost. Of course, only by coincidence will this plane also optimize performance (i.e., height of flight). However, if the combination made of the two old rockets works fairly well, even though not optimized, what would you do if you were the decision maker?

Use the old dies, of course—and this is exactly what was done. Notice how the trade-off of values was necessary. In this particular example, the trade-off was an easy one. In a real-life space application, the opposite decision—that is, to optimize performance without regard to cost below some limiting value—might have been equally easy. But the conflict is apparent: both performance and cost cannot be optimized simultaneously. And someone must decide how to decide. This kind of situation, though often with many more elements to the conflict (performance, cost, reliability, appearance, etc.), is characteristic of all engineering design. This is the reason various decision theories are being studied and used increasingly by engineers. Even though decision theory is a young science, it is advanced well beyond its use in engineering design decisions. It is very clear, however, that decision theory will, in the near future, be of paramount importance to engineers. For this reason, and despite the limited current applications, the next part of this design book deals with the process of making decisions.

10.8 SUMMARY AND PREVIEW

This chapter concludes the discussion of engineering analysis. Attention in the remainder of the book will be on the processes and skills of making decisions. Chapter 11 is an introduction and an overview of the various aspects of decision making. The remaining chapters take up specific decision processes or skills—optimization, probability, statistics, decision-utility theory—and some special applications, notably reliability. Each of these topics is the subject of whole books, so what is given here is primarily an introduction or, in some cases, a survey. The application of much decision-making theory to engineering design is not widespread. Engineers of the future, who as students learn to use these decision-making tools, will have the task and the opportunity of developing new and important design-engineering applications.

PROBLEMS

10.1 Discuss the usefulness of the ideas for getting ideas presented in Chap. 2 for a plastic-toy manufacturer who is in need of new products.

10.2 A plastic-toy manufacturer has unused production capacity. (*a*) Conduct a brainstorming session to get ideas for a new product. (*b*) Use methods other than brainstorming to get ideas for a new product.

10.3 Read "Self-Renewal: The Innovative Society," a book by John W. Gardner (see Suggested Reading, Chap. 2), and discuss how his thesis applies to (*a*) a plastic manufacturer, (*b*) a university, and (*c*) a nation.

10.4 Discuss the influence of the following factors on the value of ideas for new plastic toys:

a Trends and changes in grade school education

b Major political events
c Major scientific developments
d Major technological developments

10.5 How might the procedures discussed in Chap. 2 have been used to get the air-water concept of toy-rocket propulsion? To get the gas-generation concept?

10.6 Use the procedures discussed in Chap. 2 to obtain new ideas for types of toy rockets.

10.7 Consider each of the design decisions necessary to arrive at the drawing and specifications shown in Fig. 10.6. How was each parameter decided upon?

10.8 Suppose it was decided that the rocket of Fig. 10.6 should be bigger so that it has a length of $7\frac{1}{8}$ in. Make a drawing of the rocket you would propose as an initial design and make the specifications as shown in the figure.

10.9 Write a computer program to solve the series of Eqs. (1) to (4) and the momentum equation. Select appropriate data and find the time for the water to be ejected. How high is the rocket at this point? What is its velocity?

10.10 Write a computer program to solve Eq. (5). Also find an analytical solution. Select appropriate data and initial conditions and compare the two results.

10.11 Review the engineering analysis of the single-stage rocket in terms of the engineering analysis procedure outlined in the text. Identify the various steps. Are steps omitted? Could the procedure used be improved upon?

10.12 Conduct an inventiveness session (brainstorming or otherwise) with the objective of inventing a two-stage air-water toy rocket. How will the two stages be held and separated?

10.13 Consider the list of concepts for separating the two stages of a toy rocket given on page 235. If you were the decision maker, which one or ones would you ask the engineers in your company to pursue in more detail? Why?

10.14 Design a mechanism for holding and separating the two stages of a two-stage toy rocket, based on the fact that once the first stage has fired, there will be a large pressure difference between the two stages, and on the fact that a small amount of water can be bled from the second stage proportional to Δp and time.

10.15 Design a pump for the toy rockets discussed in this chapter.

10.16 Design and analyze a two-stage air-water toy rocket.

10.17 Make a list of the decisions that had to be made during the design of the two-stage rocket.

III
DECISION MAKING

11
INTRODUCTION TO DECISION MAKING

11.1 INTRODUCTION

This book views the engineering design activity as being composed of three fundamentally different mental processes: inventiveness, analysis, and decision making. This chapter is an introduction to the third of these: decision making.

Decision making is *compromise*. The decision maker must weigh value judgments that involve economic factors, technical practicalities, scientific necessities, human and social considerations, etc. To make a "correct" decision is to choose the one alternative from among those that are available that best balances or optimizes the total value, considering all the various factors. It is often necessary to "trade off" value in one dimension (say, reliability) in order to gain value in another dimension (say, cost). Finding the alternative that represents the optimum compromise of all the relevant factors is the task of the decision maker.

There are times when finding the optimum compromise can be done using one or more of the scientific theories of decision making: the mathematics of optimization, probability, statistics, or utility-decision theory. At other times decision making is a highly complex, subjective matter involving nonquantitative human factors and value judgments. Most often, however, decision making involves both quantitative and qualitative factors that must be considered simultaneously. In the chapters to come, specific attention will be given to the quantitative procedures for making decisions—optimization theory, probability, statistics, utility-decision theory—but in this chapter the decision process will be considered more generally. In particular, the objectives of this chapter are to consider the way in which decision making is both art and science and to present the decision process in overview.

It is said by some that decision making is essentially an art. This belief has been firmly entrenched in the thinking of many people in management, government, business, and even engineering design. The advent of the computer, however, and the corresponding growth and development of the sciences of decision making have brought about a change. Areas of decision making formerly thought to be entirely qualitative and subjective are being done quantitatively. This is especially true in military applications. In fact, there is some fear that this trend may be or become overdone. In any case, it is clear that there is a great deal of both qualitative art and quantitative science in decision making at the present time.

11.2 CHARACTERISTICS OF A DECISION PROBLEM

A decision is a solution to a particular kind of problem. What are the essential characteristics of a decision problem? Write your answer here before going on:

The essential characteristics of the decision situation are these:

1. *An Objective.* In a decision situation there is some desired goal—a task to be accomplished, a material selected, a girl dated, a new job, etc. Without some desired end, there is no need to decide anything.
2. *Alternative Courses of Action.* In a decision situation there is more than one way of proceeding to attempt to accomplish the objective. Obviously, if there is only one available course of action, then there is no choice and no decision to be made. The various alternatives may involve different costs and different probabilities of success. The costs and probabilities may or may not be known. It is for reasons such as these that decision making is often such a vague and uncertain process.
3. *Relevant Factors.* In a decision situation there are a variety of factors that are important but are different for the different alternatives. The factors may be economic, technical, personal, social, etc.

A decision-making problem exists, then, when and only when there is an objective to be reached, alternative methods of proceeding, and a variety of factors that are relevant to the evaluation of the alternatives or their probability of success. Each of these three characteristics will now be considered separately in more detail as related to engineering design decisions.

11.3 OBJECTIVES IN ENGINEERING DECISIONS

Careful attention to and awareness of the decision-making process will result in the need to clarify goals and objectives. Dean Myron Tribus of Dartmouth College has aptly described engineering design as optimization of a payoff function. The "payoff function" may be a single item (such as cost or weight) or it may be a weighted combination of two or more things. The payoff function is the goal; and before final decision making can begin objectively, it must be defined clearly. As we shall see, when the payoff function or goal can be defined simply and quantitatively, one of several decision-making theories may be applicable. However, objectives, or at least factors directly related, often involve the qualitative and subjective as well as the quantitative and objective. In these cases, decision-making theories must be applied with maturity and perspective as well as with analytical and mathematical skill.

Some of the common objectives in engineering situations are listed below, but the list should not be considered exhaustive. Remember also that an objective sometimes calls for a balance among two or more of these dimensions. Furthermore, some of these factors will appear as relevant factors or as constraints rather than as part of the objective in certain situations. That is, low weight may appear as a goal rather than a limit in such a statement as: "The weight must not exceed 40 pounds." A more detailed discussion of constraints is presented later in the chapter. Some engineering objectives are: initial cost, cost of operation or maintenance over a specified period, reliability, weight, performance, efficiency, appearance, safety, profit over a specified period, etc.

11.4 ALTERNATIVES IN ENGINEERING DECISIONS

Decision makers often (and unfortunately) are not consciously aware of the process of listing alternatives. It is obvious, however, that the final choice of an alternative can be no better than the best alternative considered. The quality of the decision is in this sense limited by the quality of the alternatives. Overt listing of the available alternatives is a great aid in making decisions. Decision making is choosing among alternatives, but generation of a list of alternatives is part of the process.

There is a sense in which listing the alternatives is directly analogous to defining the problem in engineering analysis. When the alternatives are vague, incomplete, or even unthought of, it is impossible to decide. Once the alternatives are clearly listed, however, the problem is no longer intangible. Instead, it is the very specific one of choosing one of the listed choices.

Generating a list of alternatives in a decision problem is basically an inventive step. Many of the methods of getting new and useful ideas discussed in Chap. 2 can be applied here as well.

There is one alternative that almost always appears, initially at least, on all lists. It is the alternative *not* to decide. Sometimes, but only sometimes, postponing a decision in order to allow more time to gather new facts is the optimum compromise. If the objective is to be pursued, of course, it is usual that decisions cannot be postponed indefinitely.

11.5 FACTORS RELEVANT TO ENGINEERING DECISIONS

There are actually an infinity of possibly relevant factors in any given engineering decision problem. Any attempt to catalog these completely or in detail is bound to omit some that will be encountered in real life. It is possible, however, to divide the kinds of factors that must be considered into three types and illustrate each type with the most common examples. The major types are *resource factors, technical factors*, and *human factors*.

Resource factors, as thought of here, might also be called *time, money, and availability factors*. By availability here is meant the availability of a variety of things such as materials, parts, technical or research skills, organizational capacity, etc. Figure 11.1 shows a more specific breakdown of the kinds of things that constitute resource factors. It is typical of engineering decision situations that information about the relevant aspects of such factors is not well known without special study or investigation.

Technical factors are those that are directly related to engineering analysis and design requirements. Usually the relevant technical factors are specific and quantitative. A list of the kinds of things that constitute technical factors is shown in Fig. 11.2.

When applied, these technical factors often fall into one of three categories of what are called constraints: functional constraints, regional

1. *Finances*
What will the cost and profits be? How much short term and long term money is required? Is money available? What other projects are in need of money? General business, company, and department outlook? Competition plans?
2. *Facilities for research, design, and production*
What facilities needed? Available? How long to get them and at what cost?
3. *Skills for research, design, and production*
What skills needed? Available? How long to get them and at what cost?
4. *Materials for production*
Are the necessary materials available? At what cost?
5. *Organization for research, design, production, and sales*
What supporting organization (management, office, legal, maintenance, sales, advertising, etc.) is needed? Available? If not, how long to get them and at what cost?
6. *Decision-making resources*
What will it cost to decide? How much time? Are the skills, facilities, etc., available?

Fig. 11.1 Resource factors in engineering decisions.

constraints, and extremum constraints. Functional constraints are exact specifications of performance, input, or other specific limitations. A functional constraint always expresses an equality, such as "The length must be 12 in." or "The flow rate must be 6 *gpm*."

Regional constraints are the same as functional constraints except that they express *in*equalities. Examples of regional constraints are "The length must be less than 12 in." and "The flow rate must be greater than 6 *gpm*."

Extremum constraints call for something to be as big as possible, or as small as possible, etc. They call for the item in question to take on an extreme or optimum value in a specified direction. Extremum constraints obviously lead to optimization problems. Examples of extremum constraints are "The length must be as short as possible" and "The flow rate must be as large as possible."

1. Geometric—size and shape
2. Weight—total and distribution
3. Strength—weakest link?
4. Dynamics—vibrations, natural frequency
5. First law of thermodynamics
6. Second law of thermodynamics
7. Electrical effects
8. Magnetic effects
9. Corrosion
10. Fatigue—thermal or stress
11. Creep
12. Heat transfer—conduction, convection, radiation
13. Temperature effects
14. Fluid-flow effects—drag, friction, capacity
15. Momentum
16. Wear—lubrication
17. Power—supply, output
18. Inertia
19. Etc.

Fig. 11.2 Engineering factors in engineering decisions.

1. *Ethics*

2. *Others' views of the alternative*

Can you "sell" it to the boss, to your colleagues, to your subordinates, to customers, to technicians, etc.?

3. *Resistance to change, fear of the new, and set.*

Boss, colleagues, subordinates, customers, etc.

4. *Aesthetic factors*

5. *Prestige and status*

6. *Personal preferences, tastes, and prejudices*

Self, company, society, boss, colleagues, technicians, etc.

7. *Your relations with your wife, your digestive system, your son's illness, your boss's son's illness, etc., etc., etc.*

8. *Compassion, love, hate, fear, etc.*

9. *Etc.*

Fig. 11.3 Human factors in engineering decisions.

In addition to resource factors and technical factors, human factors also are important in engineering decision making. Human factors have not only to do with the political or social practicality of carrying out or succeeding with a proposed alternative, but also have to do with personal ethics and morals. It is important to realize that these human factors must be a vital part of every significant decision. Good decision making calls not only for technical competence in weighing the resource and technical factors, but also a willingness to be influenced by the values of honesty and kindness. In today's complex and highly technical world, honesty and kindness are not enough; technical-scientific competence also is a necessity. On the other hand, neither is scientific-technical competence sufficient. Man has created something of a scientific-technical predicament for himself, and in order to ensure that he keeps his destiny under his own control, it is going to be important to keep human factors in the decision making. Honesty and kindness have been used here as illustrative of the values that are important. Such values are not based on law, reason, or logic, but on belief (or hope!) either in God or man or in oneself. As an individual decision maker you may wish to add to the list. Figure 11.3 shows some kinds of more detailed human factors that affect many decisions.

11.6 THE SCIENCES OF DECISION MAKING

There are several branches of science or engineering that may be called the study of decision making. One, probably best named decision-utility theory, is an attempt to make a unified science of decision making. But it is so new that the various individual approaches or contributions to decision making still remain rather independent, and certainly worthy of independent study. Among these more or less independent branches are optimization theory, probability, statistics, and decision-utility theory itself. Each of these is the topic of a separate chapter to follow. Here only a brief contextual description of each is presented.

Optimization involves determining the values of controllable parameters (subject to the given constraints) that result in an extreme value of the item being optimized. A function that expresses the item being optimized is called the criterion function. Thus the elements of an optimization problem are *a criterion function, constraints,* and *controllable parameters.* Mathematical techniques of optimization prescribe ways of finding the values of the parameters that maximize (or minimize) the criterion function subject to the various constraints. Some of these techniques, discussed in Chap. 12, are differential calculus, lagrangian multipliers, the calculus of variations, and a variety of numerical methods.

Probability is sometimes called the science of uncertain conclusions. Probability provides us with a way (in certain cases) of assigning a numerical value to the degree of uncertainty that may exist about a particular event under consideration. It should be clear that decisions are seldom made in the presence of complete knowledge and hence a knowledge and understanding of probability are important to modern decision making. Chapter 13 is devoted to probability.

Statistics are data or observations. The study of statistics is the study of how to make sense of, and draw rational conclusions from, the analysis of data. Probability models (theoretical statistical distributions) are used to aid in statistical decision making, and so the two fields are closely related. Statistics is the subject of Chap. 14.

A relatively new application of probability and statistics and one of great importance to engineering design is *reliability*. Reliability is becoming increasingly important as mass production of highly complex machines increases (e.g., automobiles) and as the need for highly reliable and complex systems proceeds (e.g., manned space exploration). Hence, Chap. 15 contains an introduction to the theory and practice of reliability.

Finally, the new and interesting field of study called decision-utility theory is being developed. Though applications of this work to date have been mostly in management, business, and the military, it appears to hold promise for future applications to some engineering applications. Utility theory provides a means for measuring a variety of value dimensions on a single scale. Decision theory deals with the selection of strategies to optimize the probability of obtaining maximum value on the utility scale. This subject is presented in an introductory way in Chap. 16, together with a very brief survey of several other procedures related more or less to decision making in the management of engineering design projects.

11.7 A RATIONAL DECISION-MAKING PROCEDURE

It was said earlier in the chapter that the decision-making situation involves a goal or objective, a set of alternatives, a set of relevant factors to be considered, and perhaps some uncertainty about the possible consequences of the various alternatives. In the preceding section some available scientific techniques of decision making, such as optimization, statistics, probability, and decision-utility theory, were outlined. There is still a great gap between the formulation of most decision problems and the applicability of one of the scientific techniques. There is, in other words, a great deal of "art" left in making decisions. Many factors must be considered and many alternatives eliminated before a decision problem is in a form for analysis by one of the scientific methods. In the gap, progress toward a decision can be made only subjectively or by guesses at the present time. It is valuable to be as *aware* as possible in decision circumstances, and hence an outline of a methodology for reducing a decision situation to the point where one of the various decision sciences can take over is now proposed:

1. State the objective.
2. List as many alternatives as possible. Be creative and inventive here.
3. List as many relevant considerations as possible. Some checklist such as those in Figs. 11.1, 11.2, and 11.3 may help.
4. Use the list of relevant factors to reduce the list to alternatives, taking care to note the reason for each elimination. Some alternatives will be seen to be impossible at this stage. Others will be seen to be highly impractical. This process will seem to be horribly subjective and to involve a frightening amount of guesswork in some cases. There is no alternative, however, if the decision is to be made. Making these subjective guesses and value judgments is part of the remaining art. Remember that one of the alternatives may be *not* to decide just yet, unless one of the relevant factors (such as time) eliminates *it* from the list.
5. Use the remaining list of alternatives to reduce the list of relevant factors, some of which will no longer apply now that they have been used. Others may apply equally to all the remaining alternatives and so need not be considered any further.
6. At the conclusion of the above steps, the result will be one of the following:

a. If there are *no remaining alternatives,* more creative effort is needed to list new alternatives.
b. If there are *no remaining factors,* use chance to choose among the remaining alternatives!
c. If there is only *one remaining alternative,* the decision has been made completely subjectively. If that is desirable, fine. If not, then back off a bit and don't be so particular about what *seems* to be impractical.
d. If there is only *one remaining relevant factor,* it is usually easy to pick out the best alternative. If care has been used in discarding factors, the decision is made.
e. If circumstances permit, one of the scientific decision-making techniques mentioned above and treated in more detail in the next few chapters can be used.
f. If the situation is still too complex and subjective for scientific decision analysis, then it is necessary to go back to step 2 and repeat the process, always taking care to note explicitly the reasons for elimination of the various alternatives. This list of reasons will provide an excellent list of possible personal, company, or department weaknesses.

The above outline of a methodology for decision making in engineering design recognizes the respective roles of both art and science in decision making and puts each into perspective. There is one more point to be made before going on to the next chapters, where the various sciences of decision making are discussed in detail. Engineering design calls for a *lot* of decisions. An engineer has got to be able to decide, even though uncertainty is high in some cases, and go on from there. Striking that optimum balance between too quick a decision and too much vacillation is still very much an art. Also, there is balance to strike between sticking with a decision once made and being willing to admit the need for a revision. These balances are themselves decisions, of course, but as of today they are very much a matter of personal style, temperament, and personality. Some of your success as an engineer depends on such things, however.

11.8 SUMMARY AND PREVIEW

This chapter started with the idea that all decision making is the process of finding an optimal compromise. The question of whether compromising in engineering decision making is an art or a science was discussed, and it was decided that it is both. Awareness of the qualitative factors in decision making was advocated and their importance stressed. Finally, a rational decision-making procedure was outlined, which is intended, where applicable, to be an aid in reducing complex decision-making situations to problems capable of attack by the mathematical decision theories which follow in the next chapters. The first of these techniques to be considered is optimization.

SUGGESTED READING

Ackoff, Russell L., Shiv K. Gupta, I. Sayer Minos: "Scientific Method: Optimizing Applied Research Decisions," John Wiley & Sons, Inc., New York, 1962. Another important book in decision theory.

Bronowski, Jacob: "Science and Human Values," Julian Messner, Publishers, Inc., New York, 1956. A well-known physicist discusses the relationship of science to human values and ethics.

Churchman, C. West: "Prediction and Optimal Decision," Prentice-Hall, Inc., Englewood Cliffs, N.J., 1961. An important book in modern decision theory by one of the foremost workers in the field.

Miller, David W., and Martin K. Starr: "Executive Decisions and Operations Research," Prentice-Hall, Inc., Englewood Cliffs, N.J., 1960. A good presentation of management decision making using operations research.

Simon, Herbert: "Models of Man," John Wiley & Sons, Inc., New York, 1957. More on man vis-à-vis machines.

Sorenson, T. C.: "Decision Making in the White House," Columbia University Press, New York. How President Kennedy made decisions.

Weiner, Norbert: "Cybernetics," The Technology Press of the Massachusetts Institute of Technology, Cambridge, Mass., 1948. A more technical treatment of decision making by machines and the effects on people.

———: "The Human Use of Human Beings," Houghton Mifflin Company, Boston, 1950. A discussion of cybernetics and decision making by machines and people for the layman.

PROBLEMS

11.1 Discuss the question of how to include some of the highest human values in decision processes programmed for computers.

11.2 Consider in retrospect your decision to attend your college or university. List the alternatives and factors relevant to this decision. Did you make a good decision? What can you learn from your experience?

11.3 Consider your forthcoming decision in choosing between graduate school, a job, or the Peace Corps. List the relevant factors. How will your personal values enter this decision?

11.4 Engineers planning a flood-control dam project in upper New York State decided upon a plan that would put an Indian reservation permanently under water. Alternative plans were proposed. How would you go about choosing among the various alternatives in this case?

11.5 Discuss the roles of compromise and leadership in the administration of (*a*) government, (*b*) a college or university, and (*c*) a manufacturing company.

11.6 You can choose to insulate your house with either rock wool or aluminum-foil sheets, or you can leave it uninsulated. Are there any other alternatives? What factors are relevant?

11.7 Do some reading and research so you can intelligently discuss the decision to send a man to the moon as quickly as possible.

11.8 Aluminum and plastic have been proposed as alternative materials for automobile bodies. What are some other alternatives? What factors govern this decision? Which do you think are the most important?

11.9 Discuss the decision to buy or rent a house from the viewpoint of a young family. First consider it on purely economic grounds; then include any other factors you feel are important.

11.10 Use good decision-making processes to formulate a good policy on buying and trading automobiles.

11.11 You are building a house. What kind of heating system will you install?

11.12 Suppose a reliability program costing x dollars could reduce the probability of a fatal accident in an orbital flight of an astronaut from 2 to 1 percent. What factors would you use in deciding whether to activate the program? What maximum value of x would you think reasonable?

12
OPTIMIZATION

12.1 INTRODUCTION TO THE OPTIMIZATION PROBLEM

Optimization is an important engineering design activity. Obtaining the best possible performance or the least possible cost is an integral part of every design. As noted in Sec. 11.5, the optimization task is to find the values of the controllable parameters, subject to the various constraints, that make a desired criterion function a minimum. The criterion function is the thing being optimized, and in order to apply any of the mathematical techniques of optimization, it is necessary that the criterion function be expressible in terms of the parameters. The criterion function is often cost, but may be some measure of performance, some combination, or anything else that the designer wishes to make a maximum or minimum.

An example will help clarify the meaning of the terms *parameters, criterion function,* and *constraints.* Consider the problem of insulating a home. A decision must be made about the thickness of insulation to install. Since the purpose of adding insulation is to save money, the home owner wishes to reduce his total costs as much as possible. His costs are two: fuel and insulation. Fuel cost is a per year cost and insulation is a capital cost. The latter, however, is easily converted to a per year basis by the use of simple economic principles. In this example, then, the total cost C will be the criterion function that is to be minimized.

The only controllable parameter (if different types of insulation are excluded from this example) is the thickness x. The problem now is to find the value of x such that C is a minimum. First, however, it is necessary to express C in terms of x. Assuming the cost of insulation to be proportional to the amount and the cost of fuel to be inversely proportional to the thickness, the cost can be written as

$$C = K_1 \frac{1}{x} + K_2 x$$

where K_1 and K_2 reflect fuel and insulation costs, respectively. The assumptions made here are only for example purposes, of course. Other expressions for $C(x)$ may be more realistic.

There are also constraints on x. The thickness cannot exceed the thickness of the wall space (normally $3\frac{5}{8}$ in. in existing houses) and insulation thicknesses less than 1 in. are not available. Furthermore, there may be a constraint of the total capital investment possible, due to lack of money. These constraints are all of the regional type and can be expressed as follows:

$$x \geq 1 \text{ in.} \qquad x \leq 3\tfrac{5}{8} \text{ in.} \qquad K_3 x \leq M$$

where M is the maximum money available.

It is helpful to formulate the optimization problem in a generalized way before going on to consider some of the mathematical techniques for solving certain special cases. The generalized problem is this: Find the values of the *parameters*

$$x_1, x_2, x_3, \ldots, x_n$$

that will give a maximum or minimum value of a *criterion function*

$$U = U(x_1, x_2, x_3, \ldots, x_n)$$

subject to the *functional constraints* (*equalities*)

$$\begin{aligned}
\psi_1 &= \psi_1(x_1, x_2, x_3, \ldots, x_n) = 0 \\
\psi_2 &= \psi_2(x_1, x_2, x_3, \ldots, x_n) = 0 \\
&\cdots\cdots\cdots\cdots\cdots\cdots\cdots\cdots \\
\psi_m &= \psi_m(x_1, x_2, x_3, \ldots, x_n) = 0
\end{aligned}$$

and subject to the *regional constraints* (*in*equalities)

$$l_1 = l_1(x_1, x_2, x_3, \ldots, x_n) \leq L_1$$

$$l_2 = l_2(x_1, x_2, x_3, \ldots, x_n) \leq L_2$$

$$\cdots\cdots\cdots\cdots\cdots\cdots$$

$$l_p = l_p(x_1, x_2, x_3, \ldots, x_n) \leq L_3$$

Five methods of solution of special cases of this kind of problem will be treated in this chapter: differential calculus, lagrangian multipliers, numerical methods, linear programming, and the calculus of variations. Examples of the use of each will be given.

12.2 OPTIMIZATION BY DIFFERENTIAL CALCULUS

Often there are no functional constraints, and in such cases it is sometimes possible temporarily to ignore the regional constraints. In such cases, and assuming that the criterion function U is a differential expression, differential calculus can be used to determine the optimum. The optimum is given by the solution of the n simultaneous equations found by setting the derivatives of criterion function with respect to each of the parameters equal to zero. Symbolically,

$$\frac{\partial U}{\partial x_1} = 0$$

$$\frac{\partial U}{\partial x_2} = 0$$

$$\cdots\cdots$$

$$\frac{\partial U}{\partial x_n} = 0$$

In the case of one parameter this procedure is equivalent to finding the point where the slope is zero on a plot of U vs. x. In two parameters, the geometric representation of the problem is a surface. Setting $\partial U/\partial x_1$ and $\partial U/\partial x_2$ simultaneously equal to zero gives the zero slope point and hence the maximum (or minimum) value of U.

As a simple example, consider the insulation problem introduced in the preceding section. Since fuel costs are proportional to heat loss and heat loss is inversely proportional to insulation thickness, it can be assumed that fuel costs are

$$C_{\text{fuel}} = K_1 \frac{1}{x}$$

where K_1 is factor accounting for specific fuel costs per unit heat loss, etc. Insulation costs can be considered as proportional to thickness x and so

$$C_{\text{insulation}} = K_2 x$$

The total cost is therefore

$$C = K_1 \frac{1}{x} + K_2 x$$

Since there is only one parameter, the optimum is given by

$$\frac{\partial C}{\partial x} = 0 = -K_1 \frac{1}{x^2} + K_2$$

or

$$x = \sqrt{\frac{K_2}{K_1}}$$

This is the optimum without consideration of the regional constraints. If they are satisfied, then the problem is completely solved. If they are not satisfied, then the solution will usually be given by using the limiting value of the offended constraint. Thus if $x = \sqrt{K_1/K_2}$ calls for insulation thicker than the $3\frac{5}{8}$ in. wall space, the solution will be to use the full $3\frac{5}{8}$ in. Knowledge that the optimum is $x \leq 3\frac{5}{8}$ in. indicates that $x = 3\frac{5}{8}$ in. is the best that can be obtained.

In the event there *are* functional constraints, they can usually be used to reduce the number of parameters before the differentiation begins, and thus the basic problem is not changed. Suppose, for example, a cylindrical tank of 10 ft³ is to be built with as little material as possible. The criterion function is thus the surface area, $A = 2\pi r^2 + 2\pi r$, and the functional constraint is on volume, $V = \pi r^2 l = 10$ ft³. Use the equation $V = \pi r^2 l$ to eliminate in the equation for A, and the criterion function becomes

$$A = 2\pi r^2 + \frac{2V}{r}$$

The optimization gives

$$\frac{\partial A}{\partial r} = 0 = 4\pi r - \frac{2V}{r^2}$$

or

$$r = \left(\frac{V}{2\pi}\right)^{1/3} = 1.167 \text{ ft}$$

Since $l = V/\pi r^2$, the result for l is

$$l = \left(\frac{4V}{\pi}\right)^{1/3} = 2.334 \text{ ft}$$

Notice that $l = 2r$. If a regional constraint had also been imposed, such as, say, that r cannot exceed 1 ft, then the tank would be made with $r = 1$ ft because this is as close as possible to the unconstrained optimum.

12.3 OPTIMIZING BY THE USE OF DUAL VARIABLES

The material in this section is based on a method presented by C. Zener in "A Mathematical Aid in Optimizing Engineering Design" (*Proc. Nat. Acad. Sci.*, vol. 47, pp. 537–539, 1961). Whenever it is desired to minimize a function consisting of n terms in the form $u_1 + u_2 + \cdots + u_n$, dual variables, $a_1, a_2 \cdots a_n$ can be defined such that $u_i = a_i x_i$. Then use can be made of the mathematical theorem that

$$a_1x_1 + a_2x_2 + \cdots + a_nx_n \geq x_1{}^{a_1}x_2{}^{a_2} \cdots x_n{}^{a_n}$$

This inequality becomes an equality (and hence the right side becomes the minimum value of the left) only if all the x's are equal and if $a_1 + a_2 + \cdots + a_n = 1$. As an optimization technique, the above inequality is best written as

$$u_1 + u_2 + \cdots + u_n \geq \left(\frac{u_1}{a_1}\right)^{a_1}\left(\frac{u_2}{a_2}\right)^{a_2} \cdots \left(\frac{u_n}{a_n}\right)^{a_n}$$

For example, if a criterion function $U = 10t + (20/t)$ is to be optimized, $u_1 = 10t$ and $u_2 = 20/t$. The dual function $V(a)$ is formed as follows:

$$V(a) = \left(\frac{10t}{a_1}\right)^{a_1}\left(\frac{20}{ta_2}\right)^{a_2}$$

This dual function is equal to the minimum of U when

$$a_1 + a_2 = 1 \tag{1}$$

and when simultaneously the values of the dual variables (a_1 and a_2) are such that the original variable(s) (in this case, t) vanish from the $V(a)$ expression; that is, when

$$t^{a_1}\left(\frac{1}{t}\right)^{a_2} = 1 \tag{2}$$

With this satisfied, note that

$$V(a) = \left(\frac{10}{a_1}\right)^{a_1}\left(\frac{20}{a_2}\right)^{a_2} \tag{3}$$

Now simultaneous solution of Eqs. (1) and (2) above gives $a_1 = \frac{1}{2}$ and $a_2 = \frac{1}{2}$. Using these values in Eq. (3) gives the minimum value of the criterion function $V(a) = U_{\min} = (20)^{1/2}(40)^{1/2} = 28.2$.

The minimum value of U has now been determined, but the best value for t has not yet been found. The a's have great significance in this respect, however. They give the proportion of the optimum cost associated with each term of the original expression for U. Thus in this case, each term in the original expression contributes one-half to the total cost. This makes it easy to compute the value of t:

$$10t = \tfrac{1}{2}(28.2) \qquad t = 1.4$$

or

$$\frac{20}{t} = \tfrac{1}{2}(28.2) \qquad t = 1.41$$

It should be noticed that the dual variables, and hence the proportion of the cost contributed by the separate terms, do not change if the cost coefficients (10 and 20 in this example) change.

The method of dual variables reduces the optimization of a function without regional or functional constraints (or with functional constraints incorporated into the criterion function) to solution of a set of simultaneous equations. Often this is easier and quicker than setting the various partial derivatives to zero.

When there is more than one parameter in the criterion function, the condition that each of these vanish from $V(a)$ results in an equation in the dual variables (a's). These equations, plus the condition that $\sum a_i = 1$, usually (though not necessarily) provide a solution for the a's. If the number of unknowns and the number of resulting equations is not equal, the method is not useful.

12.4 LAGRANGIAN MULTIPLIERS

In the preceding section, functional constraints were used simply to reduce the number of parameters. Sometimes this procedure and/or the differentiation that follows is impractical or impossible. The use of lagrangian multipliers is a technique that is applicable in the presence of functional constraints. Regional constraints can be used only in the same way as in the preceding section. Here is how the method works:

The criterion function $U = U(x_1, x_2, x_3, \ldots, x_n)$ is to be optimized, and so

$$dU = \frac{\partial U}{\partial x_1} dx_1 + \frac{\partial U}{\partial x_2} dx_2 + \frac{\partial U}{\partial x_3} dx_3 + \cdots + \frac{\partial U}{\partial x_n} dx_n = 0$$

This may be written more elegantly as

$$dU = \sum_i \frac{\partial U}{\partial x_1} dx_1 = 0$$

The functional constraints are

$$\psi_1 = \psi_1(x_1, x_2, x_3, \ldots, x_n) = 0$$

$$\psi_2 = \psi_2(x_1, x_2, x_3, \ldots, x_n) = 0$$

$$\cdots\cdots\cdots\cdots\cdots\cdots\cdots\cdots$$

$$\psi_m = \psi_m(x_1, x_2, x_3, \ldots, x_n) = 0$$

Each of these functional constraints may be differentiated to give

$$d\psi_1 = \sum_{i=1}^{n} \frac{\partial \psi_1}{\partial x_i} dx_i = 0$$

$$d\psi_2 = \sum_{i=1}^{n} \frac{\partial \psi_2}{\partial x_i} dx_i = 0$$

$$\cdots\cdots\cdots\cdots\cdots\cdots$$

$$d\psi_m = \sum_{i=1}^{n} \frac{\partial \psi_m}{\partial x_i} dx_i = 0$$

Now each of these m equations is multiplied by an (as yet) undetermined parameter λ different for each equation, called a lagrangian multiplier. The result is

$$\lambda_1\, d\psi_1 = \sum_{i=1}^{n} \lambda_1 \frac{\partial \psi_1}{\partial x_i} dx_i = 0$$

$$\lambda_2\, d\psi_2 = \sum_{i=1}^{n} \lambda_2 \frac{\partial \psi_2}{\partial x_i} dx_i = 0$$

$$\cdots\cdots\cdots\cdots\cdots\cdots\cdots$$

$$\lambda_n\, d\psi_m = \sum_{i=1}^{n} \lambda_m \frac{\partial \psi_m}{\partial x_i} dx_i = 0$$

Now if all these equations are added together and the dU equation also added in, there results

$$dU + d\psi_1 + d\psi_2 + d\psi_3 + \cdots + d\psi_m = 0$$

or $$\sum_{i=1}^{n} \left(\frac{\partial U}{\partial x_i} + \lambda_1 \frac{\partial \psi_1}{\partial x_i} + \lambda_2 \frac{\partial \psi_2}{\partial x_i} + \cdots + \lambda_m \frac{\partial \psi_m}{\partial x_i} \right) dx_i = 0$$

Since the parameters (x_i's) are independent, each of the n bracketed terms in the above equation must be equal to zero in order to satisfy the above equation. Hence there are n equations of the form

$$\frac{\partial U}{\partial x_i} + \lambda_1 \frac{\partial \psi_1}{\partial x_i} + \lambda_2 \frac{\partial \psi_2}{\partial x_i} + \cdots + \lambda_m \frac{\partial \psi_m}{\partial x_i} = 0 \qquad i = 1, 2, \ldots, n$$

Remember that there are also m constraint equations of the form

$$\psi_j = 0 \qquad j = 1, 2, \ldots, m$$

Thus there are $m + n$ equations and there are also $m + n$ unknowns: n parameters (x_i's) and m lagrangian multipliers (λ's). Solution of this set of $m + n$ simultaneous equations thus gives the desired optimum.

As a first example consider the tank problem worked in the preceding section by the process of differentiation. The criterion function is

$$A = 2\pi r^2 + 2\pi rl$$

and

$$\frac{\partial A}{\partial r} = 4\pi r + 2\pi l$$

$$\frac{\partial A}{\partial l} = 2\pi r$$

The functional constraint equation is

$$\psi = V - \pi r^2 l = 0$$

and

$$\frac{\partial \psi}{\partial r} = -2\pi rl$$

$$\frac{\partial \psi}{\partial l} = -\pi r^2$$

Now it is possible to form the three equations

$$\frac{\partial U}{\partial r} + \lambda \frac{\partial \psi}{\partial r} = 0$$

$$\frac{\partial U}{\partial l} + \lambda \frac{\partial \psi}{\partial l} = 0$$

$$\psi = 0$$

which determine the three unknowns r, l, and λ. They are

$$4\pi r + 2\pi l + \lambda(-2\pi rl) = 0$$

$$2\pi r - \lambda \pi r^2 = 0$$

$$V - \pi r^2 l = 0$$

Simultaneous solution gives

$$\lambda = \frac{2}{r} \qquad l = 2r \qquad r = \left(\frac{V}{2\pi}\right)^{1/3}$$

There are many engineering applications where lagrangian multipliers are a useful method of optimizing the use of resources or of minimizing cost. It is a technique which design engineers must know how to use. Problems with many parameters on constraints result in simultaneous equations that require digital computer solution.

12.5 NUMERICAL METHODS OF OPTIMIZATION

Obviously, criterion functions which involve only one parameter can be plotted graphically to determine the optimum value. It is equally easy to search out the optimum value numerically, using the digital computer if necessary. When two or more parameters are involved, however, graphical methods become impossible, and numerical methods take on a new difficulty. Two techniques will be discussed in this section: first the alternating one-dimensional-search method and then the more sophisticated method of steepest descent. Both processes are for minimizing a criterion function in the absence of constraints. Functional constraints can of course be included implicitly if they can be incorporated into the criterion function to reduce the number of parameters, and regional constraints are accounted for by limiting the parametric values to the desired region.

For discussion purposes, consider a case where there are only two parameters. (The outlines of the procedures are easily extended to three or more parameters, however.) The criterion function can now be imagined to describe a surface in space with the two parameters as coordinates. If it is a minimum that is sought, the location of the bottom of the lowest valley is the desired point. If a maximum is wanted, the top of the highest hill is desired. In both the procedures described here an initial guess must be made which is either somewhere on the *highest* hill or within the *lowest* valley as the case may be. The methods will only find the extremum of the hill or valley in which they are begun.

For the method of alternating one-dimensional search, an initial value of the two parameters is chosen. If the two parameters are x_1 and x_2, the initial values may denoted by x_{10} and x_{20}. These define an initial value of the criterion function U_{00}. Now if x_2 is kept constant at x_{20}, x_1 can be varied until the minimum (or maximum) value of U is found. Since this is a single-parameter optimization, it is easily done graphically, numerically, or perhaps differentially. Define this intermediate "optimum" value of x_1 as x_{11}. At x_{20} and x_{11} the value of U is U_{10}. Now hold x_1 constant at x_{11} and vary x_2 until a new optimum value of U (U_{11}) is found at x_{21}. Next keep

x_2 constant at x_{21} and vary x_1 to get a new optimum, and so on, until succeeding values of U are constant to the accuracy desired.

As an example, consider the following problem: A hot-water pipeline is to be designed to carry a large quantity of hot water from the heater to the point of use. The cost involved consists of four items: the cost of pumping the water, the cost of heating the water, the cost of the pipe, and the cost of the insulation which is put on the pipe. The pumping costs will be proportional to the pressure drop in the pipe, which is given by

$$\Delta p = f \frac{L}{D} \rho \frac{V^2}{2g_0}$$

For a given water flow, V can be written in terms of the pipe diameter D:

$$\Delta p = f \frac{L}{D} \frac{16\rho Q^2}{\pi^2 D^4} \frac{1}{2g_0}$$

The power consumed will be $(1/\rho)\,\Delta p = f(L/g_0)\,(8Q^2/\pi^2 D^5)$. For the purposes of this example, the cost of pumping power can thus be written as

$$C_p = K_p \frac{1}{D^5}$$

The *cost* of the heat lost from the pipe will be approximately proportional to the heat loss itself, which is given by

$$q = \frac{2\pi k L\,\Delta t}{\ln\,[(D + x)/D]}$$

D is the inside diameter of the pipe and x is the thickness of the insulation on the pipe. The cost, then, can be written simply as

$$C_h = K_h \frac{1}{\ln\,[(D + x)/D]}$$

The cost of the pipe will be said to be approximately proportional to the diameter:

$$C_{\text{pipe}} = K_3 D$$

The cost of the insulation will be approximately proportional to its thickness x:

$$C_i = K_4 x$$

The total cost, therefore, is given by

$$C = K_p \frac{1}{D^5} + K_h \frac{1}{\ln\,[(D + x)/D]} + K_3 D + K_4 x$$

The designer wishes to minimize this cost function, and he has two parameters under his control: D and x. The problem is to find the values of

D and x which minimize C. Suppose the constants have been determined as

$$K_p = 100 \qquad K_h = 1.0 \qquad K_3 = 0.50 \qquad K_4 = 1.0$$

The cost is then

$$C = 1.0 \frac{1}{\ln [(D + x)/D]} + 10.0 \frac{1}{D^5} + 0.5D + 1.0x$$

To use the method an initial guess must be made; in this case $D_0 = x_0 = 1.0$. Now holding x_0 constant, D_0 is varied until an optimum is found. It turns out to be $D_1 = 1.8$ to the nearest 0.2. (There is no point in carrying the calculations to extreme accuracy at this point.) Now with $D = 1.8$, x is varied until a new optimum is reached at $x_1 = 1.25$. And so on. The computer results for this computation are shown in Fig. 12.1. The final result

D	x	*Cost*	D	x	*Cost*
1.0	1.0	12.9427	1.83	1.25	4.5730
2.0	1.0	4.7789	1.83	1.30	4.5654
3.0	1.0	6.0172	1.83	1.35	4.5620
4.0	1.0	7.4912	1.83	1.40	4.5623
1.2	1.0	7.2686	1.83	1.31	4.5644
1.4	1.0	5.4146	1.83	1.32	4.5636
1.6	1.0	4.8234	1.83	1.33	4.5629
1.8	1.0	4.6925	1.83	1.34	4.5624
2.0	1.0	4.7789	1.83	1.35	4.5618
2.2	1.0	4.9629	1.83	1.36	4.5617
2.4	1.0	5.1967	1.83	1.37	4.5616
2.6	1.0	5.4571	1.83	1.38	4.5617
2.8	1.0	5.7326	1.83	1.39	4.5619
1.8	0.25	9.3684	1.81	1.37	4.5662
1.8	0.50	6.0088	1.82	1.37	4.5642
1.8	2.00	4.7675	1.83	1.37	4.5627
1.8	4.00	6.2839	1.84	1.37	4.5611
1.8	0.75	5.0503	1.85	1.37	4.5609
1.8	1.00	4.6925	1.86	1.37	4.5618
1.8	1.25	4.5755			
1.8	1.50	4.5790			
1.8	1.75	4.6516			
1.80	1.25	4.5755			
1.82	1.25	4.5742			
1.82	1.25	4.5734			
1.83	1.25	4.5730			
1.84	1.25	4.5731			

Fig. 12.1 Tabular results in optimization by alternating one-dimensional search.

is $D = 1.8$ and $x = 1.37$. The designer should therefore choose a pipe with inside diameter approximately 1.85 in., and he should specify a convenient insulation thickness of about 1.37 in.

The above procedure is theoretically possible for situations involving three or more variables but does not have rapid convergence properties. (Note, however, that it is easily adapted to computer solution.) It is possible to make better guesses, which involve changes in both x_1 and x_2, by computing the slope of the surface (*gradient* in vector terminology) at x_{10} and x_{20}. With the slope known it is obvious that the next guess should be in the direction of the *steepest* slope. Thus this method is called the method of steepest descent. For convenience here the method will be described in terms of the search for a minimum. The search for a maximum is exactly analogous.

Suppose the criterion function $U = U(x_1, x_2)$ is differentiated and

Fig. 12.2

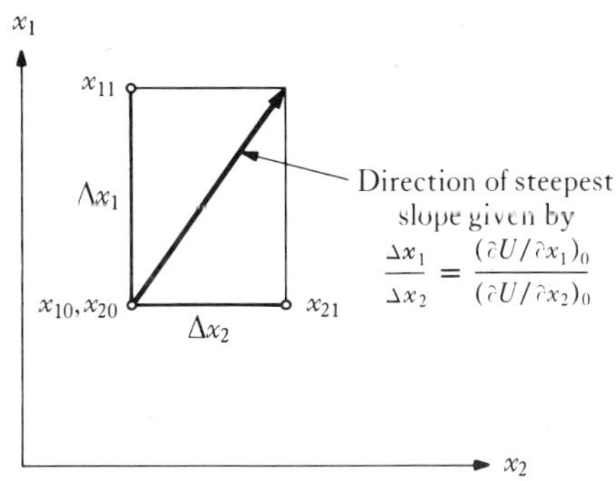

the derivatives evaluated at a starting point of x_{10} and x_{20}. Call these values $(\partial U/\partial x_1)_0$ and $(\partial U/\partial x_2)_0$. In order to move in the direction of the steepest slope then, it is necessary to move in such a direction that

$$\Delta x_1 \propto -\left(\frac{\partial U}{\partial x_1}\right)_0$$

and

$$\Delta x_2 \propto -\left(\frac{\partial U}{\partial x_2}\right)_0$$

the proportionality constant being the same in both cases. See Fig. 12.2. This indicates the *direction* in which to proceed but it does not indicate *how far* to proceed. That is, the proportionality factor in the above equations is not yet determined. It would be desirable to move in this new direction until U ceases to decrease. (At that point, a new direction of steepest descent would be sought and the process repeated.) So, let the proportionality constant be y. Now after proceeding for a distance corresponding to y, the values of the parameters are approximately:

$$x_1 = x_{10} - y\left(\frac{\partial U}{\partial x_1}\right)_0$$

$$x_2 = x_{20} - y\left(\frac{\partial U}{\partial x_2}\right)_0$$

and the criterion function is

$$U = U\left\{\left[x_{10} - y\left(\frac{\partial U}{\partial x_1}\right)_0\right], \left[x_{20} - y\left(\frac{\partial U}{\partial x_2}\right)_0\right]\right\}$$

To get the value of y where U is a minimum, set

$$\frac{dU}{dy} = 0$$

This value of y gives new values for x_1 and x_2 which can serve as starting points for another iteration. The process must be continued until a point is found where $\partial U/\partial x_1$ and $\partial U/\partial x_2$ are as close to zero as desired. Problems involving three or more parameters are done in exactly the way described here for two parameters. The method is easily adapted to computer solution.

As a simple example of the method of steepest descent, consider the function

$$U = 2x^2 + z^2$$

The minimum value of U is obviously at $x = z = 0$, but to illustrate the method an initial guess of $x = z = 1$ will be made. To use the method, the partial derivatives must be formed:

$$\frac{\partial U}{\partial x} = 4x \qquad \frac{\partial U}{\partial z} = 2z$$

Given the kth values of x and z, then, the $(k + 1)$th values are given by

$$x_{k+1} = x_k - y\left(\frac{\partial U}{\partial x}\right)_k = x_k - y(4x_k)$$

$$z_{k+1} = z_k - y\left(\frac{\partial U}{\partial z}\right)_k = z_k - y(2z_k)$$

To find y, the criterion function is evaluated at x_{k+1} and z_{k+1}:

$$U_{k+1} = 2\left[x_k - y\left(\frac{\partial U}{\partial x}\right)_k\right]^2 + \left[z_k - y\left(\frac{\partial U}{\partial z}\right)_k\right]^2$$

The best value of y is given by

$$\frac{\partial U_{k+1}}{\partial y} = 0 = -4\left[x_k - y\left(\frac{\partial U}{\partial x}\right)_k\right]\left(\frac{\partial U}{\partial x}\right)_k - 2\left[z_k - y\left(\frac{\partial U}{\partial z}\right)_k\right]\left(\frac{\partial U}{\partial z}\right)_k$$

Solving for y in this case leads to

$$y = \frac{-4x_k(\partial U/\partial x)_k - 2z_k(\partial U/\partial z)_k}{4(\partial U/\partial x)_k{}^2 + 2(\partial U/\partial z)^2} = \frac{-16x_k{}^2 - 4z_k{}^2}{64x_k{}^2 + 8z_k{}^2}$$

Thus for an initial guess of $x_0 = z_0 = 1$, there is

$$\left(\frac{\partial U}{\partial x}\right)_0 = 4 \qquad \left(\frac{\partial U}{\partial z}\right)_0 = 2 \qquad y = -\frac{5}{18}$$

This gives

$$x_1 = -\tfrac{1}{9} \qquad \text{and} \qquad z_1 = \tfrac{4}{9}$$

Using these as starting points again, there results

$$\left(\frac{\partial U}{\partial x}\right)_1 = -\frac{4}{9} \qquad \left(\frac{\partial U}{\partial z}\right)_1 = \frac{8}{9} \qquad y = 0.416$$

This gives

$$x_2 = +0.074 \qquad \text{and} \qquad z_2 = +0.074$$

Now the process is repeated until $\partial c/\partial x$ and $\partial c/\partial z$ are as close to zero as desired, indicating that the optimum has been reached. As can be seen, the answer $x = z = 0$ has been approached very closely in only two iterations. More complex problems require computer solution.

Many optimization problems result from the occurrence of criterion functions which vary both inversely and directly (in different terms) with the parameters. In one parameter a single example of this is the insulation-thickness problem, which was of the form

$$U = K_1\frac{1}{x} + K_2x$$

where x is the parameter. In two or more parameters, such functions often take the form

$$U = K_1\frac{1}{x^az^b} + K_2x^c + K_3z^d$$

There is no typical or standard form, of course, because of the infinite variety of optimization problems which occur. However, the function is typical enough, and as a second example of the steepest descent method let the constants and exponents all equal unity.

$$K_1 = K_2 = K_3 = a = b = c = d = 1$$

Thus
$$U = \frac{1}{xz} + x + z$$

To use steepest descent, the derivatives are formed:

$$\frac{\partial U}{\partial x} = -\frac{1}{x^2 z} + 1 \qquad \frac{\partial U}{\partial z} = -\frac{1}{xz^2} + 1$$

Also

$$U_{k+1} = \frac{1}{[x_k - y(\partial U/\partial x)_k][z_k - y(\partial U/\partial z)_k]} + \left[x_k - y\left(\frac{\partial U}{\partial x}\right)_k\right] + \left[z_k - y\left(\frac{\partial U}{\partial z}\right)_k\right]$$

To find y,

$$\frac{\partial U_{k+1}}{\partial y} = 0 = -\frac{(\partial U/\partial z)_k}{[x_k - y(\partial U/\partial x)_k][z_k - y(\partial U/\partial z)_k]^2} - \frac{(\partial U/\partial x)_k}{[x_k - y(\partial U/\partial x)_k]^2[z_k - y(\partial D/\partial z)_k]} - \left(\frac{\partial U}{\partial x}\right)_k - \left(\frac{\partial U}{\partial z}\right)_k$$

After an initial guess, the derivatives are easily evaluated, and the above equation has a single unknown, y. It is typical of this method that the results are complex transcendental equations at this stage, and numerical computer solution is usually required. When y is found, new values of the parameters are given, as usual, by

$$x_{k+1} = x_k - y\left(\frac{\partial U}{\partial x}\right)_k \qquad z_{k+1} = z_k - y\left(\frac{\partial U}{\partial z}\right)_k$$

and the process repeated until the desired optimum is achieved. As can be seen, the steepest-descent method is generally useful only in conjunction with the computer, and then only when the number of parameters is more than two. Two parameter problems are usually best solved by the method of alternating one-dimensional search previously outlined.

12.6 LINEAR PROGRAMMING

There are a number of circumstances which call for the optimization of a quantity which is a linear function of the parameters:

$$U = c_1x_1 + c_2x_2 + \cdots + c_kx_k$$

This problem can usually be put into a form so that only positive values of the parameters are wanted:

$$x_1 \geq 0$$
$$x_2 \geq 0$$
$$\cdots\cdots$$
$$x_k \geq 0$$

If the additional region constraints are *linear* inequalities,

$$a_{11}x_1 + a_{12}x_2 + \cdots a_{1k}x_k \leq b_1$$
$$a_{21}x_1 + a_{22}x_2 + \cdots a_{2k}x_k \leq b_2$$
$$\cdots\cdots\cdots\cdots\cdots\cdots\cdots\cdots\cdots\cdots$$
$$a_{m1}x_1 + a_{m2}x_2 + \cdots a_{mk}x_k \leq b_m$$

this problem is called a *linear-programming problem.*

When there are two or three parameters, this type of optimization problem has a rather straightforward graphical solution. As an example, consider the function

$$U = 3x_1 + 5x_2$$

which is to be maximized subject to the constraint that

$$2x_1 + 4x_2 \leq 8$$

It is assumed that the parameters have been defined so that only positive values are desired. Now the constraint equation indicates that all points outside (from the origin) the line defined by

$$2x_1 + 4x_2 = 8$$

are excluded. This line is shown in Fig. 12.3. Now the equation

$$U = 3x_1 + 5x_2 = 0$$

also defines a line shown in the figure as a dashed line passing through the origin. There is a family of lines parallel to this line, each with a different value for U. The largest value possible passes through the point ($x_2 = 0$, $x_1 = 4$), as shown in the figure. Thus the maximum value of U is 12 for this case, and the problem is solved. It should be apparent that the maximum value of U will always be found at one of the corners which define the permissible domain.

When there are three parameters, graphical solutions become difficult. When there are more than three parameters, graphical solutions are impossible. Numerous iterative schemes for solving the linear programming problem have been devised and are described in textbooks on operations research. (See the Suggested Reading at the end of the chapter.) It is more important for the engineer to recognize the situations which lead to linear programming problems, and it is also very important that he be able to formulate such problems. To this end, several such situations will now be discussed.

Example 1 A company manufactures two kinds of metal screws: type A is stainless steel and type B is mild steel. Profits on each type have been computed and are found to be \$2 per thousand of type A and \$1 per thousand of type B. Because of the hardness of the stainless steel screws, however, their production is slower. If all production units produce type B, 100,000 screws a day can be produced. If all units produce type A, only

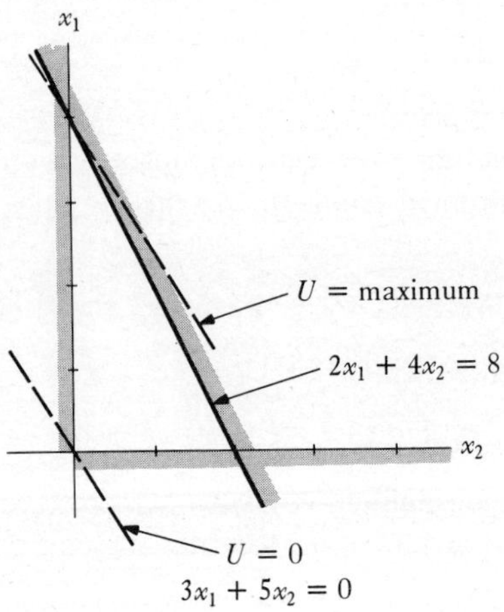

Fig. 12.3

20,000 screws a day can be produced. Furthermore, storage space, shipping problems, and supplier limitations make it impossible to exceed a total daily raw material usage of 65,000 screws. Also, stainless steel is available for only 15,000 screws daily.

This is a straightforward two-parameter linear-programming problem. Let x_A and x_B be the number of thousands of screws of type A and B respectively produced daily. Then the criterion function to be maximized is the profit:

$$U = 2x_A + x_B$$

The various constraints are as follows: on total production,

$$5x_A + x_B \leq 100$$

on the total material used,

$$x_A + x_B \leq 65$$

on the availability of stainless steel,

$$x_A \leq 15$$

Thus the problem statement has been formulated into the standard linear-programming format.

Example 2 A small aluminum-extrusion plant has two extrusion presses and produces extruded shapes in three classes: class 1 sells for 23 cents per pound, class 2 sells for 26 cents per pound, and class 3 sells for 40 cents per pound. The hourly production rate per press of the three classes is 2,000 lb for class 1, 1,500 lb for class 2, and 1,000 lb for class 3. The cost of operating the extrusion press per hour is the same, regardless of the production rate. The company must buy a special alloy *c* for class 3 extrusions at a price of 25 cents per pound. It is available in limited quantities up to 12,000 lb per day. Alloy B for producing type-2 extrusions costs 20 cents per pound and is available in any quantity desired. For type-1 extrusions, the company uses the output of its own melting and casting plant, so that the raw material for this type costs only 18 cents per pound. The melting and casting plant has a daily maximum capacity of 40,000 lb. Limitations in the heat treating and shipping departments limit the total production to 120,000 lb daily. How should this company regulate its production to maximize profit? Before going on, the reader should formulate the problem for linear-programming solution.

Let x_1, and x_2, and x_3 be the pounds of each type of extrusion produced daily. The criterion function is

$$U = 5x_1 + 6x_2 + 15x_3$$

The constraint on total hourly production per press is

$$1x_1 + \tfrac{4}{3}x_2 + 2x_3 \leq 48{,}000$$

Since there are two presses, the problem constraint is

$$x_1 + \tfrac{4}{3}x_2 + 2x_3 \leq 96{,}000$$

The constraint on melting and casting capacity is

$$x_1 \leq 40{,}000$$

The constraint on total storage capacity is

$$x_1 + x_2 + x_3 \leq 120{,}000$$

Finally, the constraint on the availability of alloy c is

$$x_3 \leq 12{,}000$$

Thus the problem is formulated for solution by one of the linear-programming techniques.

12.7 OPTIMIZING DISTRIBUTED SYSTEMS—THE CALCULUS OF VARIATIONS

So far in this chapter, lumped-parameter problems and techniques have been considered. Many optimization problems require solution for an optimum distributed function. The beam problem worked in Chap. 10 is an example. The distributed optimization problem requires solution for a continuous function, not for the values of discrete parameters (x_1, x_2, x_3, etc.). Often the constraints in such problems also involve continuously varying functions or their integrals.

The mathematical method called the calculus of variations can be used to solve certain problems of this kind. The general problem is formulated as follows. Given a twice-differentiable function F, of x, y, dy/dx and the integral I,

$$I = \int_{x_1}^{x_2} F\left(x,y,\frac{dy}{dx}\right) dx$$

Find the function $y = y(x)$ which optimizes I. It is shown in standard texts on the subject that the desired function y satisfies the following condition (known as the Euler condition):

$$\frac{\partial F}{\partial y} - \frac{d}{dx}\frac{\partial f}{\partial y'} = 0$$

where $y' = \dfrac{dy}{dx}$.

The question of constraints will be considered later. First as an example of the above, suppose it is desired to find the $y = y(x)$ such that the length of the line between point A and point B in Fig. 12.4 is a minimum. (Obviously the answer is a straight line.) For concreteness, assume point A is at $x = y = 0$ and point B is $x = y = 1$. In this case the distance to be minimized is given by

Fig. 12.4

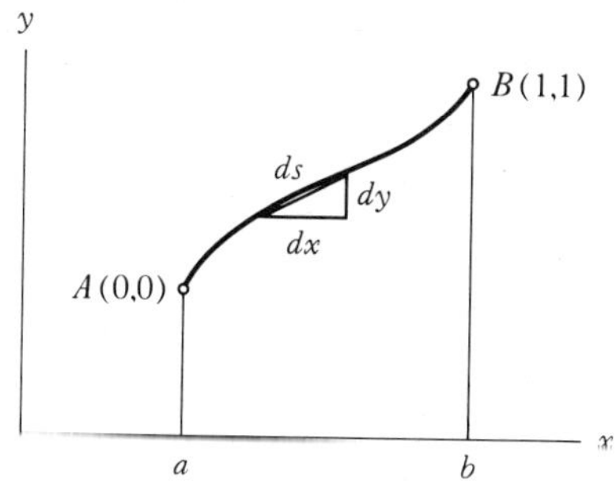

$$I = \int_A^B ds = \int_B^A \sqrt{dx^2 + dy^2} = \int_a^b \sqrt{1 + (y')^2}\, dx$$

Thus the function F is

$$F = \sqrt{1 + (y')^2}$$

The Euler condition is

$$\frac{\partial F}{\partial y} - \frac{d}{dx}\frac{\partial F}{\partial y'} = 0$$

In this case

$$\frac{\partial F}{\partial y} = 0$$

$$\frac{\partial F}{\partial y'} = \frac{y'}{\sqrt{1 + (y')^2}}$$

$$\frac{d}{dx}\frac{\partial F}{\partial y'} = \frac{d}{dx}\left(\frac{y'}{\sqrt{1 + (y')^2}}\right) = 0$$

Therefore $$\frac{y'}{\sqrt{1 + (y')^2}} = c$$

Solving for y' gives

$$y' = \frac{c}{\sqrt{1 - c^2}}$$

and so $$y = \frac{c}{\sqrt{1 - c^2}}\, x + k$$

which is the expected straight line. c and k can be evaluated from the fixed end conditions at A and B. The result is $k = 0$ and $c/\sqrt{1 - c^2} = 1$. Thus the optimum function is determined as $y = x$.

When there are constraints, a slightly different Euler condition results. Suppose it is desired to minimize

$$I = \int_a^b F(x,y,y')\, dx$$

subject to the constraint that

$$J = \int_a^b G(x,y,y')\, dx = k$$

where k is a constant. This situation is treated by finding the Euler equation for minimizing

$$\int_a^b [F(x,y,y') - G(x,y,y')]\, dx$$

which is $$\frac{\partial F}{\partial y} - \lambda \frac{\partial G}{\partial y} - \frac{d}{dx}\frac{\partial F}{\partial y'} + \lambda \frac{d}{dx}\frac{\partial G}{\partial y'} = 0$$

The λ has been introduced somewhat in the manner of an undetermined lagrangian multiplier. The two constants which result from solution of the Euler condition equation and λ are determined by satisfying the two end conditions and the value k of the constraint integral.

As an example, suppose it is desired to find the curve connecting $A(0,0)$ and $B(2,0)$ such that the length of the line is a minimum but the area under the line is $A = \pi/2$. See Fig. 12.5. The problem therefore is to minimize

$$I = \int_0^2 \sqrt{1 + (y')^2}\, dx$$

subject to
$$J = \int_0^2 y\,dx = \pi/2$$

In this case $\quad F = \sqrt{1 + (y')^2} \quad$ and $\quad G = y$

The Euler condition

$$\frac{\partial F}{\partial y} - \lambda \frac{\partial G}{\partial y} - \frac{d}{dx}\frac{\partial F}{\partial y'} + \lambda \frac{d}{dx}\frac{\partial G}{\partial y'} = 0$$

gives
$$0 - \lambda - \frac{d}{dx}\frac{y'}{\sqrt{1 + (y')^2}} + 0 = 0$$

$$\frac{d}{dx}\frac{y'}{\sqrt{1 + (y')^2}} = -\lambda$$

$$\frac{y'}{\sqrt{1 + (y')^2}} = -\lambda x + c_1$$

The solution of this differential equation can be shown to be

$$y = \sqrt{2x - x^2}$$

and the given conditions give $\lambda = c_1 = 1$. Squaring and rearranging this

Fig. 12.5

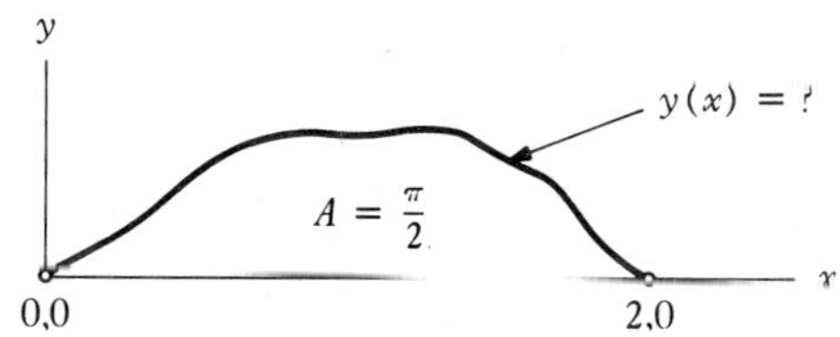

expression for y results in the familiar expression for a circle of radius 1 with center at (1,0):

$$(x - 1)^2 + y^2 = 1$$

This, then, is the desired curve.

In the event that a ratio of integrals is to be minimized (therefore, I/J) it can be shown that this is equivalent for computational purposes to minimizing I subject to J as a constraint (except, of course, that J is not a constant). The Euler condition is again

$$\frac{\partial F}{\partial y} - \lambda \frac{\partial G}{\partial y} - \frac{d}{dx}\frac{\partial F}{\partial y'} + \lambda \frac{d}{dx}\frac{\partial G}{\partial y'} = 0$$

where F and G are as defined above in I and J. Suppose it is desired to minimize the ratio

$$\frac{\int_0^{2\pi} (r^3 - R^3) r \, d\theta}{\int_0^{2\pi} (r^2 - R^2) \cos\theta \, d\theta} = \frac{I}{J}$$

where R is a constant. In this case F and G are

$$F = r^3 - R^3$$

$$G = (r^2 - R^2) \cos\theta$$

The above Euler condition (θ and r have replaced x and y, of course) gives

$$\frac{\partial F}{\partial r} - \lambda \frac{\partial G}{\partial r} - \frac{d}{d\theta}\frac{\partial F}{\partial r'} + \lambda \frac{d}{d\theta}\frac{\partial G}{\partial r'} = 0$$

$$\frac{\partial}{\partial r}(r^3 - R^3) - \lambda \frac{\partial}{\partial r}[(r^2 - R^2)\cos\theta] - 0 + 0 = 0$$

$$3r^2 - 2\lambda r \cos\theta = 0$$

Solving for r gives

$$r = \tfrac{2}{3}\lambda \cos\theta$$

A boundary condition would be needed now to give λ. The relationship between r and θ for the optimum value of I/J has been determined, however.

12.8 SUMMARY AND PREVIEW

This has been the first of a series of chapters on the sciences of decision making. When a decision situation reduces to the problem of finding the maximum or minimum value of a criterion function subject to certain given constraints, then it is a problem in optimization. The techniques of optimization discussed in the chapter are simple differentiation, lagrangian multipliers, numerical methods, linear programming, and the calculus of variations.

As has been frequently mentioned earlier in this book, probability and statistics are important fields for the modern engineering designer. Seldom are the outcomes of alternatives known for certain. Usually decisions are made in the face of uncertainty. Rational decision making, therefore, requires an ability to work with the laws of probability. Hence the next chapter turns your attention to a study of probability.

SUGGESTED READING

Asimow, Morris: "Introduction to Engineering Design," Prentice-Hall, Inc., Englewood Cliffs, N.J., 1962. This paperback has a good section on optimization techniques.

Balakrishnan, A. V., and L. W. Neustadt: "Computing Methods in Optimization Problems," Academic Press Inc., New York, 1964. A collection of papers on numerical optimization.

Johnson, Ray C.: "Optimum Design of Mechanical Elements," John Wiley & Sons, Inc., New York, 1961. This book does not explicitly treat the mathematical methods of optimization, but it does a good job of showing by example the concept and role of optimization in the design process.

Sasieni, Maurice, Arthur Yaspan, and Lawrence Friedman: "Operations Research," John Wiley & Sons, Inc., New York, 1959. A good text on operations research methods, including linear programming.

PROBLEMS

12.1 Check the dimensions of the physical equations written in this chapter.

12.2 Use the digital computer to find the optimum value of $y = 10x^2 + 5x - 2$. Work alone but try to write the most efficient program in the class.

12.3 Use the method of steepest descent to find the optimum D and x for the pipeline example described in the text.

12.4 Solve the linear-programming problems that are formulated in the chapter, using the formal technique called "the simplex method." Consult one of the books in the suggested reading to learn the method.

12.5 Consider the function $y = \sin x - \frac{1}{100} x^2$ in the region $0 \leq x \leq 10$. Find the maximum and the minimum. How can you determine *the* maximum and also distinguish it from a minimum? What implications does this example have for more complex problems?

12.6 Consider the criterion function $U = (2/x^2z) + 3x + z^2$. Find the optimum value by two numerical methods.

12.7 A five-sided box is to be built with a volume of 8 ft^3. Material costs \$10 per square foot. Joining costs \$10 per foot. What size box should be built?

12.8 The following problem appeared in Volume 7, Number 7, of the *r and d letter* of the Westinghouse Educational Department. Four hundred cubic yards of gravel must be ferried across a river. The material is to be shipped in an open box of length t_1, width t_2, and height t_3. The sides and bottom of the box cost \$10 per square yard, and the ends of the box cost \$20 per square yard. The box will have no salvage value, and each round trip of the box on the ferry will cost 10 cents. How should the contractor minimize his travel cost? Use the method of dual variables. What will the optimum cost be?

13 PROBABILITY

13.1 INTRODUCTION

The material in this chapter cannot be learned well in a cursory reading of the text. Students who wish a more thorough treatment of the subject in a completely self-teaching format should see "A Programmed Introduction to Probability," by John R. Dixon (John Wiley & Sons, Inc., New York, 1964).

Most people in the Western world have a concept of probability by the time they are in high school or college. Asked the probability that a tossed coin will come up heads, they will say "one-half." Asked the probability that a six-sided die (die is singular for dice) will come up a four, they say "one-sixth," and they will say that the probability of an even number coming up is "one-half." These responses indicate that there exists the concept of a probability scale which runs from zero, for an event which cannot happen, to one, for an event which must happen. Numbers on the scale are established as follows. Given a closed set of N possible outcomes, the probability that an even A will occur is

$$p(A) = \frac{\text{number of ways in } N \text{ that produce } A}{N}$$

The popular knowledge about probability extends a little further. Suppose a coin is to be flipped two times. What is the probability that the result will be two heads? Clearly, the answer is $\frac{1}{2} \times \frac{1}{2} = \frac{1}{4}$, and most people understand that the probability of two events occurring together is the product of the probabilities that they will occur separately.

This is about as far as popular probability goes. It is a good start, however, and in this chapter that start will be extended until a good deal more of the subject has been covered.

13.2 SYMBOLIC NOTATION

The science of probability uses a language system that must be learned before probability itself can be studied. In this language system, symbols stand for verbal statements. Thus it can be written that

A = The length of the shaft is 12 in. ± 0.01 in.

B = The diameter of the shaft is 1 in. ± 0.001 in.

To denote the denial, or negative, of a given statement B, the lowercase letter b is used. The denial of A above, for example, is

a = The length of the shaft is *not* 12 in. ± 0.01 in.

Note that the denial of a is A.

When defining symbols for use in probability work, it is important that each symbol stand for a single statement or express only a single idea. Such statements have a simple, straightforward denial. It would not be well, for example, to define a symbol, say D, as follows: "D = The length of the shaft is 12 in. ± 0.01 and the diameter of the shaft is 1 in. ± 0.001 in." Instead, each idea must be given a separate symbol, and combinations of various ideas working in various ways are made by combining the individual symbols according to the rules of the language.

For example, if it is desired to express the combined statement "Both

A and B," the notation is simply AB. That is,

AB = Both A and B.

AB = The length of the shaft is 12 in. ± 0.01 in. and the diameter of the shaft is 1 in. ± 0.001 in.

Another fundamental combined statement is "Either A or B or both." This is denoted by $A + B$. That is,

$A + B$ = Either A or B or both.

$A + B$ = Either the length of the shaft is 12 in. ± 0.01 in. *or* the diameter of the shaft is 1 in. ± 0.001 in., or both.

It should be clear from these definitions that

$$AB = BA$$

and

$$A + B = B + A$$

13.3 TRUTH TABLES

The identity of combinatory expressions can be proved with the use of so-called truth tables. The simplest possible truth table is one where only one symbol is included in the system:

Case no.	*A*
1	T
2	F

If three symbols are involved, the basic table is

Case no.	*A*	*B*	*C*
1	T	T	T
2	T	T	F
3	T	F	F
4	T	F	F
5	F	T	T
6	F	T	F
7	F	F	T
8	F	F	F

And so on. Truth tables can be used to prove the identity of combined expressions. Given two expressions [say, $A(B + C)$ and $AB + AC$], if they have the same sense in every case in the truth table, then they clearly cannot be distinguished by any test and are identical. Consider the two expressions $A(B + C)$ and $AB + AC$. The basic truth table is built up until these expressions can be compared. Thus,

Case no.	A	B	C	$B + C$	$A(B + C)$	AB	AC	$AB + AC$
1	T	T	T	T	T	T	T	T
2	T	T	F	T	T	T	F	T
3	T	F	T	T	T	F	T	T
4	T	F	F	F	F	F	F	F
5	F	T	T	T	F	F	F	F
6	F	T	F	T	F	F	F	F
7	F	F	T	T	F	F	F	F
8	F	F	F	F	F	F	F	F

Notice that AB is true only if both A and B are true but that $A + B$ is true if either A or B or both are true. In the above table, since $A(B + C)$ and $AB + AC$ are identical for every case, the two expressions are equivalent.

Students should now set up the basic truth table for two symbols, A and B, and then extend it to find the relationship (if any) between $A + B$ and ab. Do it here before going on to the next page.

Case no.	A	B	$A + B$	a	b	ab
1	T	T	T	F	F	F
2	T	F	T	F	T	F
3	F	T	T	T	F	F
4	F	F	F	T	T	T

Since $A + B$ and ab are opposites in every case in the truth table, they are opposites in meaning. Another way of saying this is to say that ab is the denial of $A + B$. This is a result that is very useful in solving many probability problems.

13.4 PROBABILITY NOTATION

The above notation conventions are used in probability work together with some additional symbols. The vertical slash line | is used to mean "given that," and p is used to denote probability. Thus if

$$A = \text{It will rain tomorrow.}$$

$$X = \text{The weather man says it will rain tomorrow.}$$

then $p(A \mid X)$ means "The probability that it will rain tomorrow *given that* the weather man says it will rain tomorrow." The probability of combined events can be expressed in this same way, such as $p(AB \mid X)$, $p(A + B) \mid X$, $p(Ab + aB)C \mid XY$, etc.

It is of the utmost importance here to note that probability has meaning only in relation to the given information. There is no such thing as the "true" probability of some event occurring. Consider, for example, the following statements:

$$A_2 = \text{An ace is drawn from a deck of 52 cards on the second draw.}$$

$$A_1 = \text{An ace is drawn from a deck of 52 cards on the first draw.}$$

$$A_n = \text{An ace is drawn from a deck of 52 cards on the } n\text{th draw.}$$

$$X = \text{The cards drawn out are not replaced before the next draw.}$$

Now it is clear that $p(A_1 \mid X) = 4/52 = 1/13$. Also note that $p(A_2 \mid X) = 1/13$. Since no information is given about the first draw, there is no alternative except to ignore it. In fact $p(A_{52} \mid X) = 1/13$!

Suppose, now, that a card is drawn from a full deck and kept hidden. Next a second card is drawn but is not looked at just yet. The probability that this second card is an ace is $p(A_2 \mid X) = 1/13$. Now the first card is examined and found to be an ace. The second card still is exactly the same card it was before, but the *probability* that a person who had not seen it would attach to its being an ace, changes when he sees the first card. The probability is now given by $p(A_2 \mid A_1X) = 3/51 = 1/17$. Notice how the additional information (from X to A_1X) changed the probability. Probability depends upon the information available and not upon any objective reality. (If this is troublesome, let it now be known that the deck of 52 cards under consideration has only two aces and recompute the probability of A_2. Now discover that the two aces are stuck together in the deck. Etc., etc., etc.) With each new piece of relevant *information* about the situation the probability changes, even though the actual physical situation does not change one bit. Finally, of course, *complete* information is available (that is, the drawn card is examined), and probability loses all significance. Probability is therefore a state of knowledge or a state of mind, and not a state of nature. It is a mental *concept* of man, not an observable fact of nature.

13.5 KNOWLEDGE OF "FUTURE" EVENTS

Return for a moment now to the deck of 52 cards:

X = The deck is standard and cards drawn are not replaced before the next draw.

A_n = An ace is drawn on the nth draw.

Several probabilities have already been discussed, such as

$$p(A_1 \mid X) = 1/13 \qquad p(A_2 \mid X) = 1/13 \qquad p(A_2 \mid A_1X) = 1/12$$

Now suppose it is necessary to evaluate $p(A_1 \mid A_2X)$; that is, it is known that the second card drawn is an ace, but the first card has not yet been examined. The answer is $p(A_1 \mid A_2X) = 3/51 = 1/17$. The knowledge that the second card is an ace excludes that card from the available N when the probabilities of earlier events are being evaluated. If this is troublesome, consider the probability $p(A_5 \mid A_2A_3A_4X)$! The reader should also verify that $p(A_1 \mid A_3X) = 1/17$, $p(a_2 \mid A_1X) = 12/13$, and $p(a_2 \mid A_1X) = 48/51 = 16/17$.

13.6 THE PRINCIPLE OF INSUFFICIENT REASON

In the event there is no relevant information about the probabilities of various outcomes, the outcomes must be considered equally likely. This is sometimes called the principle of insufficient reason. Thus if

X = There are seven balls—some red and some green—in an urn. Balls drawn out are not replaced before the next draw.

R_n = Red ball is drawn on the nth draw.

G_n = Green ball is drawn on the nth draw.

then $\qquad p(G_1 \mid X) = \frac{1}{2} \qquad$ and $\qquad p(R_1 \mid X) = \frac{1}{2}$

New information, such as

Y = There are four red and three green balls in the urn

will give new probabilities if used. Thus

$$p(G_1 \mid XY) = \tfrac{3}{7}$$

But $p(G_1 \mid X)$ is still $\frac{1}{2}$.

13.7 THE BASIC RULES OF PROBABILITY

Two basic axioms of probability can now be stated. The first is obvious:

$$p(A \mid X) + p(a \mid X) = 1$$

The second is less obvious. It has to do with combined events:

$$p(AB \mid X) = p(A \mid BX)p(B \mid X)$$

Also since $AB = BA$, it follows that

$$p(AB \mid X) = p(B \mid AX)p(A \mid X)$$

From these two rules, two additional rules can be derived. One has to do with $A + B$ type events and is

$$p[(A + B) \mid X] = p(A \mid X) + p(B \mid X) - p(AB \mid X)$$

The fourth rule is easily derived from the second one above. Since $AB = BA$ again, one can write

$$p(A \mid BX)p(B \mid X) = p(B \mid AX)p(A \mid X)$$

Rearranging,

$$p(A \mid BX) = p(A \mid X)\frac{p(B \mid AX)}{p(B \mid X)}$$

This last equation is known as Bayes' rule.

With these four rules, a great many probability problems can be solved, and in the next few sections, examples of such problems will be presented.

13.8 APPLICATIONS OF THE PROBABILITY RULES

Example 1 The design of rocket systems is such (say) that the probability of success of a moon shot is 0.20. What is the probability of at least one success if two rocket systems are sent?

SOLUTION 1 $\quad X$ = Information given in example.
A_n = Success on the nth try.
Given: $p(A_n \mid X) = 0.20$.
Find: $p[(A_1 + A_2) \mid X] = ?$

$$p[(A_1 + A_2) \mid X] = p(A_1 \mid X) + p(A_2 \mid X) - p(A_1A_2 \mid X)$$
$$p(A_2A_1 \mid X) = p(A_2 \mid A_1X)p(A_1 \mid X)$$

Since A_1 and A_2 are independent, $p(A_2 \mid A_1X) = p(A_2 \mid X)$. Therefore $p(A_2A_1 \mid X) = (0.20)(0.20) = 0.40$.

$$p[(A_1 + A_2) \mid X] = 0.20 + 0.20 - 0.40 = 0.36$$

Notice that since a_1a_2 is the negative of $A_1 + A_2$ this problem could have also been solved by

$$\begin{aligned} p[(A_1 + A_2) \mid X] &= 1 - p(a_1a_2 \mid X) = 1 - p(a_2 \mid a_1X)p(a_1 \mid X) \\ &= 1 - p(a_2 \mid X)p(a_1 \mid X) = 1 - (0.80)(0.80) \\ &= 1 - 0.64 = 0.36 \end{aligned}$$

Example 2 A designer who has the responsibility for setting reliability levels on a highly complex and sophisticated piece of apparatus wishes to know the following. Assume the factory-reliability level is set so that 70 percent of the parts are satisfactory. If a customer is sent three at random, what is the probability that at least one will be satisfactory?

SOLUTION 2 $\quad X$ = Information given in the problem statement.
A_n = nth part is satisfactory.
Given: $p(A_n \mid X) = 0.70$.
Find: $p[(A_1 + A_2 + A_3) \mid X] = ?$

$$\begin{aligned} p[(A_1 + A_2 + A_3) \mid X] &= p(A_1 \mid X) + p(A_2 \mid X) + p(A_3 \mid X) \\ &- p(A_1A_2 \mid X) - p(A_1A_3 \mid X) - p(A_2A_3 \mid X) + p(A_1A_2A_3 \mid X) \end{aligned}$$

Note that this formula can be derived by setting $B = A_2 + A_3$ and expanding $p[(A_1 + B) \mid X]$ according to the basic rule. Then substitute $A_2 + A_3$ for B and expand again. It is known that $p(A_1 \mid X) = p(A_2 \mid X) = p(A_3 \mid X) = 0.70$. To get $p(A_1A_2 \mid X)$, note that A_1 and A_2 are independent, so

$$\begin{aligned} p(A_1A_2 \mid X) &= p(A_2 \mid A_1X)p(A_1 \mid X) = p(A_2 \mid X)p(A_1 \mid X) \\ &= (0.70)(0.70) = 0.49 \end{aligned}$$

In a similar way, $p(A_1A_2A_3 \mid X) = (0.70)^3 = 0.343$. Thus

$$\begin{aligned} &p[(A_1 + A_2 + A_3) \mid X] \\ &\quad = 0.70 + 0.70 + 0.70 - 0.49 - 0.49 - 0.49 + 0.343 \\ &\quad = 0.973 \end{aligned}$$

Note that an alternative method is shorter. Since *abc* is the negative of $A + B + C$,

$$\begin{aligned} p[(A_1 + A_2 + A_3) \mid X] &= 1 - p(a_1a_2a_3 \mid X) = 1 - p(a_1 \mid a_2a_3X)p(a_2 \mid a_3X)p(a_3 \mid X) \\ &= 1 - p(a_1 \mid X)p(a_2 \mid X)p(a_3 \mid X) \\ &= 1 - (0.3)(0.3)(0.3) = 1 - 0.027 = \underline{\underline{0.973}} \end{aligned}$$

Example 3 A particular company product has been giving the customers difficulty and is to be redesigned. Data on the troubles with the product indicate that 12 percent of the products have both weak gangmops and poor crosswitches. It is also found that 24 percent have weak gangmops and 36 percent have poor crosswitches. Find (1) the probability that a product has either a weak gangmop or a poor crosswitch, or both. Also find (2) the probability that a product has neither fault.

SOLUTION 3 X = Information given in problem statement.
A = Crosswitch is weak.
B = Gangmop is poor.
Given: $p(AB \mid X) = 0.12$; $p(A \mid X) = 0.24$;
$p(B \mid X) = 0.36$.
Find: $p[(A + B) \mid X] = ?$; $p(ab \mid X) = 0$

$$\begin{aligned} p[(A + B) \mid X] &= p(A \mid X) + p(B \mid X) - p(AB \mid X) \\ &= 0.24 + 0.36 - 0.12 = 0.48 \end{aligned}$$

Since *ab* is the negative of $A + B$, $p(ab \mid X) = 0.52$.

Example 4 Your design department has sent six samples of a new experimental product to the customer for examination. It is found that two of these, sent by mistake, are early design and are unsatisfactory. What is the probability that at least one of the first two products examined by the customer will be defective?

SOLUTION 4 X = Given information.
A_n = *n*th product examined is defective.
Given: $p(A_n \mid X) = 1/3$.
Find: $p[(A_1 + A_2) \mid X] = ?$

$$p[(A_1 + A_2) \mid X] = p(A_1 \mid X) + p(A_2 \mid X) - p(A_1A_2 \mid X)$$

$$p(A_1A_2 \mid X) = p(A_2 \mid A_1X)p(A_1 \mid X) = 1/5 \times 1/3 = 1/15$$

Notice that $p(A_1A_2 \mid X) = p(A_1 \mid A_2X)p(A_2 \mid X) = 1/5 \times 1/3 = 1/15$

$$\underline{\underline{p[(A_1 + A_2) \mid X]}} = 1/3 + 1/3 - 1/15 = 9/15 = \underline{\underline{3/5}}$$

Example 5 A design engineer who wants to determine the reason for frequent failures in a system he has designed finds that there are two major sources of error: A or B. From the available data, he can determine that 36 percent have either fault A or fault B. He also finds that 12 percent have fault A but not fault B. What percent have fault B?

SOLUTION 5 X = Given information.
A = System has fault A.
B = System has fault B.
Given: $p[(A + B)\,|\,X] = 0.36$; $p(Ab\,|\,X) = 0.12$.
Find: $p(A\,|\,X) = ?$; $p(B\,|\,X) = ?$

Since ab is the negative of $A + B$, $p(ab\,|\,X) = 0.64$.

$$p(ab\,|\,X) = p(a\,|\,bX)p(b\,|\,X) = 0.64$$
$$p(Ab\,|\,X) = p(A\,|\,bX)p(b\,|\,X) = 0.12 = 1 - p(a\,|\,bX)p(b\,|\,X)$$

The above two equations contain two unknowns, $p(b\,|\,X)$ and $p(a\,|\,bX)$. Simultaneous solution gives

$$p(a\,|\,bX) = \tfrac{16}{19} \qquad p(b\,|\,X) = 0.76$$

Thus $p(B\,|\,X) = 0.24$.

Example 6 Customer complaints indicate that there is something radically wrong with the company's new snooper-scope. The data received shows that 75 percent have either a good snooper or a good scope or both. Also, it is found that of those with a good snooper, 80 percent have a good scope. Furthermore, 65 percent have both a good snooper and a good scope. Should you give first priority to redesigning the snooper or the scope?

SOLUTION 6 X = Given information.
A = Product has a good snooper.
B = Product has a good scope.
Given: $p[(A + B)\,|\,X] = 0.75$; $p(B\,|\,AX) = 0.80$;
$p(AB\,|\,X) = 0.65$.
Find: $p(A\,|\,X) = ?$; $p(B\,|\,X) = ?$

$$\begin{aligned} p[(A + B)\,|\,X] &= p(A\,|\,X) + p(B\,|\,X) - p(AB\,|\,X) \\ 0.75 &= p(A\,|\,X) + p(B\,|\,X) - 0.65 \\ 1.40 &= p(A\,|\,X) + p(B\,|\,X) \\ p(AB\,|\,X) &= p(B\,|\,AX)p(A\,|\,X) \\ 0.65 &= 0.80\,p(a\,|\,X) \\ p(A\,|\,X) &= 13/16 = 0.81 \\ p(B\,|\,X) &= 1.40 - p(A\,|\,X) = 1.40 - \tfrac{13}{16} = 0.59 \end{aligned}$$

Thus the percentage of bad snoopers is 19 percent and the percentage of bad scopes is 41 percent. Better redesign both, but the scopes are twice as critical!

13.9 EXPECTED VALUE, VARIANCE, AND STANDARD DEVIATION

It was pointed out earlier in this chapter that probability is an abstract mental concept, not to be confused with any physical reality. In a collection of data, suppose there are N total pieces of data and let N_i be the number of outcomes of the ith type. Thus $\sum_i N_i = N$. Suppose further that V_i is the value of the ith outcome. In this notation, then, the familiar average of the data is given by

$$V_{\text{av}} = \frac{1}{N} \sum_i N_i V_i = \bar{V}$$

The average is itself a datum. It is an observation. Suppose, however, that it is desirable to *predict* the average. The predicted average is called the *expected value*, and if p_i is the probability of getting the ith outcome on any trial, the expected value is given by

$$V_{\substack{\text{expected}\\ \text{value}}} = \sum_i p_i V_i = \langle V \rangle$$

The expected value is a mental concept. When N is large, however, the average value and the expected value become essentially equivalent numerically.

Knowledge of the average value or expected value is important when dealing with probabilistic situations. However, the average value of 0.48, 0.49, 0.50, 0.50, and 0.53 is 0.50, and the average value of 0.20, 0.30, 0.50, 0.60, 0.90 is also 0.50. Another quantity, called the variance, is used to help indicate the *spread*, or the expected spread, of the data. The variance of a collection of data is

$$\sigma^2(V) = \frac{1}{N} \sum_i N_i (V_i - \langle V \rangle)^2$$

The predicted variance is given by

$$\sigma^2(V) = \sum_i p_i (V_i - \langle V \rangle)^2$$

To understand the usefulness of knowledge about the variance, notice that the variance of the closely packed data above is 0.00028, whereas for the much spread-out data the variance is 0.060.

The standard deviation, which is simply the square root of the variance, is a more descriptive term than the variance because it has the same dimensions as the average. In the example, the standard deviation of the closely packed data is 0.017. This indicates that most of the data lie between 0.50 ± 0.017. The standard deviation of the spread data is 0.244. These parameters become increasingly meaningful when the size of the data increases.

13.10 ESTIMATING PROBABILITIES FROM AVERAGE AND STANDARD DEVIATION DATA

Though it is often necessary to use the N_i data or p_i estimates to compute or predict averages or standard deviations, it is often necessary or important to be able to do the reverse. That is, suppose the average and standard deviation are known, then what are the probabilities of each of the ith states?

Consider a specific example. Suppose it is learned that there are four salary brackets in an engineering design department: \$7,000, \$9,000, \$11,000, and \$15,000. Furthermore, it is learned that the average salary is \$9,000. What is the best estimate of the percentage of men in each bracket?

Obviously, in problems of this kind, there are an infinite number of answers which satisfy the given average value. Of these infinite number, which one is best?

For the answer to this question, one must turn to the science of information theory. For more complete derivations, see the Suggested Reading at the end of the chapter. Information theory shows that the least biased estimate of the probabilities is given by maximizing the function

$$S = -k \sum_i p_i \ln p_i$$

subject to the given constraints, in this case $\sum_i p_i = 1$ and $\sum p_i V_i =$ \$9,000 $= \bar{V}$. The function S is called the entropy and represents the average uncertainty about the system.

To maximize S, use is conveniently made of lagrangian multipliers. Differentiate S and set the result equal to zero for an optimum. Then differentiate the constraint equations and multiply each by a lagrangian multiplier.

$$dS = -k \sum_i (\ln \uparrow_i + 1)\, dp_i = 0$$

$$\lambda \sum_i dp_i = 0$$

$$\beta \sum_i V_i\, dp_i = 0$$

Adding these three expressions and noting that $dp_i = 0$:

$$\sum_i (\ln P_i + \lambda + \beta V_i)\, dp_i = 0$$

Since this must be satisfied always, the P_i's are given by

$$\sum_i \ln P_i + \lambda + \beta V_i = 0$$

$$P_i = \exp\,(-\lambda - \beta V_i)$$

The lagrangian multipliers λ and β are now found by using the two original constraint equations. Solution for them gives

$$\lambda = \ln \sum_i \exp(-\beta V_i)$$

$$V = \frac{\sum_i V_i \exp(-\beta V_i)}{\sum_i \exp(-\beta V_i)}$$

And hence $$p_i = \exp(-\lambda - \beta V_i) = \frac{\exp(-\beta V_i)}{\sum \exp(-\beta V_i)}$$

It can also be shown that

$$\frac{\partial \lambda}{\partial \beta} = -\langle V \rangle \qquad \frac{\partial^2 \lambda}{\partial \beta^2} = \sigma^2(V) = -\frac{\partial \langle V \rangle}{\partial \beta}$$

applied to the specific example defined above, is found from

$$\bar{V} = \frac{\sum_i V_i \exp(-\beta V_i)}{\sum_i \exp(-\beta V_i)} = 9 = \frac{7x^7 + 9x^9 + 11x^{11} + 15x^{15}}{x^7 + x^9 + x^{11} + x^{15}}$$

where $x = e^{-\beta}$ and the V's are written in thousands of dollars. The only meaningful root of this equation is found by dividing first by rearranging to give

$$6x^{15} + 2x^{11} - 2x^7 = 0$$

Since $x = 0$ is meaningless, division by x^7 leads to

$$6x^8 + 2x^4 - 2 = 0$$

Solution of the quadratic gives $y = 0.433$ or $x = 0.809$ if only real, positive roots are retained. The P's are then

$$P_i = \frac{x^{V_i}}{\sum_i x^{V_i}}$$

$$P_1 = \frac{0.809^7}{0.809^7 + 0.809^9 + 0.809^{11} + 0.809^{15}} = 0.432$$

$$P_2 = 0.302 \qquad P_3 = 0.186 \qquad P_4 = 0.085$$

The best estimate, therefore, that can be made based on the given information is that 43 percent make \$7,000, 30 percent make \$9,000, 19 percent make \$11,000, and 8 percent make \$15,000.

The procedure outlined above works equally well when more data, such as the standard deviation, are given, though the computation is more cumbersome. The procedure is the same, however: to find the least biased (i.e., entropy $S = -k \sum_i P_i \ln P$) subject to the given constraints.

13.11 SUMMARY AND PREVIEW

Not all decisions can be made with certain knowledge of the outcome. Often only partial knowledge of the actual situation is available and there is uncertainty involved. Probability is a means of quantifying uncertainty, and as such it is a very useful technique for the scientific decision maker. This chapter has been a relatively nonmathematical introduction to the concept of probability, its basic rules, and some of its applications. The student or engineer, of course, must recognize those situations where probability analysis will be appropriate and useful.

Probability is a mental concept. A statistic, on the other hand, is a datum. Statistics, then, is the analysis of data, whereas probability is a mathematical theory. But probability and statistics are sister sciences, for statistics uses probability models in its analyses. Probabilities need not, of course, be discrete ones, as they have been in this chapter. It is common also to think of a continuous probability distribution. It is in this continuously distributed form that probability theory is most useful to statistical analysis.

The next chapter deals with both probability and statistics. The objective is to find ways of rationally answering such questions as this: A sample of 100 parts out of a total lot of 10,000 was tested and found to have an average life of 100 hr with a standard deviation of 10 hr. How certain can one be from these data that 90 percent of the total have a life of at least 95 hr? Areas intimately related to design, such as quality control, reliability, etc., depend greatly upon statistical analysis and prediction. It is also increasingly common to find statistical tolerances (specification of an average and a standard deviation) replacing the conventional limit tolerances (specification of a maximum and a minimum).

SUGGESTED READING

Dixon, John R.: "A Programmed Introduction to Probability," John Wiley & Sons, Inc., New York, 1964. A teach-yourself book, written by the author of this design book, which will strengthen and extend the probability presented in this chapter. Many problems are worked. Strongly recommended, of course!

Feller, William: "Introduction to Probability Theory and Its Applications," John Wiley & Sons, Inc., New York, 1950. An outstanding mathematical treatment of probability.

Tribus, Myron: "Thermostatics and Thermodynamics," D. Van Nostrand Company, Inc., Princeton, N.J., 1961. Though this is a book on thermodynamics, it has a good introduction to probability and information theory for engineers.

PROBLEMS

13.1 Z = There are 15 balls in an urn: six red, five green, and four blue. X = Balls drawn out are replaced before next draw. Y = Balls drawn out are not replaced. Evaluate the following probabilities.

a $p(R_1 \mid Z)$
b $p(R_4 \mid ZX)$
c $p(R_4 \mid ZY)$
d $p(R_2 \mid R_1 ZX)$
e $p(R_2 \mid R_1 ZY)$
f $p(r_2 \mid R_1 ZY)$
g $p(R_1 R_2 \mid ZX)$
h $p(R_1 R_2 \mid ZY)$
i $p(r_1 r_2 \mid ZY)$
j $p[(R_1 + R_2) \mid XZ]$
k $p[(R_1 + R_2) \mid XY]$
l $p(R_1 \mid R_2 XY)$
m $p(R_2 \mid r_1 XY)$
n $p(R_2 \mid g_1 XY)$

13.2 Another symbolic language, used expecially in set theory, is sometimes used in place of the one presented in this chapter. In this other notation

$\bar{A}$ = Denial of A.
$A \cup B$ = Either A or B or both (A union B).
$A \cap B$ Both A and B (A intersection B).

Rewrite the following expressions in this language:

$$A(B + C) =$$
$$A + a =$$
$$Bb =$$
$$(A + b)(DE) =$$
$$aB(A + B + E) =$$

13.3 Rewrite all the expressions in this chapter, using the set theory notation defined in Prob. 13.2.

13.4 Experience establishes that the probability of a hole in one for a certain golfer on a certain hole is 1/10,000. What is the probability he will make a hole in one in two tries? three tries?

13.5 A missile probability of success is 0.6. How many of these missiles must be fired to have a 0.95 probability of success of hitting the target?

13.6 Forty percent of the delicate instruments shipped by a company are being rejected for quality reasons by the customers. In order to try to stop the shipment of inferior items, a quality control procedure is established. To test the quality control procedure, a test is run on all items shipped and the results recorded. Sixty-five percent of all intruments passed the inspection process. Sixty percent, as usual, were accepted by the customers. Of those that were accepted, 85 percent have passed the inspection test. Evaluate the inspection procedure by computing the probability that an instrument that passes the inspection will be accepted by the customers.

13.7 In a study of congenital malformations in a deprived community, records gave the following information: two percent of the births had deformed arms and deformed legs. Thirty percent had neither deformed arms nor deformed legs. Of those who had deformed arms, three percent had deformed legs. Find (*a*) the percentage that had deformed arms, (*b*) the percentage that had deformed legs, and (*c*) the percentage of those with deformed legs that had deformed arms.

13.8 In a certain manufacturing operation, it is known that when the abolotox exceeds 10 at the same time the boktometer exceeds 20, the product will fail to meet the design specifications. During one month the following data were taken: (*a*) on 40 percent of the production runs, the abolotox exceeded 10; (*b*) on 30 percent of the runs in which the abolotox was less than 10, the boktometer exceeded 20 or the abolotox was less than 10, or both. What percentage of subspecification product would you expect? On what percentage of the runs does the boktometer exceed 20?

14 PROBABILITY AND STATISTICS

14.1 INTRODUCTION

It is common practice to approximate actual statistical distributions by probability models and then to draw conclusions from the analysis of the model. For this reason, this chapter begins with the presentation of some common probability distributions: the binomial and multinomial distributions, and the so-called standard normal distribution. Once these are developed, most of the remainder of the chapter deals with their use in solving statistical problems, which may be of use to designers.

14.2 THE BINOMIAL AND MULTINOMIAL DISTRIBUTIONS

Suppose there are only two possible outcomes, a or b, on each of a series of trials. The outcomes might be, say, heads and tails, or success and failure. Let n be the total number of trials, n_a be the number of outcomes of type a, and n_b be the probabilities of outcomes a and b respectively on any one trial. Under these conditions, it is clear that

$$n_a + n_b = n \qquad \text{and} \qquad p_a + p_b = 1$$

It can be shown that the probability of obtaining n_a outcomes and n_b outcomes is

$$p(n_a,n_b \mid n_x) = \frac{n!}{n_a!n_b!}(p_a)^{na}(p_b)^{nb}$$

Sometimes this equation, called the binomial formula, is written in slightly different form. Suppose one outcome, n_a, is considered to be "success"; n_b is thus "failure." Let r be the number of successes, replacing n_a. Also let p be the probability of success on any trial. Then the binomial formula becomes

$$p(r \mid nx) = \frac{n!}{r!(n-r)!}p^r(1-p)^{n-r}$$

As an example of the use of these equations, consider the following problem. An honest coin is tossed six times. What is the probability of getting four heads? In this case, $n = 6$, $n_a = 4$, $n_b = 2$, and $p_a = p_b = \frac{1}{2}$. Thus

$$p(n_a{=}4,\ n_b{=}2 \mid 6x) = \frac{6!}{4!2!}\left(\frac{1}{2}\right)^4\left(\frac{1}{2}\right)^2 = \frac{15}{64}$$

Another example: Suppose that there are five balls in an urn, three of which are red and two of which are blue. Balls drawn out are replaced before the next draw. What is the probability of drawing one red and three blue balls in four tries? Here $n = 4$, $n_a = 1$, $n_b = 3$, $p_a = \frac{3}{5}$, and $p_b = \frac{2}{5}$. Hence

$$p(n_a{=}1,\ n_b{=}3 \mid 4x) = \frac{4!}{3!1!}\left(\frac{3}{5}\right)^1\left(\frac{2}{5}\right)^3 = \frac{96}{625}$$

Another kind of question which can be answered using the binomial formula is this: What is the probability of getting *at least* x successes? In the preceding example, for instance, what is the probability of getting at least one red ball in four tries? The answer, of course, is the sum of the probabilities of getting one, two, three, and four. The probability of getting one is 96/625, as has been seen. The other probabilities are

$$p(n_a{=}2,\ n_b{=}2 \mid 4x) = \frac{4!}{2!2!}\left(\frac{3}{5}\right)^2\left(\frac{2}{5}\right)^2 = \frac{216}{625}$$

$$p(n_a{=}3,\ n_b{=}1 \mid 4x) = \frac{4!}{1!3!}\left(\frac{3}{5}\right)^3\left(\frac{2}{5}\right)^1 = \frac{216}{625}$$

$$p(n_a{=}4,\ n_b{=}0 \mid 4x) = \frac{4!}{4!0!}\left(\frac{3}{5}\right)^4\left(\frac{2}{5}\right)^0 = \frac{81}{625}$$

(Note that $0! = 1$ by convention.) The probability of at least one red is therefore $(96 + 216 + 216 + 81)/625 = 609/625$. Alternatively, one might have calculated the probability of no successes, which in this case is

$$p(n_a{=}0,\ n_b{=}4 \mid 4x) = \frac{4!}{0!4!}\left(\frac{3}{5}\right)^0\left(\frac{2}{5}\right)^4 = \frac{16}{625}$$

Then the probability of one or more (i.e., at least one) is $1 - 16/625 = 609/625$.

The *bi*nomial formula is so named because it deals with situations where there are only *two* outcomes (e.g., heads or tails). More than two outcomes are handled by an extension of the formula, called the *multi*nomial distribution. Now the outcomes are of type a, b, c, . . ., k, and

$$n_a + n_b + n_c + \cdots + n_k = n$$

$$p_a + p_b + p_c + \cdots + p_k = 1$$

as before. The probability of a particular combination of outcomes (n_a of type a, etc.) is then

$$p(n_a,n_b,n_c,\cdots,n_k \mid nx) = \frac{n!}{n_a!n_b!n_c!\cdots n_k!}\,(p_a)^{na}(p_b)^{nb}(p_c)^{nc}\cdots(p_k)^{nk}$$

As an example, consider a roulette wheel which is one-half red, one-third white, and one-sixth blue. In six spins, what is the probability of getting two reds, two whites, and two blues? Here $n = 6$, $n_r = 2$, $n_w = 2$, $n_b = 2$, $p_r = \frac{1}{2}$, $p_w = \frac{1}{3}$, and $p_b = \frac{1}{6}$. Thus

$$p(n_r{=}2,\ n_w{=}2,\ n_b{=}2 \mid 6x) = \frac{6!}{2!2!2!}\left(\frac{1}{2}\right)^2\left(\frac{1}{3}\right)^2\left(\frac{1}{6}\right)^2 = \frac{5}{72}$$

Returning to the binomial distribution again, it is useful to look at graphs of the probability of r successes plotted vs. the number of successes. A number of these are shown in Fig. 14.1 for the case where $p = \frac{1}{2}$. Notice that the length of the vertical lines denotes the probability of success at that point. Thus the probability of three successes in four trials is $\frac{1}{4}$. Also the sums of the probabilities can be added to give the probability of achieving a number of successes within specified limits. Thus the probability of

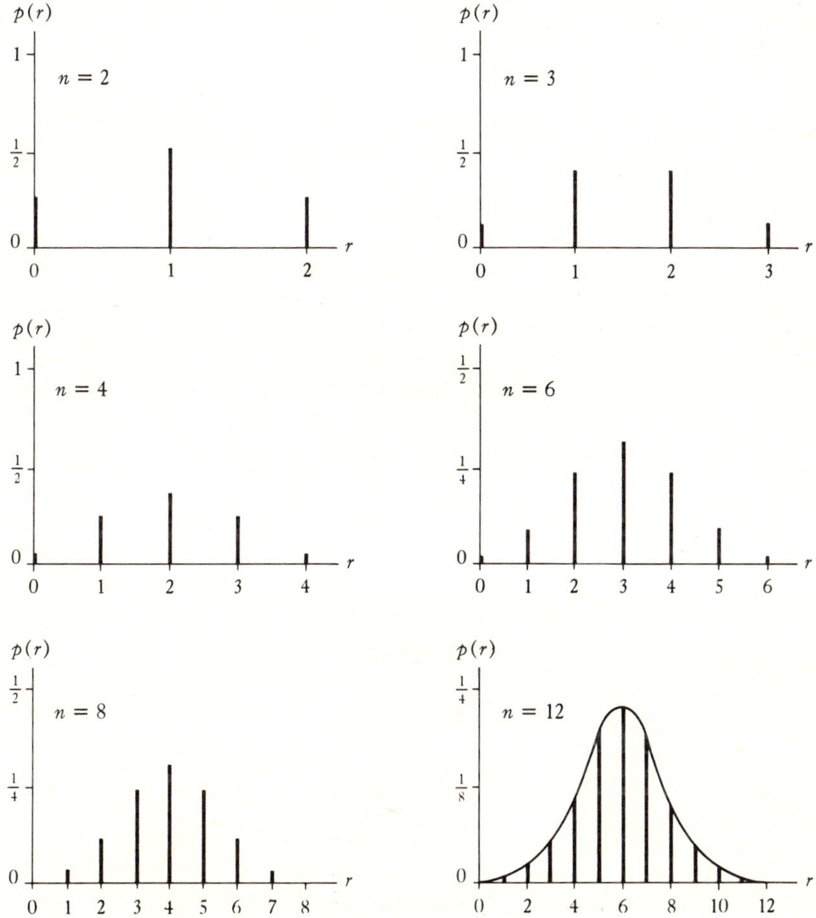

Fig. 14.1 Probabilities of r successes in n trials when probability of success in any given trial is one-half.

getting either two or three successes in four tries is $\frac{1}{4} + \frac{3}{8} = \frac{5}{8}$. With data of this type, statistics books often provide tables for various values of p and for n's up to 20 or 25. Above these values for n, the standard normal distribution is used.

14.3 THE NORMAL-DISTRIBUTION CURVE

As shown in Fig. 14.1, when the number of binomial trials n becomes large, the binomial distribution looks increasingly like a smooth, continuous function. In fact, when $n \to \infty$, it can be shown that the curve produced is given by

$$y = \frac{1}{\sqrt{2\pi}} \exp \frac{-x^2}{2}$$

This curve is shown plotted in Fig. 14.2. The total area under this curve is unity. That is,

$$\int_{-\infty}^{+\infty} y\, dx = \int_{-\infty}^{+\infty} \frac{1}{\sqrt{2\pi}} \exp \left(\frac{-x^2}{2}\right) dx = 1$$

Areas under the curve therefore represent probabilities. The variable x

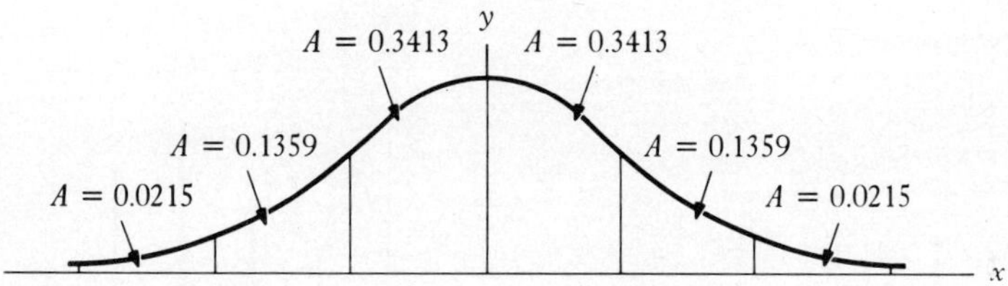

Fig. 14.2 Normal-distribution curve.

given by the curve is called a *normal random variable*. The probability, for example, of finding a normal random variable x within, say, $x = 0$ and $x = 1$ is

$$\int_0^1 \frac{1}{\sqrt{2\pi}} \exp \left(\frac{-x^2}{2}\right) dx = 0.34$$

Other areas—i.e., probabilities—are shown in Fig. 14.2. The standard deviation of a normal variable x can be shown to be 1. The mean value μ of the distribution is 0 (the area to the right of $x = 0$ equals the area to the left of $x = 0$). By analogy to the discrete case where the variance is given by

$$\sigma^2(v) = \sum_i p_i(v_i - \bar{v})^2$$

there is

$$\sigma^2(x) = \int_{-\infty}^{\infty} \frac{1}{\sqrt{2\pi}} x^2 \exp \frac{-x^2}{2} = 1$$

Thus $\sigma(x) = 1$. Figure 14.2 shows, therefore, that for a standard normal variable x, 68 percent of the outcomes fall between $\pm$ one standard deviation of the mean, 96 percent fall within two standard deviations, etc. Statistics books also contain tables listing areas under various portions of this curve.

When dealing with a random curve derived from actual data, of course, the data are not normalized with a mean of zero and a standard deviation of 1. It can easily be converted to this form, however. Suppose, for example, data on the heights of men are collected, and the mean is found to be 70 in. with a standard deviation of 2 in. If it is assumed that the data can be approximated by a random curve, then it can be converted to normal form by defining a new variable Z such that

$$Z = \frac{x - 70}{2}$$

where x is the height of a man in inches. Or, in general notation, if μ is the mean of the nonnormalized random curve for x, and σ is its standard deviation, then the new variable Z which will give a standard normal curve is

$$Z = \frac{x - \mu}{\sigma}$$

Note that the percentage of men with heights between 68 in. and 72 in. (one standard deviation) will be 68 percent. At $x = 68$, $Z = -1$ and at $x = 72$ in., $Z = +1$, showing that the Z curve is indeed normalized as desired.

In the preceding discussion, it was assumed that the distribution of heights was random about the average. This was, of course, an approximation. The assumed random curve is actually a *probability model* of the actual statistical data. It is a very good model, however, for many kinds of circumstances. Whenever a reasonably random sample is under discussion and there are no effects which would tend to bias or truncate the data, something close to a standard random distribution can be expected, and the use of the corresponding probability model will give valuable results.

As an example of the use of the above subject matter, suppose it is learned that the average length of a throw of a baseball by major league outfielders in a contest was 250 ft and that the standard deviation *of the data* was 25 ft. Thus

$$\bar{x} = 250 \qquad \sigma = 25$$

Now if it is assumed that the data can be approximated with a standard curve, then

$$\mu = 250 \qquad \sigma = 25$$

To determine anything about the distribution is now easy. For instance, the percentage of throws exceeding 300 ft (two standard deviations above the mean) is 2.28 percent. Though this can be seen by inspection to be two standard deviations above the mean, it is found formally by converting to the *normal* standard variable:

$$Z = \frac{x - \mu}{\sigma} = \frac{300 - 250}{25} = 2$$

The tables now give the data that the percentage of area in excess of two standard deviations is 2.28 percent. See Fig. 14.2.

Finally, suppose an automatic machine turning out thousands of parts is found to be producing small shafts with an average diameter of 0.2521 in. and a standard deviation of 0.0136 in. What percentage of these parts would be expected to be less than 0.2450 in.? Again, assuming that the

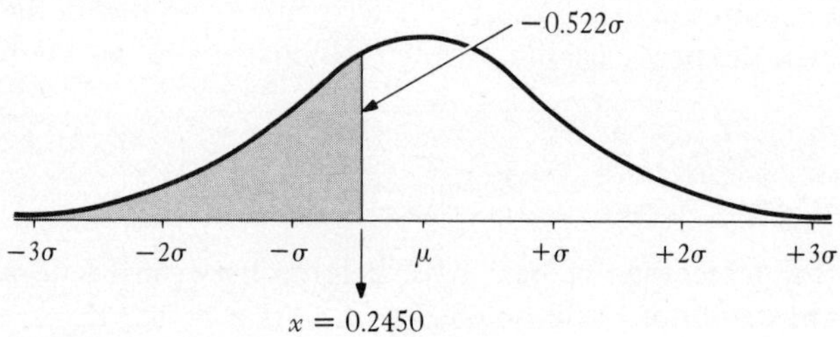

Fig. 14.3

actual distribution corresponds with the normal one, this question can be quickly answered. Converting to a standard distribution,

$$Z = \frac{0.2450 - 0.2521}{0.0136} = \frac{-0.0071}{0.0136} = -0.522$$

The situation is therefore as shown in Fig. 14.3. The shaded area can be found in the tables to be 0.7008. That is, 70 percent of the parts will be less than 0.2450.

14.4 SAMPLING

In the preceding discussion it was assumed that the average values and standard deviations reported applied to the entire population. Usually, of course, this is far too expensive and sometimes impossible because certain tests are destructive of the part. Suppose, then, that out of a large population, a sample of size n is selected at random and the average and standard deviation of the sample determined. (These will be called the sample average and sample standard deviation.) The problem then is to determine the characteristics of these data (the sampling problem) and what these data imply about the original population (the *estimation problem*).

It is naturally possible to select several random samples of size n and to determine the sample average for each of these samples. This set of sample averages now itself has a distribution. That is, one can now meaningfully discuss the *distribution of sample averages.* There is now a key mathematical theorem (not to be proved here) that says: *If a population has a mean μ and a standard deviation σ, then the sample mean $\bar{x}$, based on a random sample of size n, will have approximately a normal distribution with the same mean μ and with a standard deviation $\sigma_{\bar{x}}$ equal to $\sigma/\sqrt{n}$.* Notice that the population need *not* be a standard random distribution, and yet the distribution of the sample averages will be approximately standard. The approximation becomes increasingly accurate as n increases.

In practice, it is not necessary to take a large number of samples. It is enough to know that *if* a large number were taken, the averages would yield a standard distribution with a mean equal to the population mean and a standard deviation equal to the population standard deviation divided by the square root of the sample size $(\sigma/\sqrt{n})$. A single sample can then be used to estimate information about the population—in particular, the mean and standard deviation of the population.

It is easiest to show how this is done by first considering an example. Suppose a manufacturer of light bulbs assumes that the life of his bulbs is normally distributed and from experience knows that, though the average life varies, the standard deviation remains reasonably constant at 96 hr. From a large population he now selects a sample of 36 and finds that the average life is 1,000 hr. What does this imply about the average life of the bulbs in the population?

Clearly, he must conclude that the average life of a bulb in the population is also 1,000 hr. But there is some uncertainty about this conclusion, and now he can quantify his uncertainty. If he had taken a large number of samples of size 36, the sample means would be distributed about the population with a standard deviation of $\sigma/\sqrt{n} = 96/6 = 16$. The standard deviation of the distribution of sample means is 16. Therefore, there is a 68.26 percent chance that the given sample mean is within ± 16 of the actual population mean. Or, this can now be stated within any degree of accuracy desired. For example, there is a 95.44 percent chance that the given sample mean is within ± 32 of the population mean. The tables can be used to compute other values.

The problem can also be worked in reverse order. The necessary sample size can be computed to give a particular accuracy. Suppose the engineer involved in this problem wishes to be 90 percent certain that the sample mean life is within ± 10 hr of the mean population life. The tables show that in a normal random distribution, 90 percent of the area is enclosed between values of $\sigma = \pm 1.65$. Thus 1.65 standard deviations should not exceed 10 hr. Or

$$1.65\sigma = 10$$

Now since $\sigma_{\bar{x}} = \sigma/\sqrt{n}$, there results

$$1.65 \frac{96}{\sqrt{n}} = 10 \qquad n = 251$$

A sample size of 251 is required to give the desired degree of certainty. (When the cost of this is evaluated, the engineer may decide he can live with greater uncertainty! Note how it is necessary to pay for reduced uncertainty—that is, for information.)

It is common, in problems of estimation using a sample from a population, to talk of "confidence limits." The term is used as follows: If one is, say, 95 percent certain that a property of the population lies between x_1 and x_2, then the interval x_1 to x_2 is called the 95 percent confidence interval. Suppose in the preceding example, it is desired to find the 95 percent confidence interval for the population mean based on the original sample size of 36. Remember that the sample mean was 1,000 hr and the standard deviation of sample means was 16. Since 95 percent of the normal curve area is contained within ± two standard deviations (actually it is 95.44 percent), it is clear that with a sample mean of 1,000 there is a 95 percent chance that the population mean lies between $(1{,}000 - 32)$ and $(1{,}000 + 32)$; that is, between 968 and 1,032. This is thus the 95 percent confidence interval.

All of the above discussion has been based on an assumed knowledge of the population standard deviation. If this is not known, more samples must be taken until a sample mean distribution is established. Then of course the population standard deviation is $\sigma_{\bar{x}}\sqrt{n}$.

The above discussion applies strictly only when large samples (n greater than about 25) are employed and when the sample size is small compared with the population. Students are referred to books on statistics for methods of handling small samples and/or small populations.

14.5 QUICK AND DIRTY ESTIMATION OF A MEDIAN[1]

There is a simple scheme for estimating the median (not necessarily the mean) of a population using a small sample. Suppose from a large population, five samples are drawn. What is the probability that none of the five samples is below the median?

[1] I am indebted to my colleague Frank A. McClintock, professor of mechanical engineering, M.I.T., for calling this neat technique to my attention.

The answer is $(\frac{1}{2})^5 = 1/32$. What is the probability that none of the samples is above the median? Again, the answer is 1/32. Now, what is the probability that *at least* one sample is above the median *and* at least one is below the median?

The answer is $1 - \frac{2}{32} = 15/16$. That is, if a sample of size five is drawn from a population, one can be 15/16 confident that the median of the population lies between the extremes of the sample. 15/16 is about 94 percent. Roughly, then, such a test gives the 95 percent confidence interval neatly and quickly. For example, if five shafts from a large shipment are tested and found to have the following diameters: 0.512, 0.508, 0.507, 0.503, and 0.499, it can be said with only 5 percent chance (actually $6\frac{2}{3}$ percent!) of error that the median diameter in the shipment lies between 0.499 and 0.512.

14.6 HYPOTHESIS TESTING

How should one reach conclusions based on statistical information? Suppose, for example, that shafts for a certain delicate assembly are being supplied by two firms. It is known from past experience that shafts from supplier A have a mean diameter of 0.502 in. with a standard deviation of 0.015 in., whereas those from supplier B have a mean diameter of 0.490 in. also with a standard deviation of 0.015 in. This diameter variation is acceptable normally, so that shafts from the two suppliers are used interchangably. Bent shafts present a problem at the testing stage, however, and a collection of bent shafts has been made. The design department does not wish to have to redesign to account for the possibility of bent shafts, so it is important to identify the supplier who is providing them. Those that are bent represent a small percentage, so it is inconvenient to test incoming shafts. Besides, the delicacy of the equipment is such that special and expensive instrumentation would be required to detect the unstraight shafts prior to assembly.

In the past, supplier B has been generally more reliable than supplier A. It is also known that supplier B has a more efficient quality-control department.

For the purposes of this example, let it be assumed that there are two possible hypotheses:

H_A = The population from which the bent shafts come has a mean diameter of 0.502 in.

H_B = The population from which the bent shafts come has a mean diameter of 0.490 in.

There are also two errors which can be made.

Error 1 = H_A is accepted, and H_B is true.
Error 2 = H_B is accepted, and H_A is true.

Now if a number of samples ($n = 9$) of bent shafts were available, and *if* they come from supplier A, the distribution of sample means would be as shown by curve H_A in Fig. 14.4. Note that the standard deviation of the distribution of sample means is $\sigma/\sqrt{n} = 0.005$. The curve labeled H_B in the figure is the distribution of sample means from the B population.

The problem now is to decide how to decide. That is, a strategy—a plan for deciding—is needed. Suppose the nine bent shafts are measured and the average computed, what can be concluded? Suppose their average is 0.485 or 0.494 or 0.498 or 0.504, what is to be the conclusion?

It is to be emphasized that there is no theoretically objective way to select a strategy. Strategies—that is, plans for deciding—are mostly subjective and are chosen because they seem to make sense or because they conform to convention. Sometimes, though, they are a matter of personal

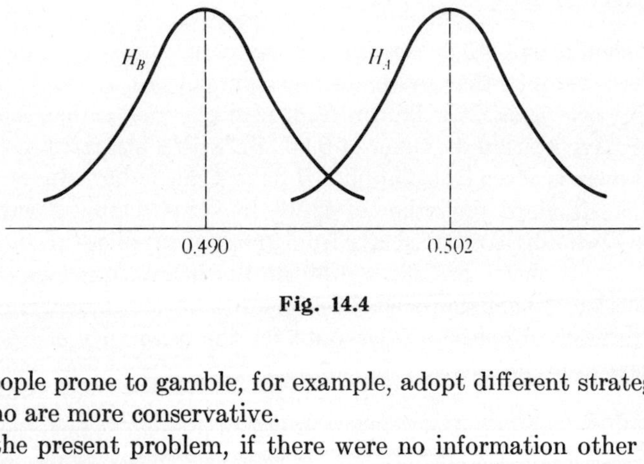

Fig. 14.4

style! People prone to gamble, for example, adopt different strategies from those who are more conservative.

In the present problem, if there were no information other than the distributions themselves, it would be most reasonable to select a diameter of 0.496 as the decision point. If the sample average falls above 0.496, H_A would be accepted. Possible errors can be computed here. 0.496 is 1.2 standard deviations away from each mean. The area under a normal curve from $\sigma = 1.2$ to ∞ is 0.1151. Thus the possibility that H_B will be accepted when H_A is true (error 2) is 11.51 percent. What is the possibility of error 1?

In the present case more information is available, and it is of such a nature that the engineer is rather reluctant to reject hypothesis H_A. He will therefore probably wish to move the decision point to the left. Notice what happens when he does so by referring to Fig. 14.5. The area representing the possible errors changes as the point is moved. At 0.496 the possible errors were equal at 11.51 percent. As the decision point moves left, the area under H_B from the decision point to ∞ (A_B) increases while

A_A decreases. The engineer must adjust the decision point mentally until he is satisfied with the resulting areas or possible errors.

For example, if a decision point of 0.495 is chosen, A_B will be 15.87 percent and A_A will be 8.08 percent. At this point, then, the probability of making error 1 is 15.87 percent, whereas the probability of error 2 is only 8.08 percent. If the decision point is moved to 0.494, the resulting probabilities are 21.19 percent for error 1 and 5.48 percent for error 2.

What point should be chosen as the decision point? There is no way to select this strategy scientifically! Once it is chosen, however, the analysis can give the likelihood that future predictions will be wrong.

Suppose the engineer feels most comfortable with a decision point of 0.494 and upon measuring the sample finds an average of 0.495. His conclusion is that the bent shafts come from supplier A. That is, he accepts hypothesis H_A and knows that there is 21.19 percent chance that, based

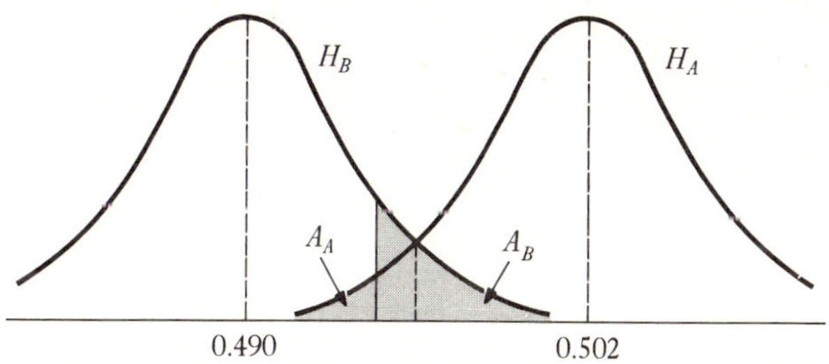

Fig. 14.5

on diameter measurement alone, H_B is true. On the other hand, if the sample mean turns out to be 0.493, he will accept H_B, knowing that there is 5.48 percent chance that H_A is true on this basis.

The above discussion has, for purposes of illustration, assumed that the two hypotheses must be compared—that one of the two hypotheses must be accepted as true and the other false. Often this situation arises when one of the hypotheses is the so-called null hypothesis. The null hypothesis always says that nothing significant has happened or changed. As an example, suppose the automatic production of parts normally yields an average length of 1.000 in. with a standard deviation of 0.010 in. Say a sample of 25 parts turns out to have an average length of 1.006 in. The null hypothesis says that this sample has actually come from the acceptable normal production output. In this case, the probability that the null hypothesis is not true can be found as follows: The distribution of sample means with $n = 25$ from the normal output will have a mean of 1.000 in. and a standard deviation $\sigma = 0.010/\sqrt{25} = 0.002$ in. Thus a sample mean with $n = 25$ of 1.006 in. is three standard deviations away from the mean. The probability that this could take place is 0.003 or 0.3 percent. In this case, the null hypothesis could be rejected with great confidence. The machine should definitely be checked.

It is common practice to reject the null hypothesis when the probability of error is less than either 5 percent or 1 percent, but the selection of these values is simply an arbitrary custom. Often the term "5 percent (or 1 percent) level of confidence" is used. This simply means that one can be 95 percent (or 99 percent) certain that the conclusion given is correct. These are the usual levels demanded of a result that is to be called "statistically significant." Thus in testing, the null hypothesis cannot be rejected unless the results show that there is only a 5 percent (or 1 percent) chance that it is, in fact, not correct.

14.7 CORRELATION

A great deal of the value of probability and statistics is that it quantifies ignorance or subjective feelings. For example, study the two plots in Fig. 14.6. (The data for the plots are shown in Fig. 14.7.) It appears to the eye

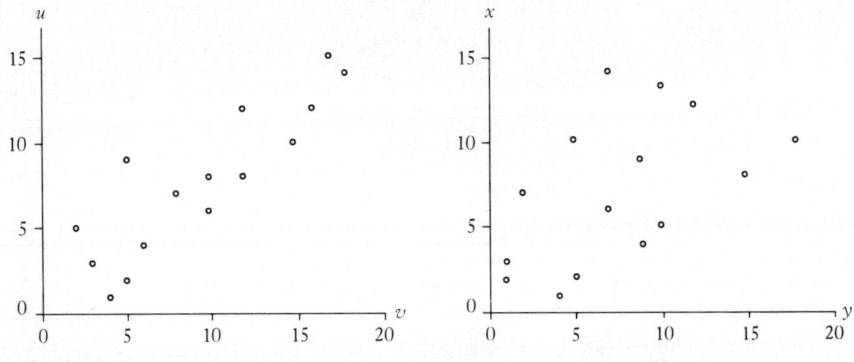

Fig. 14.6

that u and v correlate better with each other than do x and y. It is often helpful, however, to have a quantitative measure of what is meant by "correlation." For this purpose, theoretical statistics has developed a parameter called the *correlation coefficient*. The correlation coefficient r has been devised so that when the relation between the two variables under consideration falls on a straight line, the value of r is 1. When the data are completely random, the correlation coefficient r will be 0. Values between

0 and 1 then give a quantitative indication of the strength of the correlation between the variables.

To obtain the correlation coefficient between two variables, the mean and standard deviation of each must be computed. Let these be denoted

u	v	x	y
1	4	1	4
5	2	2	1
7	8	2	6
2	5	4	9
9	5	3	1
8	10	5	10
8	12	7	2
10	15	8	15
12	12	10	18
12	16	10	5
15	17	12	12
4	6	14	7
3	3	9	9
6	10	13	10
14	18	6	7
116	143	106	116
$\bar{u} = 7.73$	$\bar{v} = 9.53$	$\bar{x} = 7.07$	$\bar{y} = 7.73$

Fig. 14.7

by $\bar{u}$, $\bar{v}$, σ_u, and σ_v for the u and v data given in Fig. 14.6. The individual data points are u_i and v_i. Remember that

$$\bar{u} = \frac{1}{n} \sum_i^n u_i \qquad \sigma_u = \left[\sum \frac{1}{n-1} (u_i - \bar{u})^2 \right]^{1/2}$$

where n is the size of the sample. Now the correlation coefficient is given by

$$r = \frac{\sum_{i=1}^{n} (u_i - \bar{u})(v_i - \bar{v})}{(n-1)\sigma_u \sigma_v}$$

Working out the numbers in this case gives $r = 0.84$, which indicates a rather strong correlation, as expected. By contrast the correlation between x and y is (see Prob. 14.4), indicating the weaker correlation as expected.

Quantifying the concept of correlation is very useful in studying test data, reliability information, production records, etc. Engineers must be very careful, however, to distinguish between the concept of correlation and the ideas of cause and effect. The fact that two variables correlate does not necessarily mean that they are cause or effect to each other. Smoking correlates with lung cancer in people, for example, but it also correlates with several other trends in our society. Before researchers could conclude that smoking can cause cancer, considerable more research, both statistical and medical, was necessary.

14.8 LINEAR REGRESSION

Regression is the term used by statisticians to denote the process of finding the best single curve that can be drawn through a set of data points. *Linear* regression denotes the process of finding the best *straight line* through the points. Thus if the correlation coefficient r is reasonably large, it makes sense to apply linear-regression techniques to the problem

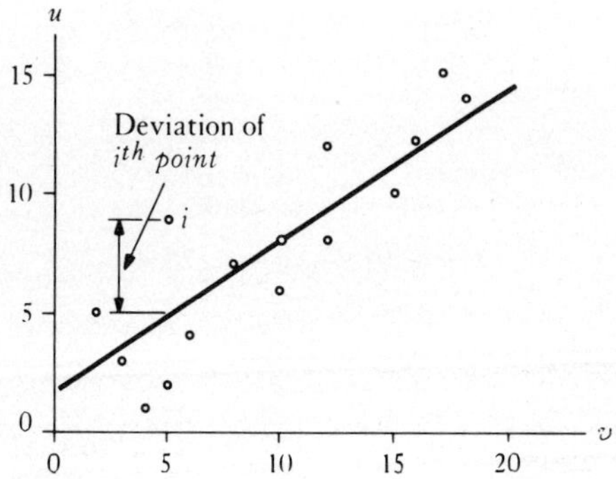

Fig. 14.8

to find the line that best fits the data. The reason for wanting a single-line representation is usually that the equation of the line can then be used as an empirical relation in analytical work.

There are a number of methods for determining the "best" straight line through the points. As was the case in the selection of a strategy, there is no known way to decide objectively which method is *really* best. The most common method, however, and one which gives quite satisfying results, is the least-squares method. The concept of the method is this: The line is chosen which gives a least or minimum value to the sum of the squares of the deviations of the data points from the line. Suppose, for example, that a trial line is drawn through the data, as shown in Fig. 14.8 (These are the same data as shown in Fig. 14.6*a*.) This line can be represented by an equation of the form

$$u = av + b$$

It turns out, however, to be more convenient to write it in the form

$$u = a(v - \bar{v}) + b$$

where $\bar{v}$ is the average value of v in the data. For a given point, say the ith point (see the figure), the deviation from the line is

$$u_i - [a(v_i - \bar{v}) + b]$$

The least-squares method calls for finding the line that results in a minimum when all the deviations are computed, squared, and added.

It can be shown that the equation which satisfies this requirement is given by

$$b = \bar{u}$$

$$a = \frac{\sum_{i=1}^{n} (u_i - \bar{u})v_i}{\sum_{i=1}^{n} (u_i - \bar{u})^2}$$

For the data in Fig. 14.7, these formulas give the results:

$$b = 7.73 \qquad a = 0.626$$

This is the line shown in the figure.

14.9 THE CHI-SQUARE DISTRIBUTION

The binomial distribution can deal only with two possible outcomes. In many applications, however, there are more than two categories of outcomes. In a quality-control situation, for example, there may be grades of acceptable (excellent, good, fair) instead of merely acceptable or unacceptable. There is a theoretical distribution, called the chi-square distribution, analogous to the binomial distribution, which treats this situation. And just as the discrete binomial distribution has the continuous normal distribution as a limiting case, so too the chi-square distribution has a continuous counterpart.

As an introduction to the chi-square distribution, consider the following example. Grades received by students in an engineering design course when averaged over the past 15 years are A = 10 percent, B = 30 percent, C = 40 percent, D = 10 percent, and F = 10 percent. The normal class size is 50, so that the distribution of grades in any one year if it conforms to the average is: A = 5, B = 15, C = 20, D = 5, and F = 5. Suppose now that a new teacher is assigned to the course and gives the following distribution: A = 2, B = 7, C = 18, D = 15, and F = 8. The question is whether this distribution is due to reasonably probable variations in class ability from year to year, or whether the new teacher has graded improperly or taught poorly. (Note how a similar problem could have been conceived using the quality of machined parts as the variable. The question then would be to determine whether the production was out of adjustment or whether this sample was just an unusual one.)

There are five possible categories or outcomes in this example. If the outcomes are denoted by the subscript i, then i varies from 1 to k, where $k = 5$. The expected outcomes e_i are 5, 15, 20, 5, and 5, whereas the actual outcomes a_i are 2, 7, 18, 15, and 8. A measure of the extent to which the

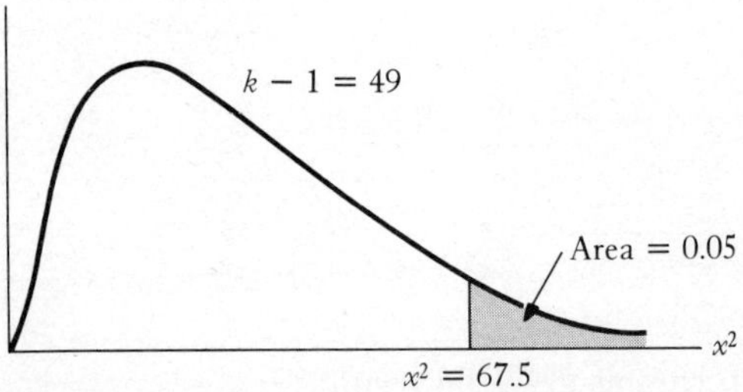

Fig. 14.9 Chi-square distribution curve.

expected and actual outcomes agree is obtained by computing χ^2 ("chi-squared") as follows:

$$\chi^2 = \sum_{i=1}^{k} \frac{(a_i - e_i)^2}{e_i}$$

For this case

$$\chi^2 = \frac{(2-5)^2}{5} + \frac{(7-15)^2}{15} + \frac{(18-20)^2}{20} + \frac{(15-5)^2}{5} + \frac{(8-5)^2}{5} = 28.1.$$

Note that for perfect agreement, χ^2 equals zero, and that the larger the disagreement, the larger will be the value of χ^2.

Using theoretical considerations beyond the scope of this text, it can be shown that if a large number of samples of size 50 were chosen from a population which has the expected outcomes given, and χ^2 computed for each, the distribution of χ^2 values would be as shown in Fig. 14.9. A study of the figure shows that the null hypothesis must be accepted in the grading example—that is, the class could *easily* vary as much as the actual grades indicate—because the probability that $\chi^2 = 67.5$ or greater is 5 percent, as shown. If χ^2 were greater than 67.5, it could be concluded that there was less than 5 percent chance that the class actually varied, as the new teacher's grades indicate.

Tables of the chi-square distribution are nearly as common as those for the normal distribution. The chi-square curve, however, is a function of the number of possible outcomes. That is, there is a different distribution for each different value of k. (Often $k - 1$ is the parameter used in tables.)

14.10 THE POISSON DISTRIBUTION

The binomial distribution presented in Sec. 14.2 is suitable for approximating cases where the size of the sample n and the probability p of the event in question are any given values. If n is large, the normal distribution discussed in Sec. 14.3 is appropriate and very useful, since computations with the binomial for large n are lengthy. If n is large *and* if p is small, another approximation to the binomial can be developed. It is known as the Poisson distribution and is useful in engineering reliability and in certain areas of operations research.

The derivation of the Poisson distribution will not be shown here (see the Suggested Reading). The result of taking the limit of the binomial as $n \to \infty$ and $p \to 0$ is

$$p(x_i) = \sum_i \frac{m^{x_i}}{x_i!} e^{-m}$$

where m is the average or expected number of occurrences. For example, suppose the average number of failures per hour in a reliability test were 3. Then the Poisson distribution can be used to predict the probability of $x_i (x_i = 0,\ 1,\ 2 \cdots)$ failures per hour. The probability of 0 failures is $3^0/0!\,e^{-3} = 0.050$. The probability of 1 failure is $3^1/1!\,e^{-3} = 0.150$, and so on, as shown below:

x_i	$p(x_i)$	
0	$\frac{3^0}{0!} e^{-3} = 0.050$	
1	$\frac{3^1}{1!} e^{-3} = 0.150$	
2	$\frac{3^2}{2!} e^{-3} = 0.225$	Poisson distribution $m = 3$
3	$\frac{3^3}{3!} e^{-3} = 0.225$	
4	$\frac{3^4}{4!} e^{-3} = 0.169$	
....		
	1.000	

The Poisson distribution will be used again in the next chapter, where reliability is discussed.

14.11 SUMMARY AND PREVIEW

This chapter, together with Chap. 13, has been an introduction to probability and statistics. Both probability and statistics are powerful tools in scientific decision making. These chapters are by no means a substitute for a course or for study of one or more complete books on probability and statistics. See the Suggested Reading.

In the next chapter the subject of reliability is taken up. Reliability theory is a direct application of probability and statistics to engineering design, and it is becoming increasingly important in both space and earth-bound design problems.

SUGGESTED READING

Brownlee, Kenneth A.: "Statistical Theory and Methodology in Science and Engineering," John Wiley & Sons, Inc., New York, 1960. A good book—most mathematical of those listed here.

Fraser, D. A. S.: "Statistics: An Introduction," John Wiley & Sons, Inc., New York, 1958. Slightly more mathematical than Hoel's book (below).

Gotkin, Lassar G., and Leo S. Goldstein: "Descriptive Statistics," John Wiley & Sons, Inc., New York, 1964. Two volumes of a popular programmed text in statistics. Good way to teach yourself statistics.

Hoel, Paul H.: "Elementary Statistics," John Wiley & Sons, Inc., New York, 1960. A very good introduction to classical statistics. Little mathematics needed.

PROBLEMS

14.1 *a* What is the probability of getting three heads in four tosses of a coin? Do this two ways—once using binomial theorem and once using basic reasoning.

b What is the probability of getting at least three heads in four tosses of a coin?

14.2 A shipment, though randomly mixed, is known to contain 20 percent of type A parts, 30 percent of type B parts, and 50 percent of type C parts. How many must you select to be 95 percent certain of getting a type A part?

14.3 In the above shipment, what is the probability of getting one of each type of part in a sample of three?

14.4 A population of shafts is known to have a mean diameter of 0.750 in. and a standard deviation of 0.050 in. What percent of the shafts

a exceeds 0.800 in.?

b is less than 0.650 in.?

c is less than 0.600 in.?

d is less than 0.620 in.? (You may need to consult an outside reference for this one.)

14.5 For the shaft population in Prob. 14.4, what is the 95 percent confidence interval?

14.6 For the shaft population in Prob. 14.4, plot the distribution of sample means to be expected if a sample size of 16 is used.

14.7 From an unknown population a sample of 10 shafts is selected and is found to have a mean of 1.920 in. and a standard deviation of 0.020 in. What can you conclude about the population, and with what confidence?

14.8 In 120 rolls of a six-sided die, the following results were obtained: one came up 10 times, two came up 15 times, three came up 20 times, four come up 25 times, five came up 30 times, and six came up 40 times. Do you think this is an "honest" die? Support your conclusion quantitatively.

14.9 It is sometimes said that experimental data in scientific research ought to be good enough that statistical analysis using correlation and regression should be unnecessary. Comment on this. Also, do you see any difference between the need for statistical analysis in scientific research and the need in production or design work?

15 RELIABILITY

15.1 INTRODUCTION

Astronaut Walter Shirra was thinking subjectively about the importance of reliability when he said that it gives one an uneasy feeling to have one's life depend upon the proper functioning of thousands of parts, each produced by the lowest bidder! As technical systems become more complex, consideration of operating reliability becomes more crucial. The need for concern with design reliability in a manned satellite is apparent, but it is no less important in the mass production of items for the civilian economy. Reliability is an important quality of all technical systems, subsystems, and components.

Reliability engineering became an identifiable field in the early 1950s. The success of the field has been due partly to its ability to make the concept of "reliable performance" a quantitative parameter. Various definitions have been proposed, and the following states the generally accepted idea: Reliability is the probability that a device will perform its specified function for a specified time under specified operating conditions. Note that reliability is a probability. It is also important to note that satisfactory performance, the time involved, and the operating conditions must be clearly and objectively stated. Otherwise, reliability has no useful, quantitative meaning.

Quality and reliability should be distinguished from one another. In nontechnical terms, reliability is one aspect of quality. For design-engineering purposes, quality is defined as the degree to which a device meets the various standards specified for it. Quality standards can include, in addition to reliability, geometric tolerances, capacity of one kind or another, strength, etc. It is, however, something of an oversimplification to say that reliability is merely one factor of quality, because all the other factors of quality also affect the reliability. Thus the two are related intimately, but remain unique items, each with its own measure.

The field of reliability engineering has many facets. Engineering design for reliability has its special interests and concerns. The measurement of reliability is another area. Quite a lot has been written about the management of and organization for the design process for highly reliable systems. The prediction of reliability, by means of mathematics, is also an important area of reliability engineering. Finally, many reliability engineers are concerned with the problems of "trading off" reliability with other factors, such as cost or inventory levels. In one chapter, there is no hope of covering all of these facets thoroughly. Whole books are being written on reliability. (See Suggested Reading.) It will be possible in this chapter, however, to introduce the subject rather completely and to cover its major principles and methods. Students interested in a more thorough treatment should first study probability and statistics and then read one of the advanced texts on the subject. Some engineering graduate schools are offering advanced courses in reliability.

15.2 A RELIABLE EXAMPLE

In order to illustrate the basic concepts of reliability, the following example is presented. Suppose that the reliability test data of 400 parts have given the results shown in Fig. 15.1. With these data, several important reliability parameters can be computed as follows.

The reliability R is clearly a function of the time the part is to operate. If the time for operating performance is 1 hr, then the reliability is $300/400 = 0.75$. On the other hand, if the 6-hr reliability is wanted, the result is $176/400 = 0.44$. A plot of reliability vs. time is shown in Fig. 15.2.

Time interval t	*Number of failures* f	*Cumulative failures*	*Number of survivors* s
0		0	400
1	100	100	300
2	46	146	254
3	23	169	231
4	20	189	211
5	18	207	193
6	17	224	176
7	15	239	161
8	14	254	146
9	13	267	133
10	12	279	121
11	11	290	110
12	10	300	100
13	9	309	91
14	7	316	84
15	12	328	72
16	16	344	56
17	20	364	36
18	18	382	18
19	10	392	8
20	6	398	2
21	2	400	0

Fig. 15.1

Another parameter that is useful and that is often used in reliability work is the failure rate F. The failure rate is the number of failures per unit time expressed as a percentage of the existing population. Since in this example there are only data by hours, the increment of time available for

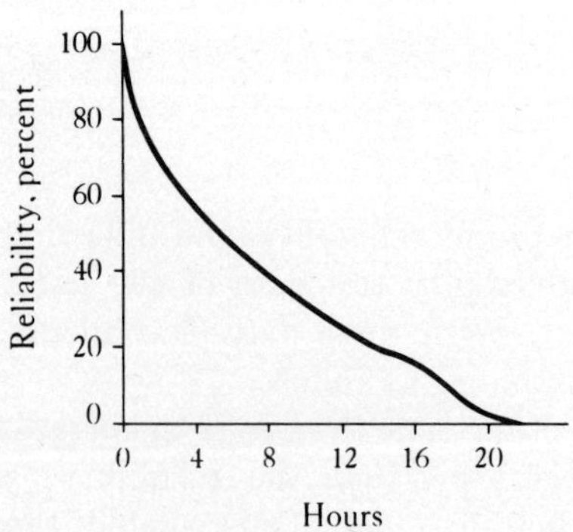

Fig. 15.2

use is 1 hr. More formally, however, failure rate is defined as

$$F = \frac{1}{N} \lim_{t \to 0} \frac{\Delta f}{\Delta t} = \frac{1}{N} \frac{df}{dt}$$

where f is the number of failures. The failure rate for the data of Fig. 15.1 is shown in Fig. 15.3. In obtaining data for Fig. 15.3, the average population during each time interval was used for N. Thus, during the interval 0 to 1 hr, there were 100 failures, and the average population was $(400 + 300)/2 = 350$. Hence the failure rate for that period is $100/350 = 0.286$.

The failure-rate curve shown in the figure is typical and shows three

Fig. 15.3

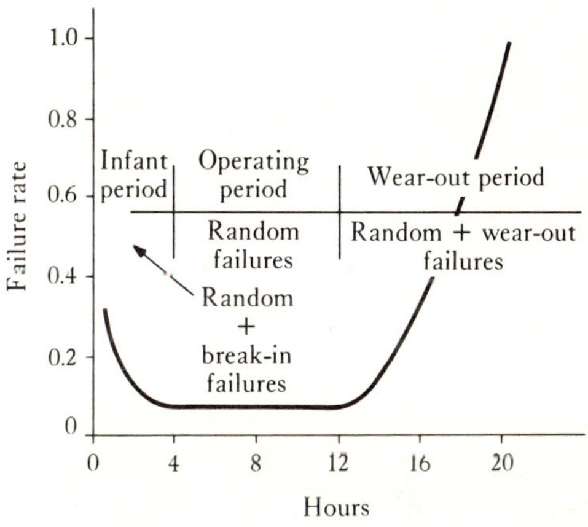

distinct regions. Early failures (sometimes called "infant mortality") are due not only to the inherent and inevitable random failures but also due to errors in manufacture, poor materials, etc. Careful manufacture, quality control, and an adequate breaking-in period should eliminate these errors from marketed products. Late failures are due to wear-out failures and indicate that the device is on its last legs. In between is the normal operating period for the system, and a constant failure rate during this period is usually encountered.

Another term frequently computed and reported in reliability work is the *mean time between failures*. Using M for this parameter, it is defined simply as the reciprocal of the failure rate:

$$M = \frac{1}{F}$$

During the operating period, then, for the example under consideration, the mean time between failures is approximately $M = 1/0.09 = 11$ hr.

15.3 THE EXPONENTIAL RELIABILITY CURVE

In Sec. 14.9, the Poisson distribution was introduced. This theoretical distribution has direct application to reliability analyses. Recall that the Poisson distribution is

$$1 = e^{-a} + ae^{-a} + \frac{a^2e^{-a}}{2!} + \frac{a^3e^{-a}}{3!} + \cdots + \frac{a^ne^{-a}}{n!}$$

where e^{-a} = probability of 0 events per interval or unit

ae^{-a} = probability of 1 event per interval or unit

$\frac{a^2e^{-a}}{2!}$ = probability of 2 events per interval or unit

etc.

and where a is the average number of events per interval or unit.

The applicability of this type of theoretical distribution to reliability is as follows. The event under consideration is a failure. If the failure rate is F and a time interval of t is considered, then the average number of failures in an interval t is $f = Ft$. Now the probabilities of 0, 1, 2, . . . failures in an interval t are given by the individual terms in the distribution. In the example described in the preceding section, if $t = 1$ hr is used and the failure rate F is taken as 0.09, then the probability of 0 failures in an hour is $e^{-1\times0.09} = 0.914$. The probability of 1 failure is $0.09\ e^{-1\times0.09} = 0.082$, and so on. Note that $t = 0$ must correspond to the beginning of the time period for which F was computed. Otherwise the fact that the total lot size is being depleted by failures is not accounted for. The failure rate F depends upon the initial lot size.

Use of this distribution leads to what is known as the exponential failure law. In the example, the probability of 0 failures during the first hour ($t = 1$) is equivalent to the reliability for a 1-hr life. Numerically this is $R_1 = e^{-0.09} = 0.914$. The 2-hr reliability is the probability of 0 failures for $t = 2$ or $R_2 = e^{-0.18}$. Thus the reliability can be written in general as

$$R_t = e^{-Ft}$$

This is known as the exponential failure law or exponential reliability law for obvious reasons. Note that it is based on the existence of a constant failure rate F. For the example, the theoretical exponential reliability based on $F = 0.09$ is plotted in Fig. 15.4 together with the actual curve from the data. The deviation is due to the inclusion of the infant mortality and wear-

Fig. 15.4

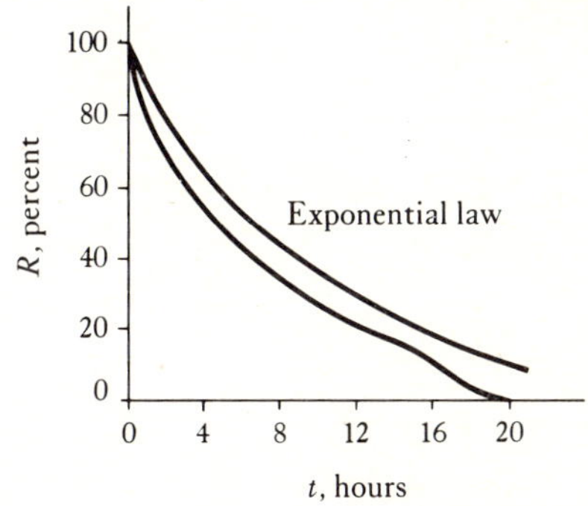

out failures in the actual data. If these are excluded by proper break in tests and replacements, the exponential curve will fit the data very well.

15.4 THE WEIBULL DISTRIBUTION

Another theoretical reliability curve that is widely used is the Weibull distribution. The Weibull distribution as used in reliability is given by

$$f(t) = 1 - \exp(-t^m/M)$$

where m is called a shape factor and M is the mean time between failures. The factors M and m are determined experimentally, using actual data. In using this theoretical distribution, $f(t)$ is interpreted as the probability of failure at time t. Thus the reliability at time t is given by $1 - f(t)$:

$$R_t = \exp(-t^m/M)$$

When sufficient data have been accumulated to determine m and M for the product or item in question, this equation can be used to predict reliability.

It should be noticed that when $m = 1$, the Weibull reliability curve is the same as the exponential curve. The Weibull curve thus is more general and contains the exponential law as a special case. As m increases to about three or more, the Weibull curve becomes a normal curve. Fig. 15.5 shows Weibull probability curves for a given value of M and for various values of m.

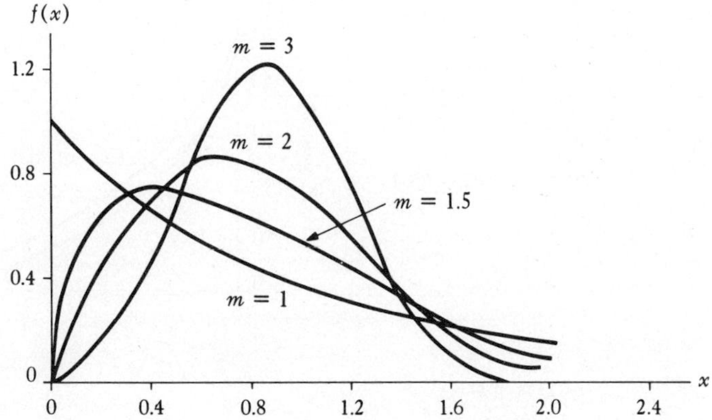

Fig. 15.5 Weibull distribution.

15.5 COMBINED FAILURE RATES AND REDUNDANCY

What happens to reliability when a system depends upon the successful operation of two parts having reliabilities R_1 and R_2 respectively? Clearly the reliability of the combined system is the product R_1R_2. When the number of parts involved is in the hundreds or thousands, even though each has a very high reliability, the resulting system reliability can become very low. For example, a 20-component system with each component having $R = 0.99$ has a reliability of only $0.99^{20} = 0.82$.

Some care must be used when multiplying component reliabilities in order to obtain a system reliability. Often the system may continue to function satisfactorily, though not perfectly, even though one or more of its components has fallen below its individual standard or specified performance level. Thus in some cases, the system reliability may be greater than that predicted simply by multiplication of component reliability. There are no general methods for handling these complex cases, however, and each one must be considered individually.

Reliability engineers have devised a scheme for improving the reliability of complex systems. The term used is *redundancy*. An easy way to illustrate the ideas of redundancy in design is with an example from electric circuits. Suppose that a system consists of two switches, both of which must function properly in order to close a circuit. If the reliability of each one closing properly is R, then the reliability of thc system comprising the two switches is R^2.

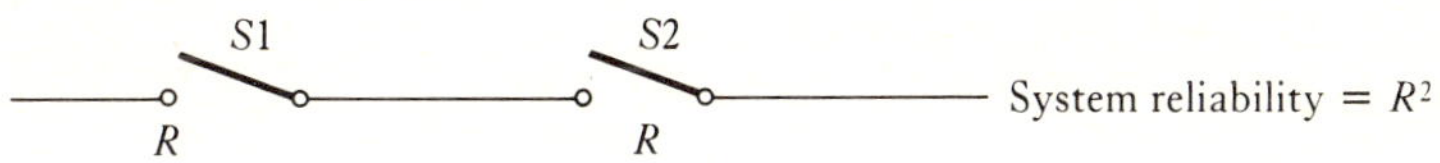

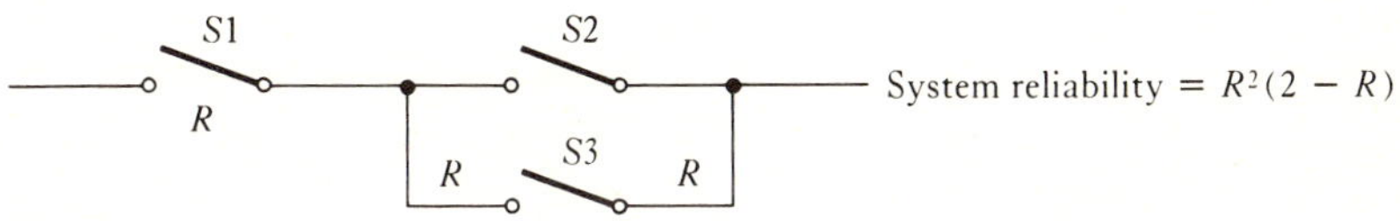

Suppose now that another switch $S3$ with reliability R is placed in parallel with $S2$ and controlled to close whenever $S2$ closes. Now the system will function properly when either $S2$ or $S3$ or both operate properly. The reliability of the combination of $S2$ and $S3$ (remember that reliability is a probability) is given by

$$\begin{aligned} R(S2 + S3) &= R(S2) + R(S3) - R(S2)\,R(S3) \\ &= 2R - R^2 \end{aligned}$$

The reliability of the whole system is thus

$$R[S1(S2 + S3)] = R(2R - R^2) = R^2(2 - R)$$

Since $R < 1$ always, it can be seen that the system reliability has been improved by the addition of the third switch. The added switch is redundant in the sense that it is not *necessary* for the operation of the system.

As can be seen above, the usefulness of logical symbolic notation in designing and analyzing system reliability is great. As a more complex example, consider the circuit shown below:

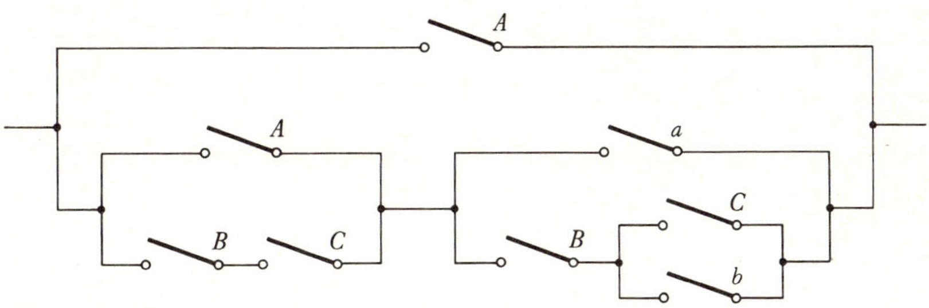

This circuit system will be closed and will conduct current when the following statement is true:

$$A + (A + BC)\{a + [B(C + b)]\}$$

This is a function of the form $A + f(A,B,C)$. It can be shown that logical functions of the form $f(X_1,X_2,\ldots)$ can be expanded and written as

$$f(X_1,X_2,\ldots) = X_1 f(I,X_2,X_3,\ldots) + X_1 f(\phi,X_2,X_3,\ldots)$$

where I and ϕ are called the universal set and the null set respectively. In circuit terms I is always closed and ϕ always open. In $f(I,X_2,\ldots)$, I has replaced X_1 and ϕ has replaced x_1. In $f(\phi,X_2,\ldots)$, ϕ has replaced X_1 and I has replaced x_1. Applying this expansion to the above equation gives

$$\begin{aligned}
&(A + BC)[a + B(C + b)] \\
&\qquad = A(I + BC)[\phi + B(C + B)] + a(\phi + BC)[I + B(C + b)] \\
&\qquad = A(I + BC)[B(C + b)] + a(BC)[I + B(C + b)] \\
&\qquad = AB(C + b) + ABCB(C + b) + aBC + aBCB(C + b) \\
&\qquad = ABC + ABC + aBC + aBC \\
&\qquad = ABC + aBC \\
&\qquad = (A + a)BC \\
&\qquad = BC
\end{aligned}$$

Combining this result with the A in the original expression gives an equivalent circuit expression:

$$A + BC$$

Diagrammatically, the equivalent circuit is

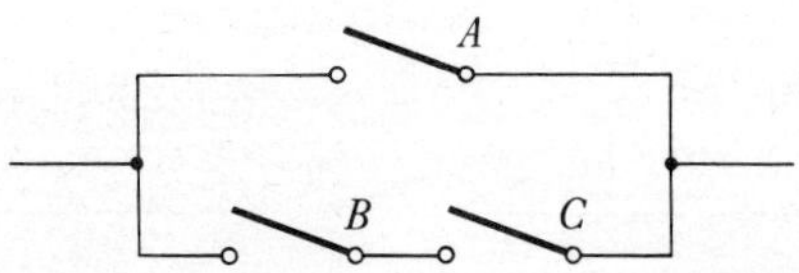

Students may be able to see by inspection of the original circuit that this result is correct, but in more complex cases, the method presented makes simplification possible. The above circuit is logically equivalent to the original one, but it is considerably more reliable because it is simpler. The expansion of $f(X_1,X_2,\ldots)$ is a technique from boolean algebra. Other boolean-algebra techniques are also useful in analyzing systems for unnecessary complication or redundancy.

Adding redundancy to a system adds cost and often affects other design constraints, such as weight or size. In any given design, the value of the added reliability must be balanced against the value of the added cost, weight, etc. Studies of this nature are often called *trade-off studies*. Actually, it is mostly simply a matter of minimizing total cost. Designing redundancy into a component or system is one of a number of ways of increasing reliability, all of which add to the initial cost but reduce the operating cost. Thus qualitatively the situation is as shown in Fig. 15.6. R_1 represents the design reliability for minimum total cost. Of course, considerations other

Fig. 15.6

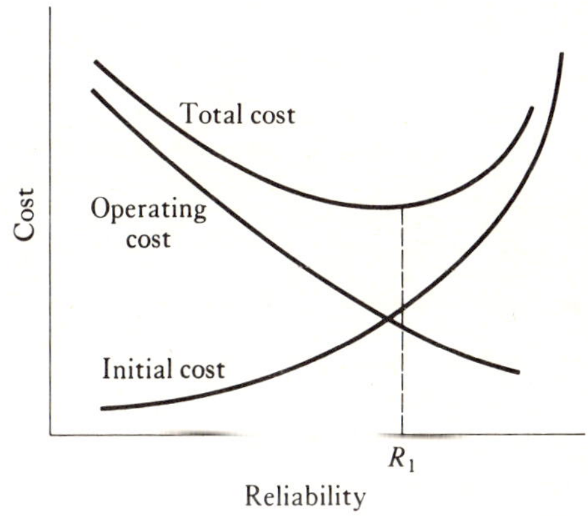

than economic may be significant in certain cases; as in a manned satellite, for example.

15.6 MECHANICAL RELIABILITY AND FACTORS OF SAFETY

It should be clear (but often is not) to engineering students that material properties for the "same" materials vary from piece to piece. Assuming a normal distribution curve, the situation is something like that shown in Fig. 15.7*a*. Note that if the design load requires strength S_1, then the shaded area represents the probability of failure.

The actual loading situation may itself be a distributed probability function, resulting in the situation represented by Fig. 15.7*b*. The shaded area is the probability of failure. The probability-of-failure curve is found by multiplying the probability that a given load will be exceeded (i.e., the area to the right of the point under the load-distribution curve) by the probability that the strength will be below that load (i.e., the area to the left of the point under the strength-distribution curve).

This discussion, using probabilities, is considerably different from one which takes only a factor of safety into account. Referring to Fig. 15.7*b*, if the means are used for design purposes, then the factor of safety is

S_1/L_1. The probability of failure can be quite different in cases with the same factor of safety depending on the shape (or standard deviation) of the material-strength curves. A significant new feature of design with reliability in mind is the need to be concerned with the *distribution* of material properties, load factors, etc.

Of course, the mean load and mean strength need not necessarily be chosen for design criteria. The design load can be taken as, say, one or

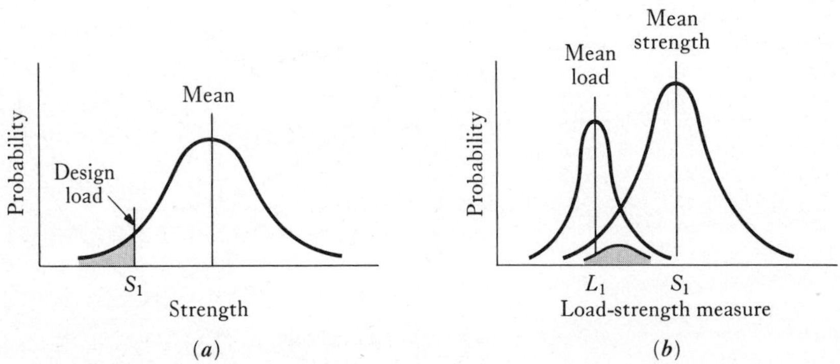

Fig. 15.7

even two standard deviations above the mean and the design strength as one or two standard deviations below the mean strength. Such decisions are arbitrary or based on experience or legal or professional codes. All of this discussion is based on properties that do not change with time.

15.7 REMARKS ON DESIGNING FOR RELIABILITY

A reliability constraint is one of those that is now common in engineering design situations, together with such familiar constraints as cost, weight, capacity, and strength. In a sense then, it is not possible to discuss designing for reliability *alone* any more than it is to discuss designing solely for any of the other constraints. The engineering designer–decision maker must find ways of meeting all the minimum specifications and must compromise (trade off) when optimizing beyond the minimums. The good designer will be aware of reliability as he designs, therefore, just as he is aware of cost, ease of production, etc. In general, there are four methods of building in greater reliability in the design of components and systems. One of these, *redundancy*, has already been mentioned. Another is obviously *simplicity*—the fewer parts the better for reliability. A third method is to *use standardized parts of known and proved reliability*. Finally, the designer can prescribe *derating;* that is, operation of parts and systems at only partial full-load capacity, a practice which usually adds to the life of equipment.

15.8 SUMMARY AND PREVIEW

This chapter has been a very brief introduction to the theory of engineering reliability. Reliability is an example of the use of probability and statistics in the making of engineering decisions. The reliability of engineering systems has become increasingly important as the systems have become more complex and as people—individually and collectively—have become more dependent upon them.

In the next chapter a survey of several fields which roughly can be called decision science is presented. None of these is as important as reliability to the engineering designer at the present time. However, they are growing and promising fields and deserve to be watched by those engineers who wish to keep abreast of all the possible tools of engineering decision making.

SUGGESTED READING

Bazovsky, Igor: "Reliability Theory and Practice," Prentice-Hall, Inc., Englewood Cliffs, N.J., 1961. A good standard book on reliability.

Calabro, S. R.: "Reliability Principles and Practices," McGraw-Hill Book Company, New York, 1962. This is an excellent treatment of the subject of reliability and well within the mathematical grasp of undergraduate engineers.

Haviland, R. P.: "Engineering Reliability and Long Life Design," D. Van Nostrand Company, Inc., Princeton, N.J., 1964. This book concentrates more on the specific reliability of mechanical designs than on the general subject of reliability. It is well written, and designers of mechanical equipment should consider this "must" reading.

Mechanical Reliability Concepts, United Engineering Center, 395 East 47th St., New York City, N.Y., ASME Design Engineering Conference, New York, May, 1965. Can be ordered directly from ASME. A good introduction to concepts of reliability, mostly nonmathematical. Highly recommended as supplementary reading.

Roberts, N. H.: "Mathematical Methods in Reliability Engineering," McGraw-Hill Book Company, New York, 1964. This is another excellent reliability book, and though the mathematics is more advanced and sophisticated than in the Calabro book, it is not beyond advanced undergraduate or first-year graduate students.

PROBLEMS

15.1 Discuss, in terms of failure-rate curves, an auto guarantee (*a*) of six months or 6,000 miles, (*b*) of five years or 50,000 miles. What is the purpose of each guarantee?

15.2 Draw a failure-rate curve for the "one-horse shay."

15.3 Refer to Fig. 15.6. Sometimes instead of designing for R_1, a design reliability is selected that gives the optimum ratio of reliability to total cost. Draw a line from the origin of the figure that is tangent to the total-cost curve. Show that the point of tangency defines the desired reliability.

15.4 List as many ways as possible that a mechanical structure can fail.

15.5 Discuss the relationship between factors of safety and design for reliability in mechanical systems.

15.6 The failure rate for an electronic missile component is 0.025 failures per hour. Compute the mean time between failures. What is the probability of a component failing in 1 hr? in 3 hr?

15.7 During the reliability test of 600 parts, the number of failures between 3 and 4 hr. was 34. The number of survivors at 3 hr was 356 and at 4 hr was 322. What is the failure rate?

15.8 A system consists of two components connected in series. Each component has 50 parts. In one component the failure rate of each part is 0.0001 per hour. In the other, it is 0.0002 per hour. Find the reliability of the system for a 10-hr run.

15.9 For electron tubes, it has been found that a good value for m in the Weibull distribution equation is 1.7 (Prof. J. H. Kao, Cornell University). Suppose that for a certain type of tube, M is 17.8 kilohours. What is the expected value of the reliability at the end of 1,000 hr? 600 hr?

15.10 $M = 100$ kilohours for a certain set of tubes ($m = 1.7$). If the reliability is to be 95 percent what tube life can be claimed?

15.11 Consider the systems below. Write logical expressions for the function of each. (*a*) If each component (X,Y,Z) has a reliability R, compute the reliability of each system. (*b*) If AX has a reliability e^{-Xt}, Y has a reliability e^{-Yt}, etc., find the reliability of each system.

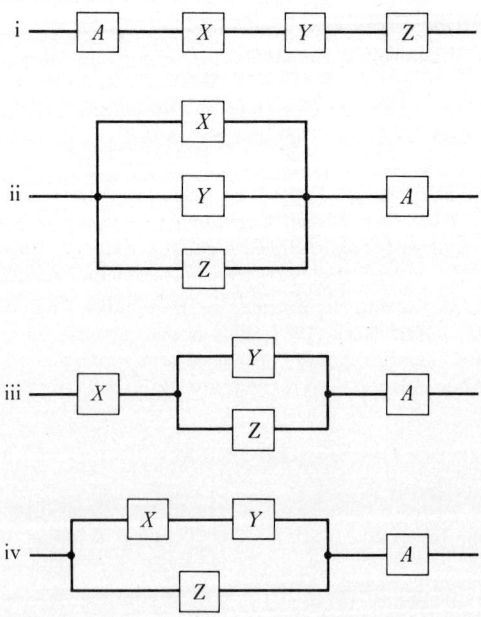

16
SOME MANAGEMENT METHODS RELATED TO ENGINEERING DESIGN

16.1 INTRODUCTION

In the preceding chapter, probability and statistics were applied to the subject of engineering reliability. In this chapter, the general subject of management of the design process is considered from a decision-theory point of view. Some of the topics covered, such as the PERT system, will have evident design-management applications. Other topics, such as decision-utility theory, will have less obvious current usefulness. For the most part the topics introduced in this chapter are recently developed and growing sciences.

Their application to engineering design and to the management of engineering design has not yet been made. Some appear to be more applicable to business management than to design management. They are, however, general decision-making theories. They are included here because their increased use and future applicability to the design process seem assured.

16.2 DECISION THEORY

As was pointed out in Chap. 11, an essential feature of a decision situation is the existence of alternative courses of action. Decision theory requires that some additional concepts be defined so that the elements of the decision situation can be made more explicit and operational.

Alternative Courses of Action Let the possible actions be denoted by a_1, a_2, a_3, . . . , a_n. In a homely example, a man walking to work has a choice of taking his raincoat (a_1) or not taking it (a_2). Or an engineer designing a satellite has a choice of using, say, either aluminum, steel, or titanium for the outer shell material. These could be denoted by a_1, a_2, and a_3, respectively.

States of Nature Usually there are conditions which are out of the control of the decision maker. These are referred to as states of nature, and the symbol θ will be used here. It may rain (θ_1) or not rain (θ_2).

In the case of the satellite shell, the states of nature may have to do with the size and frequency of the meteorites which are encountered. This kind of "nature" is continuous rather than discrete. (Actually, the rain example is continuous, too. It cannot rain, sprinkle lightly, rain a little, etc.) Normally such situations are handled by dividing the continuous distribution of nature into arbitrary but convenient discrete states. Suppose, therefore, that the meteorite showers to be encountered by the satellite are described by θ_1 = very light, θ_2 = light, θ_3 = average, θ_4 = heavy, and θ_5 = very heavy. In practice there would have to be operational definitions of these states, but these qualitative definitions will serve the purposes of this example.

Losses or Gains—Utility A decision situation resulted from there being an objective or goal which could be accomplished by varying degrees of certainty and desirability by the various alternative courses of action. A particular course of action may have a probability of success somewhat less than unity, of course. It is possible now to develop what is called a loss table (or matrix). For each given combination of action and state of nature, there is an associated cost. Now the "cost" may be money, but it also may include many other items such as time, prestige, loss of irreplaceable resources, etc. All of these factors have to be included in a single measure called *utility*. Utility can be thought of as a kind of generalized loss or gain where all values have been converted to a single scale. The subject of utility is taken up in Sec. 16.3. For now, it can be assumed that utility scales have been developed for this problem situation and that they have led to the

		States of nature	
	Courses of action	*Rain* θ_1	*No rain* θ_2
Take coat	a_1	1	2
Do not take coat	a_2	7	0

Fig. 16.1 Loss table for raincoat decision.

table of losses shown in Fig. 16.1 for the raincoat example and Fig. 16.2 for the satellite design.

If the utilities lead naturally to gains rather than losses, these can easily be converted to losses by simply subtracting the value of the largest gain from each of the quantities. All entries will then be either negative or

	States of nature				
Courses of action	θ_1	θ_2	θ_3	θ_4	θ_5
a_1	2	7	12	12	12
a_2	5	5	10	15	15
a_3	6	6	6	11	16

Fig. 16.2 Loss table for satellite-material decision.

zero. Since a negative gain is a loss, the minus signs can be dropped, and the result is a loss table.

Probabilities on States of Nature As can be seen, the situation can be greatly complicated if probabilities are attached to the various losses; e.g., if θ_2 occurs, a_1 may lead to loss L_1 with probability p_1 or to L_2 with probability p_2. However, it serves the purposes of this introductory example to consider only probabilities attached to the various states of nature. Suppose, for example, that the weather bureau establishes the probability of rain $p(\theta_1) = 0.30$ and of no rain $p(\theta_2) = 0.70$, and that previous satellite experience has shown:

State of Nature	*Probability*
θ_1	0.10
θ_2	0.20
θ_3	0.40
θ_4	0.20
θ_5	0.10

Expected Loss It is now possible to compute an expected value of loss for each of the possible actions. Remember that the expected value of a quantity x is given by

$$\text{Expected value of } x = \sum_i p_i x_i$$

so that the expected losses in the simple rain case are

Action a_1: Expected loss $= 0.30(1) + 0.70(2) = 1.7$
Action a_2: Expected loss $= 0.30(7) + 0.70(0) = 2.1$

Thus to minimize expected loss, this man should follow action a_1 and take his raincoat.

In the example of the satellite material, the expected losses are

a_1: Expected loss $= 0.1(2) + 0.2(7) + 0.4(12) + 0.2(12) + 0.1(12) = 10.0$
a_2: Expected loss $= 0.1(5) + 0.2(5) + 0.4(10) + 0.2(15) + 0.1(15) = 10.0$
a_3: Expected loss $= 0.1(6) + 0.2(6) + 0.4(6) + 0.2(11) + 0.1(16) = 8.0$

Thus the decision-theory model indicates that the engineer should adopt action a_3. (The numbers in this example are contrived and do not necessarily correspond to a real situation in detail.)

Minimax Though expected loss seems to be a common sense criterion to use in arriving at decisions, other criteria are possible and are used occasionally. One of the common alternatives is called *minimax*. This method calls for *minimizing the maximum loss*. It is thus a conservative criterion which says, "Pick the strategy which takes the least risk of a large loss." In the rain example, the maximum losses are 7 for a_2 and 2 for a_1. Thus minimax calls for action a_1. Note that it would call for a_1 regardless of the weatherman's prediction, an illustration of its conservativeness.

In the satellite-material example, the maximum losses are 12 for a_1, 15 for a_2, and 16 for a_3. The minimum of these maximum losses is 12 for action a_1. Minimax says select action a_1, whereas expected loss says that a_3

is best. There is no objective way of choosing between the criteria. Earlier it was noted that there was room for personal style in decision making. This is true even in decision theory because the choice of criteria—expected loss, minimax, or some other—is subjective.

Strategies Depending on the circumstances of the particular problem at hand, there are many variations on the above procedure for arriving at a decision. Often, for example, the situation is such that a *strategy* can be employed. Usually this is the case when there is a series of similar decisions to be made. A strategy is a *plan for deciding.*

The little example of the raincoat decision is good for illustrating strategies. Each day the man hears the weather forecast and each day he

	States of nature	
	Rain	*No rain*
Forecast observation	θ_1	θ_2
$p(\theta_1) = 0.25$	0.10	0.20
$p(\theta_1) = 0.50$	0.20	0.15
$p(\theta_1) = 0.75$	0.25	0.10
$p(\theta_1) = 1.00$	0.40	0.05

Fig. 16.3 Observation probabilities for raincoat decision.

must decide whether or not to take his raincoat. With his years of experience he has learned that the weather forecaster makes one of the following predictions: $p(\theta_1)$ rain $= 0$, 0.25, 0.50, 0.75, or 1.00. He also has learned that when the forecaster says $p(\theta_1) = 0$, it actually rains anyway 5 percent of the time. His other information of this sort is shown in Fig. 16.3. This information should be thought of as experimental data which have been obtained to help with the decision. With this data at his disposal, strategies can now be formed. Figure 16.4 shows a number of possible ones. Strategy S_1, for example, says that when the forecaster predicts $p(\theta_1) = 0.0$, 0.25, or 0.50, take action a_2; otherwise take action a_1. Strategy S_2 says to take the raincoat (a_1) all the time. And so on. Only four have been included here, but many others could be added.

Once these strategies have been enumerated, *action probabilities* can be determined. Notice in Fig. 16.3 that where it rains (θ_1), the frequencies of the various forecasts are given. Thus on a θ_1 day, there is an 0.05 chance that the forecast has been $p(\theta_1) = 0.0$, and so on. Now S_1 calls action a_2 for $(\theta_1) = 0.0$, 0.25, and 0.50, which according to Fig. 16.3 will occur 35 percent of the time. Hence for strategy S_1 and for state of nature θ_1, the

	Observations				
Strategy	$p(\theta_1) = 0.0$	$p(\theta_1) = 0.25$	$p(\theta_1) = 0.50$	$p(\theta_1) = 0.75$	$p(\theta_1) = 1.00$
S_1	a_2	a_2	a_2	a_1	a_1
S_2	a_1	a_1	a_1	a_1	a_1
S_3	a_2	a_2	a_2	a_2	a_2
S_4	a_2	a_2	a_2	a_2	a_2

Fig. 16.4 Strategies for raincoat decision.

probability of action a_2 is 0.35 and of action a_1 is 0.65. Other action probabilities are found in the same way and are shown in Fig. 16.5.

With the action probabilities determined, the *average loss* for each

	Actions							
	S_1		S_2		S_3		S_4	
Observations	a_1	a_2	a_1	a_2	a_1	a_2	a_1	a_2
θ_1	0.65	0.35	1.00	0	0.40	0.60	0	1.00
θ_2	0.85	0.15	1.00	0	0.05	0.95	0	1.00

Fig. 16.5 Action probabilities for raincoat decision.

state of nature for the various strategies can be determined. For strategy S_1, when θ_1 is the case, the probability of action a_1 is 0.65 and the loss for (a_1, θ_1) is, from Fig. 16.1, 1. The probability of action a_2 is 0.35, and the corresponding loss is 2. Thus the average loss is

$$0.65(1) + 0.35(2) = 3.10$$

For state of nature θ_2, the average loss is

$$0.85(2) + 0.15(0) = 1.70$$

3.10 and 1.70 represent the average losses for their respective states of nature θ_1 and θ_2 for strategy S_1. Average losses for the other strategies are shown in Fig. 16.6.

Notice that Fig. 16.6 is very much like Fig. 16.1 in content. In fact, S_2 and S_4 correspond to Fig. 16.1 exactly. S_1 and S_3 are different, however, and the figure gives losses obtained on the average with this strategy over a period of time.

Minimax can be used directly in conjunction with Fig. 16.6 to select a best strategy. It calls for selecting the strategy that has the minimum maximum loss, in this case S_2, where the maximum loss is 2.0.

Before expected loss can be used here, data on the average number of θ_1 and θ_2 days are needed. Suppose it is established by a study of the records

	S_1	S_2	S_3	S_4
θ_1	3.10	1.00	4.60	7.00
θ_2	1.70	2.00	0.10	0.00

Fig. 16.6 Average losses in raincoat decision.

that there is rain on 10 percent of the days in the locality under consideration. Then the expected loss can be computed for each strategy as follows:

S_1: Expected loss $= 0.10(3.10) + 0.90(1.70) = 1.84$
S_2: Expected loss $= 0.10(1.00) + 0.90(2.00) = 1.90$
S_3: Expected loss $= 0.10(4.60) + 0.90(0.10) = 0.55$
S_4: Expected loss $= 0.10(7.00) + 0.90(0.00) = 0.70$

Computed in this way, the best strategy is S_3, again a somewhat less conservative choice than that given by minimax. Significantly, however, by either method, one of the strategies S_1 or S_2 is better than either of the rigid no-choice plans, S_2 and S_4.

Mixed Strategies Strategies which rely on chance can also be developed. For example, a possible strategy is: "If the weatherman predicts $p(\theta_1) = 0.25$, I will use a random means to select action a_1 or a_2 such that $p(a_1) = 0.25$." Such strategies will not be discussed here in detail but are quite frequently encountered in game theory, where their dependence on randomness prevents an opponent from discovering a plan for a given day, since none exists.

The preceding discussion has developed the main ideas of decision theory. As was noted in the beginning, applications of this method of making decisions have not yet been made in any significant ways in engineering design or in the management of engineering design. It has been used somewhat in business management and in the analysis of strategic military decisions. Its future application to engineering is left as a long-range exercise for the reader.

16.3 UTILITY

When the loss table was presented in the last section in connection with decision theory, it was noted that the "losses" represented more than money. Such items as prestige, good will, reputation, time, anxiety, and others are also potential costs in many situations. Even in the case of money, its *value* or *usefulness* is not directly proportional to the amount that one has. It is easy to illustrate this. Consider the value of $1,000 to a millionaire compared with its value to you (millionaires excepted!). In order to make a loss table or to make much sense with decision or game theory, the concept of utility has first to be accepted. The idea is that there exists a single scale called *utility* upon which any expressed event or outcome can be given a value. That is, utility theory is based upon the idea that there is a number, usually designated as $U(P)$, which can be attached to any possible event (prospect P), that gives the *utility* of the event.

The utility scale is based upon personal preference. That is, if one prefers P_1 to P_2 (e.g., $10 to $5), then $U(P_1) > U(P_2)$. It is also assumed that mixed prospects are meaningful and can be meaningfully considered. That is, P_3 may be a mixture of P_1 and P_2 such that the probability of P_1 is p and the probability of P_2 is $1 - p$. Then

$$U(P_3) = pU(P_1) + (1 - p)U(P_2)$$

These are powerful assumptions. It is often very difficult to decide things in the way that utility theory demands. Do you prefer $1,000 or a 1 percent chance at $10,000? Do you prefer a month in Paris or a 0.001 percent chance of death in an airplane crash? In addition to the difficulty of deciding, another difficulty with utility theory is that deciding and doing are often different things. Thus, one may *say* he prefers this to that, but if action were required, behavior and words would not necessarily correspond. It is therefore important in utility theory that decision makers be willing to support verbal decisions with behavior.

Utility theory assumes that people can decide their preferences when confronted with prospects or possible outcomes, some of which may be mixed as P_3 above, and that these decisions are valid and reliable indicators of behavior. To illustrate how a utility scale is developed, a student's preference for a grade in his design course was explored. Since the absolute values of the numbers on the scale are arbitrary, a value of 10 can be assigned to an A and a value of 0 to an F. These values are chosen simply for their convenience. Thus

$$U(\text{A}) = 10 \qquad U(\text{F}) = 0$$

Now to get the utility of the intermediate grades, the student must state his preference for some mixed prospects as follows. To find the value of a B grade the student is asked to determine the value of p that satisfies the relation

$$U(\text{B}) = pU(\text{A}) + (1 - p)U(\text{F})$$

That is, the student is asked: "Would you prefer a B grade, or an A grade with probability, say, of 0.1 and an F grade with probability 0.9?" Here p is assumed to be 0.1. To this question the student answered easily that he would prefer the B grade. Now p is changed. "Would you prefer a B grade,

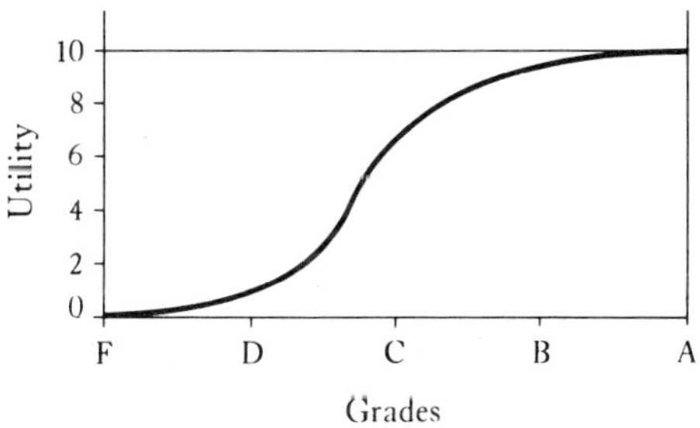

Fig. 16.7 One utility curve for grades.

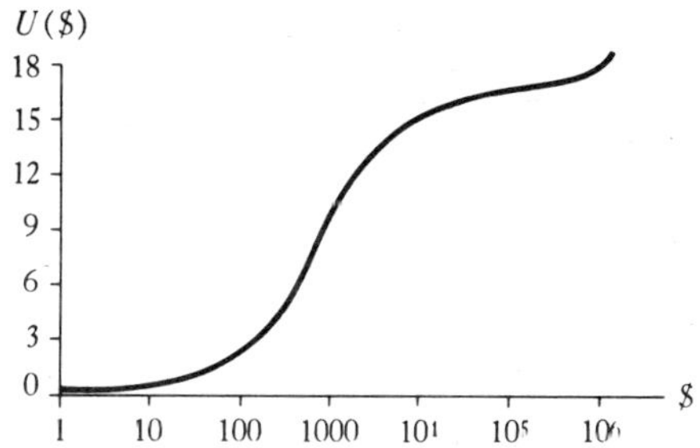

Fig. 16.8 One utility curve for money.

or an A grade with probability 0.5 and an F grade with probability 0.5?" Again the student prefers B. This process is repeated with as many values of p as needed to find the place where the student prefers both prospects (i.e., the B grade or the mixture of A and F for the given p) equally. In the example case being reported, the student was not much of a gambler and disliked F grades very much, so the value of p he finally decided upon was 0.99. Solving for $U(\text{B})$ with $p = 0.99$ gives

$$U(\text{B}) = 0.99(10) + 0.01(0) = 9.9$$

In a similar way, $U(\text{C})$ and $U(\text{D})$ can be found. The results are shown in Fig. 16.7. Though a curve is meaningless in this case because the grades are discrete, not continuous, the curve is drawn because it is characteristic of the shape of utility curves. A utility curve for money for the same student is shown in Fig. 16.8. For obtaining this curve (note that the abscissa is a log scale) arbitrary utility values were established as 0 and 10 at dollar values of 1 and 1,000. Note the characteristic shape of the curve, and the second jump which sometimes occurs at very large values of a commodity such as money. Clearly the value of money is not linearly proportional to the amount of it.

The connection of utility with gambling and investments like insurance should be evident. In making a bet, it is expected *utility* that should be optimized. It is then possible to make a *good* bet at *bad* mathematical odds. Property insurance is like that. The insurance company makes money, and so those who buy insurance clearly do not get back all that they put in plus appropriate interest. But because the utility curve for money for most individuals reflects the great hardship that would be caused by a large loss and because the companies have huge financial resources, both individuals and the companies gain *utility* by their cooperation. Generally

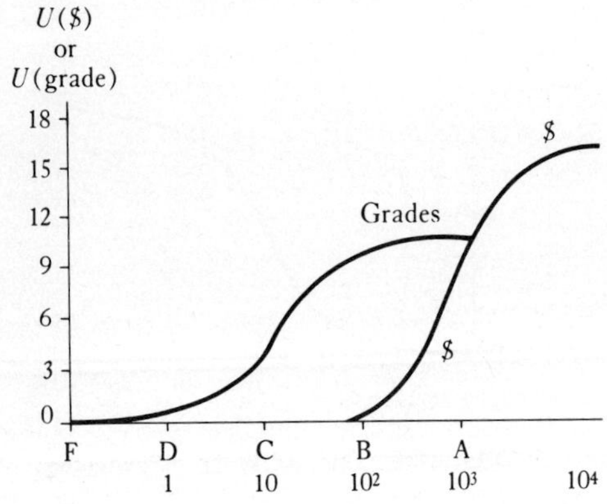

Fig. 16.9 Combined utility curves.

speaking, people will make bets at poor mathematical odds when the sums are small, but will insist upon better than fair mathematical odds when the sums become large.

Often a single utility curve can be developed for a given situation, but sometimes it is necessary to draw separate curves for the separate factors (time, money, prestige, etc.) that are involved. For use in a decision- or game-theory loss table, a single scale is of course needed, so the separate curves must be combined onto a single scale. This is easily done if corresponding points can be found. For example, if the student whose utility curves are shown in Fig. 16.7 and 16.8 wanted grades and money on a single scale, he would have to make some decisions on his preferences for money and grades. He might, for example, decide that an A grade is worth $1,000 but that a B grade has zero monetary value to him. Now a single scale could be drawn as shown in Fig. 16.9. Now both factors are referred to a single comparable utility scale and can be used to compile loss tables. Note that the curve for $U(\$)$ had to be linearly extended to fit the new scale.

16.4 GAME THEORY

Game theory and decision theory are closely related, but there is a significant difference. In decision theory as outlined in Sec. 15.2, there was a "state of nature" included. Knowledge of the state of nature was not necessarily complete, but the circumstances of nature were assumed to be unalterable. The decision maker in a decision-theory situation has a strategy, but the state of nature just is what it is. In game theory, an intelligent opponent who may also have a strategy is included in the situation. Sometimes the decision-theory model is called a *game against nature* and nature is thought of as an *opponent*. Generally, however, it is just as well to restrict game theory to those situations where there are two or more intelligent opponents.

This section will present only a very brief introduction to the jargon and principles of game theory. It is the business manager who is most concerned with game-theory situations. The design engineer is more often confronted by nature as an opponent and is hence more likely to use decision theory.

The simplest of all games is the two-person zero sum game. Suppose that two persons, A and B, playing a game, can see the *payoff matrices* shown below but cannot communicate with each other:

	b_1	b_2
a_1	+2	+1
a_2	+3	−2

Payoff for A

	b_1	b_2
a_1	−2	−1
a_2	−3	+2

Payoff for B

Each must then choose a course of action independently. That is, A must choose either a_1 or a_2 and B must choose either b_1 or b_2. It is called a zero-sum game when the payoff for A is always the negative of that for B. In a zero-sum game there is no "banker" or source or sink for utility. That which one player loses, the other gains. Note that the values entered in the payoff matrices are, strictly, utilities. Often the matrix for the opponent is not known, and this makes a different (and more complex) game. In the simple example at hand, the matrices are considered known.

Now A has two choices, a_1 and a_2, and B has two choices, b_1 and b_2. How will A reason? The first thing that comes to mind is that he will have to consider what B might do. He will note, for example, that B is better off with choice b_2 no matter what A does. That is, if A chooses a_1, B will do best by choosing b_2. If A chooses a_2, B will do best by choosing b_2. Hence an intelligent B will choose b_2, and A will obviously select a_1. This game thus has a unique and obvious solution (a_1, b_2). The point (a_1, b_2) is called a *saddle point*, and there are a variety of ways that even more complex games can have a saddle point. The significance of a saddle point is that, if both players are rational and assume rationality in the other, a saddle point will be an obvious solution which both will recognize. The game is said to be "stable."

There is a way out of the dilemma of the zero-sum no-saddle-point game if the game is to be played a large number of times. It can be shown, however, that for each player there exists a mixed strategy (each player chooses his alternatives a specified percentage of the time) which if played will result in the best possible return for each player. Readers interested in a more detailed study of game theory should consult the Suggested Reading.

Non-zero-sum games can come in many varieties, but the classical example of the prisoners' dilemma will serve to illustrate the type. Two men suspected of having jointly committed a serious crime are being kept separated for questioning. Since they actually have committed the crime, there are two alternatives open to each: confess or keep quiet. Since the police do not have proof of the serious crime, but do have proof against both of a lesser offense which would result in a short sentence, they offer complete freedom to the one who confesses. The other would then get convicted of the serious crime and receive a long sentence. Of course, if both confess, they both get long sentences but somewhat reduced because of their confession. The utility payoff matrix now looks something like the following for this problem:

	b_1	b_2
a_1	$-5, -5$	$10, -10$
a_2	$-10, 10$	$5,5$

In the entries, the payoff for A is listed first; the actions a_1 and b_1 mean confession, whereas a_2 and b_2 mean to remain silent.

This presents a keen dilemma for both prisoners. (Remember that they cannot communicate with each other.) If they choose to cooperate and both remain silent (a_2,b_2), they each get $+5$. But if A, for example, were sure that B would remain silent, he could then choose a_1 and thereby increase his return to $+10$. B, seeing this is possible, will not want to suffer the -10 loss and so may well select b_1. The result will then be (a_1b_1), which is a -5 loss for each man. It appears that cooperation will lead to a $+5$ gain for both, whereas if each pursues his own self-interest independently, they will both lose!

There are no solutions to such dilemmas. Two "ethical" men (in this case, group ethics means obedience to the criminal code not to "talk") will both win, whereas two self-centered men will both lose. If one is ethical and one self-centered, the self-centered one will win big and the ethical one will lose big. Readers should not draw generalities about how to behave from this example, because different payoff matrices can be concocted that lead to different conclusions. The point is that game theory can provide a model with which to study certain decision or conflict situations.

Such is the world of game theory. Its possible application to business situations should be obvious, though, as with decision theory, present knowledge does not allow great accuracy in assessing probabilities or in obtaining utility data.

16.5 CRITICAL-PATH METHOD (CPM) AND PROGRAM-EVALUATION-AND-REVIEW TECHNIQUE (PERT)

Two methods helpful in the planning and direction of engineering design, development, and other projects have been devised in recent years. One of these, the Critical-path Method (CPM), will be discussed in this section. The other, Program-evaluation-and-review Technique (PERT), will be discussed in the next section.

Neither CPM nor PERT are decision-making techniques. They are aids to the decision maker, especially by calling attention to places where decisions are necessary, but the decisions themselves remain the function of the person in charge of the project. CPM and PERT help to point out needs regarding time and resources that are involved in meeting project-completion schedules.

Much of the jargon used by CPM and PERT is identical. PERT, in fact, can be considered to be an extension or refinement of CPM. Both methods are based on the concept of an *event*, defined as the start or completion of a task. An event, as used by these methods, is *not* the task or phase of the project; an event is the *start* or *completion* of the task. Note that an event itself does not require either time or resources.

The first step in setting up a CPM or PERT program is to compile a list of the events that will occur during the project and to arrange these events in a logical sequence. As an example, suppose a toy company is looking for a new product and the idea of a toy rocket has been conceived. Then the events for the first phase of the project might be the following:

A. Market survey completed
B. Idea for toy rocket presented
C. Program approval made
D. Engineering feasibility study completed

Note that these are not listed in logical sequence. The events are placed in logical sequence in a *network* as follows:

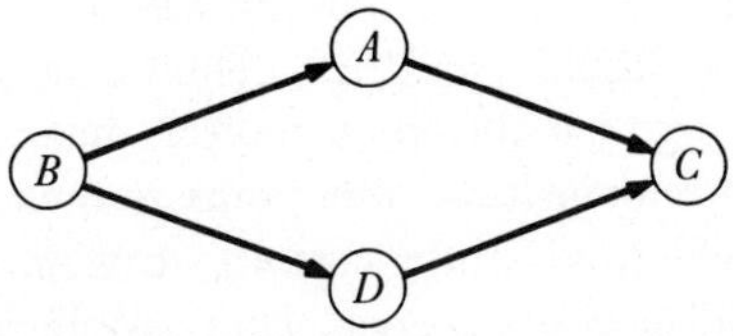

In the network the circles represent *events* and the arrows represent *activities*. Activities are *time-consuming*, whereas events are not. The network is drawn in such a way that the necessary sequence of events is clearly indicated. Thus event C cannot take place until both event A and event D take place. Similarly, the network shows that neither A nor D can take place until B has occurred. In order to draw the network, the manager must consciously consider the logical sequence and interdependence of events in the project. This in itself is one of the most valuable results of employing both CPM and PERT.

These methods do more, however. They enable the manager to determine the *critical path* through the network. This is possible as a result of the manager's having estimated the time necessary for the activities that make up the project. For the network above, suppose that the activities and their time estimates are as shown below:

Activity	*Predecessor event*	*Successor event*	*Best estimate time t_e, days*
Market survey	B	A	28
Engineering study	B	D	21
Market report preparation	A	C	4
Engineering report preparation	D	C	2

This information would be shown on the network as follows:

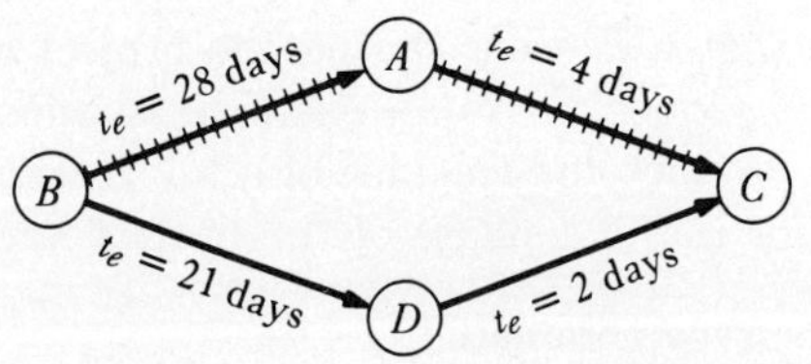

The *critical path* is that sequence of events that requires the most time for completion of the project. In the simple network above, then, the critical path is B–A–C, and is shown cross-hatched. The other path, B–D–C, is said to contain *slack*. The manager now knows to concentrate his efforts along the critical path in attempting to reduce the completion time of the project. He might, for example, plan to transfer men or resources from a slack portion to the critical portion in order to shorten the critical path. In any event, he now knows where his biggest problems lie if he is to meet a schedule. Of course, his information is no more accurate than his own estimates of the time consumed by the activities. As was noted earlier, CPM and PERT are management aids, but *the decisions still remain with the manager*. It should be apparent that skill and experience are necessary in order for a manager to be able to make the necessary estimates accurately.

Networks are usually much more complicated than the one used above as an example. Suppose the network was the one below. Which is the critical path?

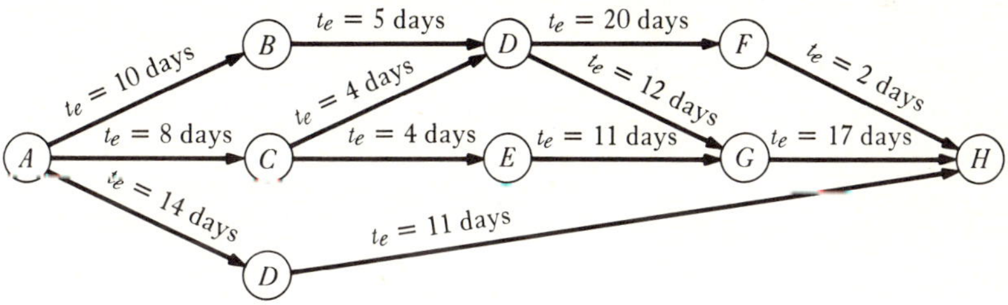

16.6 PROGRAM-EVALUATION-AND-REVIEW TECHNIQUE—PERT

The critical path in the above network is *A–B–D–G–H*. The total time is 44 days.

PERT is a somewhat extended version of CPM. The procedure when using PERT is the same as when using CPM until the time when time estimates for the activities are to be made. Thus CPM and PERT consider events and activities in the same way and begin by drawing an event-activity network for the project. At this point PERT becomes more sophisticated because it makes use of *probabilities* in connection with the activity-time estimates. In particular, whereas in CPM only one time estimate for an activity was recorded—the manager's best estimate—in PERT *three* time estimates are made for *each* activity. The times recorded are the following:

t_1: an optimistic time—the shortest possible period which the activity can require

t_2: the best estimate—the same time used in CPM

t_3: a pessimistic time—the longest possible time which the activity can require

The usual procedure at this point is to assume a distribution function covering a total span of six standard deviations (6σ). That is, it is assumed that $t_3 - t_1 = 6\sigma$, and so the standard deviation and variance are

$$\sigma = \tfrac{1}{6}(t_3 - t_1) \qquad \sigma^2 = \left(\frac{t_3 - t_1}{6}\right)^2$$

To find the mean time that the activity will consume, a distribution is assumed such that the mean time t_e is given by

$$t_e = \tfrac{1}{6}(t_1 + 4t_2 + t_3)$$

The next step in a PERT analysis of a project is to compute t_e for each activity and, using these, to determine the critical path. Referring again to the simple example of the toy company, suppose the time estimates are the following:

Activity	*Optimistic time* t_1	*Best estimate time* t_2	*Pessimistic time* t_3	*Average time* t_e	*Variance* σ^2
Market survey	21	28	35	28	5.5
Engineering study	17	21	28	21.5	3.4
Market report	2	4	5	3.8	0.3
Engineering report	1	2	3	2	0.1

If the earliest expected completion date for an event is t_E, then t_E for event C by path B–A–C is $28 + 3.8 = 31.8 = 32$. By path B–D–C it is $21.5 + 2 = 23.5 = 24$. Hence, B–A–C is the critical path, and $T_E = 32$ is the earliest possible completion day expected for this phase of the project.

For the network shown at the end of Sec. 16.6, what are the critical path and the earliest possible expected completion date for this network if the following times are the manager's estimates?

Preceding event	*Succeeding event*	t_1	t_2	t_3	t_e	t^2
A	*B*	8	10	12	10	0.44
A	*C*	7	8	9	8	
A	*D*	10	14	18		
B	*D*	4	5	7		
C	*D*	2	4	5		
C	*E*	3	4	5		
D	*F*	15	20	30		
D	*G*	10	12	15		
E	*G*	10	11	12		
D	*H*	9	11	13		
F	*H*	1	2	3		
G	*H*	14	17	20		

PERT is helpful in planning to meet target dates or in estimating the probability of meeting a target date. The latest possible completion times are denoted by T_L. Target, or scheduled completion, times are denoted by T_S. Suppose that the toy company's little project must be completed 40 days from today. Then $T_L = 40$ and the latest possible starting date for the market survey is $40 - 32 = 8$ days from now. Note that because of the slack in the other path, the latest possible starting date for the engineering study is $40 - 24 = 16$ days from now. Notice how this kind of information can aid a project manager in the efficient use of his men and resources.

The information on variance is useful in estimating the probability of meeting a scheduled completion date T_S. If a normal distribution of completion dates is assumed, it will have a mean of T_E and its standard deviation can be estimated from

$$\sigma_{T_E} = \left[\sum_i (\sigma_i^2)\right]^{1/2}$$

where σ_i^2 is the variance of the ith activity in the critical path of the project. σ_i^2 is thus the sum of the variances along the critical path. Now the distribution of T_E's will look like this, with a particular T_S also shown:

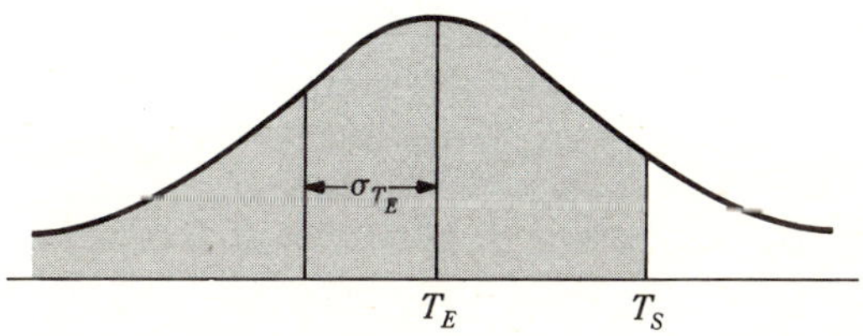

The area to the left of T_S, the cross-hatched area in the diagram, can be seen to represent the probability that the event will be completed by the time T_S. The table of areas for the normal distribution will give the area (i.e., probability). Suppose, for example, that $T_E = 32$ and that $\sigma_{T_E} = [\sum_i (\sigma_i^2)]^{1/2} = (5.5 + 0.3)^{1/2} = 2.4$ as above and that the probability of meeting a scheduled completion date of $T_S = 34$ days is desired. In this case T_S is $2.0/2.4 = 0.833$ standard deviations to the right of T_E. From the tables, then, the area to the right of T_S is 0.797. Note that if $T_S = T_E$, the probability is 0.500. If T_S is less than T_E, then the probability of meeting the date is less than 0.500. What is the probability of meeting a scheduled completion time of 29 days in the example?

Since 29 days is 3 less than the mean of 32 and the standard deviation is 2.4, this T_S is $3.0/2.4 = 1.25$ standard deviations to the left of T_E. The area to the left of this T_S is, from the normal distribution tables, 0.1. Hence the probability of being completed in 29 is only 0.1.

These are the major principles of PERT. As projects and their networks become larger and more complex, the computational work becomes longer but remains as straightforward as the above. The results are no better than the manager's estimates, but the method focuses his attention on the critical questions and provides him with a rational way of planning based on his own estimates. In many industries, standard computer programs are being used or prepared to do the computational work associated with complex CPM or PERT problems.

16.7 THE ASSIGNMENT PROBLEM

Assignment problem is the term used to describe that problem from operations research in which n items are to be distributed among n assignments in such a way that the return is optimized. Problems of this type have many applications, as in dispatching trucks from city to city, assigning salesmen to territories, and the like. It will be discussed here as a means for assigning men to various projects—with engineering design in mind, of course.

Suppose that a project manager has four engineers and four design tasks to be accomplished. The engineers differ in ability, training, and motivation, and the tasks differ in nature and difficulty. The assignment problem is to find the optimum way of assigning the men to the tasks so as to minimize cost.

To begin with, the manager must estimate the costs for each man to do each job. This is best shown in matrix form. With three men and three tasks, the cost matrix may look something like this:

	Man		
Task	*A*	*B*	*C*
1	9	12	13
2	5	3	3
3	10	11	15

One way to solve such a problem is simply to write out all the alternative arrangements. There are $3! = 6$ in this case, so it is quite feasible. Thus the cost of assignment A to 1, B to 2, and C to 3 is $9 + 3 + 15 = 27$. The cost of assigning A to 2, B to 1, and C to 3 is $5 + 12 + 15 = 32$. Etc.

A more practical approach is needed, however, whenever the number of men and assignments is above about four ($5! = 120$). Techniques for doing this are discussed fully in the operations-research literature and will not be repeated here.

When there are more men than tasks, the problem can be formulated to look like the so-called transportation problem in operations research. The reader will recognize, however, that in any case it is the cost matrix that is difficult to come by. Solving the problem for the optimum is not difficult once the cost matrix is accurately and reliably obtained. To date, however, people have remained sufficiently unpredictable and complicated that reduction of their abilities and talents for achievement to a single number in a matrix is a fairly unreliable task. Nevertheless, the methods of operations research have proved useful in managing materials, and as time goes on, they may well prove useful in the management of people—even engineers themselves!

16.8 SUMMARY AND PREVIEW

As noted at the outset, this chapter is a survey and an introduction to some schemes now used effectively in the management of business and materials that may eventually have use in the management of the design activity. Decision and game theory are fascinating fields and have been only briefly introduced here. In utility theory, or something similar, lies the key to the application of these methods to engineering design, because that theory makes it possible to express worth or value of a variety of things on a single scale—a necessary prerequisite to the optimization of value. CPM and PERT are tools to help managers plan and keep track of their chances of success in meeting a schedule. The assignment problem requires a quantitative knowledge of men and tasks that is not now generally available, though it should be noted that design project managers solve assignment problems subjectively all the time.

This chapter completes Part III of the book, the decision-making part. In the beginning, the design activity was separated into the three mental processes it uses: inventiveness, analysis, and decision making. Each of these has been considered in some detail. In the epilogue that follows, and completes the book, the three sections are reviewed briefly, and the problem of how to integrate these skills into an inspired and competent whole is posed.

SUGGESTED READING

Chernoff, Herman, and Lincoln E. Moses: "Elementary Decision Theory," John Wiley & Sons, Inc., New York, 1959. The best introduction to decision-utility theory now available.

Federal Electric Corporation: "A Programmed Introduction to PERT," John Wiley & Sons, Inc., New York, 1963. A good, teach-yourself elementary introduction to PERT.

Fishburn, Peter C.: "Decision and Value Theory," John Wiley & Sons, Inc., New York, 1964. A more advanced and more mathematical study of decision theory.

Rapoport, Anatol: "Fights, Games, and Debates," The University of Michigan Press, Ann Arbor, Mich., 1960. A very interesting nonmathematical review of game theory as applied to life.

PROBLEMS

16.1 Draw your own utility curve for money.

16.2 Draw your own utility curve for grades.

16.3 Discuss the relative merits of minimax and expected loss as criteria for decision making.

16.4 Plan your next group design project, using PERT.

16.5 In planning your next group design project, use the methods of the assignment problem to assign tasks.

16.6 Consider again your choice of where to go to college. Use decision theory and utility to make the decision.

16.7 You have to decide whether to make or buy a workbench for your home basement. Apply decision theory to decide.

16.8 Discuss the usefulness of decision-utility theory in make-or-buy decisions.

16.9 Where do you think decision-utility theory will be most useful in engineering in the future?

EPLIOGUE— THE DESIGN PROCESS RECONSIDERED

In this book, engineering design has been divided into the three mental activities which it involves: inventiveness, analysis, and decision making. Each of these mental processes is fundamentally different from the other two. For this reason each has been discussed separately. Before the integration of these processes is discussed briefly in this closing chapter, a very short summary of each is in order.

Inventiveness is the ability to get good, useful ideas for solving engineering problems. Inventive mental activity is characterized by its open-endedness, its dependence on perception and breadth of experience, and its tendency to be inhibited by constraints and mental set. Though people tend to be limited naturally in inventive ability, nevertheless most people do not perform near their natural limit. There are both general and specific methods discussed in Chap. 2 for improving inventiveness.

Analysis is the ability to get meaningful answers to engineering questions within reasonable time and cost. Analytical activity, though not without its creative qualities, is characterized by its converging nature and its dependence upon judgment and special knowledge. Analysis involves the construction of a physical or mathematical model, application of fundamental physical principles to the model, and interpretation of the results. There is some evidence that strong analytical skill or training interferes with inventive performance. Inventiveness and analysis are clearly different, in the sense that the object of inventive effort is *many possible answers*, whereas the object of analytical effort is *one actual answer*.

Still another way to distinguish these mental processes is the following. Inventiveness is the generation of alternatives and analysis is the study of one alternative. In this context, decision making is the process of choosing among alternatives. Decision making is characterized by its involvement with information, value assessments, and optimization. Thus whereas inventiveness seeks many possible answers and analysis seeks one actual answer, decision making seeks to choose the *one best answer*.

Much less has been done by psychologists on decision-making ability than on inventiveness, so that little is known about possible interference between the skills of decision making and either inventiveness or analysis. What is clear, however, is that inventiveness, analysis, and decision making are three quite different activities—all of which are required of the engineering designer.

The difficulty for the engineering designer is that not only are all three skills part of the design process but they are all part of almost every step! Certain ones are emphasized at various stages, of course, and in special cases, the activity may become almost purely one of the three. In general, however, the engineer is confronted with the need to be *simultaneously* an inventor, analyst, and decision maker. At the very least, quick and repeated changes of roles are called for by the good engineering designer.

Very little attention has been given by engineering educators to the problem of how to integrate these three mental processes in the designer. It has been suggested that inventiveness and analysis can be turned on and off consciously, to help avoid the inhibiting effects they seem to have on each other. This same technique can perhaps be extended to the deliberate decision-making stages, providing of course people can be made analogous to three-way switches! But this does not aid in handling the myriad of tiny—and not so tiny—decisions that permeate the design process.

One final point on the problem of integrating inventiveness, analysis, and decision making. A designer is often required to decide and move on, even though he may not know that the decision is the best. That is, decisions often *precede* analysis. An alternative, perhaps even one generated for the occasion, must be selected before it can be analyzed and optimized. Such decisions need not be final, they can be reconsidered when the analysis is in, but *they must be made!* This is the source of the frequently heard remark that "design is decision making." Actually, of course, design is also inventiveness and analysis (including communication), but there is no denying the critical role that the *ability to decide* plays in the necessary skills of the excellent design engineer.

A simple example may help to bring this point home. Suppose a heat exchanger is to be designed to do a specified job. Tube length, diameter, spacing, material, etc., are all design parameters. Before any progress at all can be made, some initial decisions must be made. One cannot analyze an infinite number of combinations. Such decisions (circumstances may lead to many occasions for them in one project) to limit consideration to one or a few specific alternatives ought to be made with good engineering sense, but they must be made.

Design involves inventiveness, analysis, and decision making. It also involves the above-mentioned capacity to decide and move on, as well as skill in the management of one's own time, energy, and ability. Such management is simply *planning*. These extra skills—planning, the willingness to decide, the integration and utilization of inventiveness, analysis, and decision making—make up another dimension in the design problem-solving process. At the present time this fourth dimension can be learned only by practice on actual design projects. It is customary that *something* be left as an exercise for the students.

SOME IDEAS FOR ENGINEERING DESIGN PROJECTS

1 Design a toy or game that illustrates or utilizes an engineering skill, principle, or concept. Specify an age group at the outset. A game might be called "Optimization" or "Estimation" or a toy called "Feedback System."

2 Design a container with specified weight and size constraints to prevent an enclosed fresh egg from breaking when dropped from a third-story window.

3 Design a blood heat exchanger for minimum volume that will cool a dog from body temperature to 40°F in 20 min. Assume that the heat exchanger can be put in series with the heart so that the cooled blood is circulated throughout the dog's system.

4 Design a system for providing a variable quantity of constant-temperature air for psychophysical experiments.

5 Design an "effective temperature" meter. See a text on air conditioning to learn the definition of effective temperature.

6 Design a new kind of handcart carrier for factories. Use pressurized air as the method of support and lubrication.

7 Design a record-player system that will skip bands on a record as desired.

8 Design an experimental apparatus for measuring surface tension.

9 Design a toilet that uses less water.

10 Design an automatic control for a lawn-sprinkler system.

11 Design an air pump for fish bowls.

12 Design a system for producing and throwing ice pigeons. Note that the field never has to be cleaned up as with clay pigeons.

13 Design, construct, and test a structure made of balsa wood and glue to support a specified load. Also specify geometrical constraint as desired, to make the problem interesting. Make the optimization criterion (*a*) minimum weight, (*b*) maximum ratio of strength to weight, or (*c*) maximum load for a given weight.

14 Design and construct a fuel cell for classroom-demonstration purposes.

15 Make a paper-and-pencil design of a chair.

16 Design a new and better teaching machine for programmed instruction.

17 Select a particularly bad traffic situation in your community and recommend a solution.

18 Design a small supersonic wind tunnel for laboratory-demonstration purposes.

19 Design a walk-in plant environment chamber.

20 See the August, 1965, issue of *Scientific American* and discuss the transportation system for the East Coast of the United States presented there. Design such a system for your local city.

21 What could be done with the Sahara Desert?

22 Suppose a society on earth consisted of collie dogs with man-like intelligence. Design their "automobiles" and public transportation systems.

23 Design a system to read numbers printed on a page.

24 Design a moon landing gear. Test your concept by building a model and using it to land a 10-lb weight dropped from 10 ft high onto an irregular surface.

25 Design a decelerometer for use on the weight in Prob. 24.

26 Select three hammers of different design and evaluate the design of each.
Note: Instead of hammers, one can use another tool, appliances (e.g., vacuum cleaners), or anything else.

27 Make up your own design project!

INDEX